£18.50

REPRODUCTION OF MARINE INVERTEBRATES

Volume II

Entoprocts and Lesser Coelomates

REPRODUCTION OF
MARINE INVERTEBRATES

Volume II

Entoprocts and Lesser Coelomates

Edited by

Arthur C. Giese

*Department of Biological Sciences and
Hopkins Marine Station
Stanford University
Stanford, California*

John S. Pearse

*Division of Natural Sciences
University of California
Santa Cruz, California*

ACADEMIC PRESS   New York   San Francisco   London   1975

A Subsidiary of Harcourt Brace Jovanovich, Publishers

ACADEMIC PRESS, INC.
111 Fifth Avenue, New York, New York 10003

United Kingdom Edition published by
ACADEMIC PRESS, INC. (LONDON) LTD.
24/28 Oval Road, London NW1

Library of Congress Cataloging in Publication Data

Giese, Arthur Charles, Date
    Reproduction of marine invertebrates.

    1.   Reproduction.   2.   Marine invertebrates—
Physiology.   I.   Pearse, John S., joint author.
II.   Title.   [DNLM: 1.   Invertebrates—Physiology.
2.   Marine biology.   3.   Reproduction.   QL364 G455r]
QP251.G437 vol. 3   592'01'6   72-84365
ISBN 0-12-282502-0 (v. 2)

PRINTED IN THE UNITED STATES OF AMERICA

CONTENTS

CHAPTER 1   ENTOPROCTA

RICHARD N. MARISCAL

CHAPTER 2   TARDIGRADA

LELAND W. POLLOCK

CHAPTER 3   PRIAPULIDA

JACOB VAN DER LAND

CHAPTER 4   SIPUNCULA

MARY E. RICE

CHAPTER 5 POGONOPHORA

Eve C. Southward

CHAPTER 6 CHAETOGNATHA

M. R. Reeve and T. C. Cosper

CHAPTER 7 HEMICHORDATA

Michael G. Hadfield

CHAPTER 8 CHORDATA: TUNICATA

N. J. Berrill

CHAPTER 9   CHORDATA: ACRANIA (CEPHALOCHORDATA)

JOHN H. WICKSTEAD

LIST OF CONTRIBUTORS

Numbers in parentheses indicate the pages on which the authors' contributions begin.

N. J. Berrill (241), *410 Swarthmore Avenue, Swarthmore, Pennsylvania*

T. C. Cosper (157),* *Division of Biology and Living Resources, University of Miami, Miami, Florida*

Michael G. Hadfield (185), *Pacific Biomedical Research Center, University of Hawaii, Honolulu, Hawaii*

Richard N. Mariscal (1), *Department of Biological Science, Florida State University, Tallahassee, Florida*

Leland W. Pollock (43),† *Systematics-Ecology Program, Marine Biological Laboratory, Woods Hole, Massachusetts*

M. R. Reeve (157), *School of Marine and Atmospheric Science, University of Miami, Miami, Florida*

Mary E. Rice (67), *Department of Invertebrate Zoology, National Museum of Natural History, Smithsonian Institution, Washington, D. C.*

Eve C. Southward (129), *The Laboratory, Citadel Hill, Plymouth, England*

Jacob van der Land (55), *Rijksmuseum van Natuurlijke Historie, Leiden, The Netherlands*

John H. Wickstead (283), *Ministry of Overseas Development, London, and The Marine Biological Association, Citadel Hill, Plymouth, England*

---

* Present address: Office of Technological Development, QL&M Laboratories, Hudson River at Burd Street, Nyack, New York.

† Present address: Department of Zoology, Drew University, Madison, New Jersey.

PREFACE

This volume brings together information on many of the smaller groups of marine metazoans, most of which are difficult to relate to any other group of animals. Each group, however, is of both intrinsic and general interest and should be considered in any broad review of animal reproduction. Some provide favorable material for special studies, for example, the entoprocts with their remarkable asexual capacities. Others, such as the carnivorous pelagic chaetognaths and the interstitial tardigrades, are especially important in particular marine environments. The priapulids, sipunculans, and pogonophorans appear to be related to the annelids and echiurans, and may provide perspective in viewing these important phyla which are considered in Volume III. The hemichordates, and particularly the tunicates and acraniates, are especially interesting in view of their presumed position as predecessors to the vertebrates.

We are indebted to our Advisory Board for suggestions on the scope and organization of the treatise, to the Board and to a larger community of biologists for encouragement and suggestions for additional prospective authors, and to all the authors who enthusiastically assumed responsibility for chapters which required of them much effort and time. We are indebted to Ms. Jean McIntosh for the preparation of the Subject and Taxonomic indexes. Finally, we are indebted to Dr. Vicki Buchsbaum Pearse for her painstaking editorial assistance and to the staff of Academic Press for their help with the development of the treatise.

ARTHUR C. GIESE

JOHN S. PEARSE

CONTENTS OF OTHER VOLUMES

Chapter 1

ENTOPROCTA

Richard N. Mariscal

1.1 Introduction

The Entoprocta (Kamptozoa, Calyssozoa, Endoprocta) comprise a little-known phylum of about 120 species by my count (1971). They are nearly all microscopic, being only a few millimeters in length and are easily overlooked both in the field and in the laboratory.

In spite of their pseudocoelomate nature, for many years they have been included together with the Ectoprocta in the phylum Bryozoa. Because of the presence of a large number of extant species and the extensive ectoproct fossil record, most of the attention paid to the entoprocts has been systematic and by bryozoologists primarily interested in the ectoprocts. For this reason, relatively little modern work has been

1

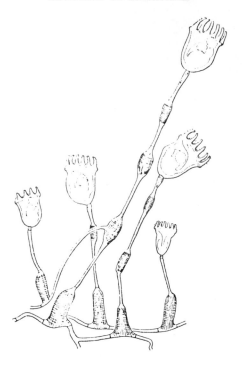

Fig. 1.   A portion of a colony of *Barentsia* (probably *B. benedeni*) showing the characteristic muscular swelling at the base of the stalk which distinguishes this genus. This species also may have one or more muscular enlargements along the stalk which allow for the bending of the individual polyp. Note the development of a stolon from one of the muscular nodes on the stalk (from Cori, 1936).

done on entoprocts and much of what we know of their biology comes from several lengthy papers and monographs from the nineteenth and early part of the twentieth centuries.

The basic body plan is similar throughout the phylum. The adult polyps or zooids all consist of a tentacle-bearing, bowl-shaped calyx perched on a long thin stalk which may have bulbous muscular enlargements along its length (Fig. 1). Both mouth and anus are connected by the characteristic **U**-shaped gut, opening within the circlet of tentacles, hence, the name of the phylum (Fig. 2). However, there have been two rather divergent lines of evolution within the group. The point of divergence lies in the nature of the base of the stalk, i.e., whether it is attached to the substrate directly or whether it joins with a ramifying stolon which, at periodic intervals, gives rise to other similar polyps. If the polyp is attached directly to the substrate (i.e., solitary), the species is placed in the family Loxosomatidae (Fig. 3). If the polyps are colonial

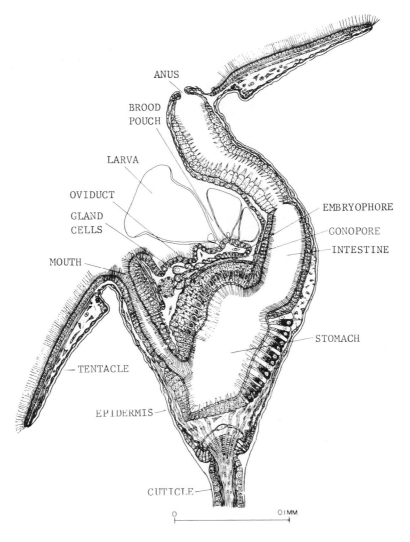

ANUS

BROOD
POUCH

LARVA

OVIDUCT

GLAND
CELLS

MOUTH

EMBRYOPHORE

GONOPORE

INTESTINE

STOMACH

TENTACLE

EPIDERMIS

CUTICLE

0   0.1MM.

Fig. 2. Median sagittal section of the calyx of an adult female pedicellinid entoproct (*Barentsia benedeni*), showing the relationship of the oviduct, embryophore,
brood pouch, and the developing larvae (from Mariscal, 1965).

and interconnected by branching stolons, they belong to the family
Pedicellinidae (Fig. 1). A third family is also recognized, the Urnatellidae,
which is restricted to fresh water and contains only two species (Fig. 4).
The base of the stalk of this form is somewhat intermediate between those
of the other two families in that several stalks arise from a single basal

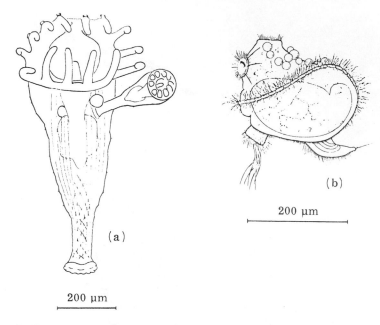

200 μm

FIG. 3. *Loxosoma jaegersteni.* (a) Adult individual in frontal view. Note the large bud extending from the side of the calyx. (b) *Loxosoma jaegersteni* larva in side view containing an internal bud (from Nielsen, 1966a).

attachment disc. Although higher taxa can be devised, it is convenient and natural to go directly from the phylum to the family level and for that reason the characteristics of the major families and genera will be briefly reviewed.

The solitary Loxosomatidae constitute by far the largest single natural grouping among the Entoprocta. The family is composed of a remarkable collection of largely symbiotic entoprocts which are distinguished by the nature of the foot or attachment organ and the unique development of buds from the calyx. Nearly half of the Loxosomatidae have been found in close association with various families of polychaetes or their tubes. Nielsen (1964a) has summarized much of this information and has listed ten families of Polychaeta with which loxosomatids are known to associate. Although little enough is known about the entoprocts themselves, essentially nothing is known of the biology of this association or the possibilities of any reproductive interaction. In addition, loxosomatids have been found living symbiotically with ectoprocts, sipunculans, sponges, echinoderms, and ascidians, among others. Most of these associations appear to be commensal in nature with the entoproct utilizing

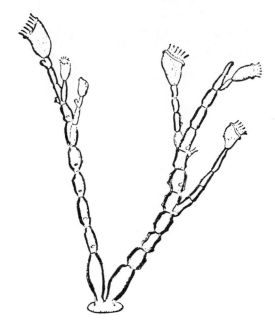

Fig. 4. *Urnatella gracilis* colony showing the beaded stalk and small attachment disc (from Cori, 1936).

the feeding currents and substrate provided, but apparently neither benefiting nor harming the host.

At most, five genera of loxosomatids are currently recognized, the major distinguishing feature being the nature of the foot or pedal attachment organ:

Loxosoma. Basal portion of the stalk of both buds and adults consists of a muscular sucking disc containing scattered glandular cells. The animals are capable of movement on the host throughout life (Fig. 3).

Loxocalyx. Basal portion of the stalk contains a pedal gland, pedal groove, and accessory glandular cells, which are found both in the buds and adults throughout life. Some species, at least, appear capable of movement on the host (Fig. 5).

Loxosomella. Basal portion of the stalk of the buds consists of a foot with a pedal gland and pedal groove with accessory glandular cells. After the bud detaches from the parent it generally becomes permanently cemented to its host and the foot gland degenerates in the adult and may be entirely lost (Fig. 6). Because of the difficulty in clearly distinguishing between the characteristics separating *Loxocalyx*

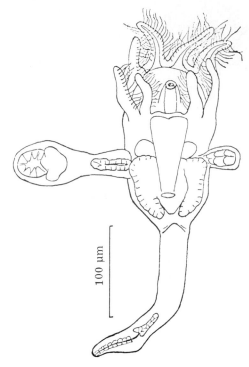

100 μm

Fɪɢ. 5. *Loxocalyx cochlear* adult with two buds (after Schmidt, 1876, from Prenant and Bobin, 1956).

and *Loxosomella,* Nielsen (1964a) has abolished the genus *Loxocalyx* and set up two subgenera, *Loxosomella* and *Loxomitra,* which are separated by the nature of the attachment of the buds to the parent.

Loxomespilon. No stalk; the calyx is attached directly to the host by means of a small adhesive gland. Only one species described.

Loxostemma. No stalk; the calyx is attached directly to the host either by means of four large muscular suckers which lie just beneath the aboral tentacles, or less commonly, by means of four small suckers at the base of the calyx. Only one species described.

The colonial Pedicellinidae contain only three genera about which much is known of the biology:

Barentsia. The basal portion of the stalk forms an enlarged muscular swelling which abruptly narrows down to give rise to the much thinner stalk. The stalk may be of the same diameter throughout or it may possess one to many muscular enlargements along its length (Figs. 1 and 2).

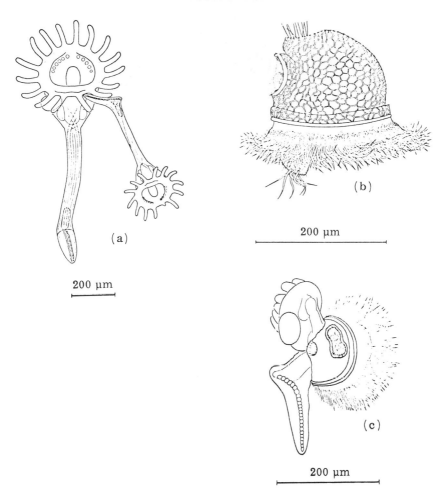

Fɪɢ. 6. *Loxosomella vivipara.* (a) Frontal view of adult with one large bud. (b) *Loxosomella vivipara* larva in side view. (c) The liberation of the bud from the larva by rupture of the larval body wall (from Nielsen, 1966a).

Pedicellina. The stalk lacks any distinct muscular enlargement at its base and is of approximately the same diameter and muscularity throughout (Fig. 7).

Myosoma. The stalk, like that of *Pedicellina,* also lacks muscular enlargements along its length, but both the stalk and calyx exhibit a much greater development of muscular fibers which are continuous from one to the other. The calyx is situated at an oblique angle to the stalk.

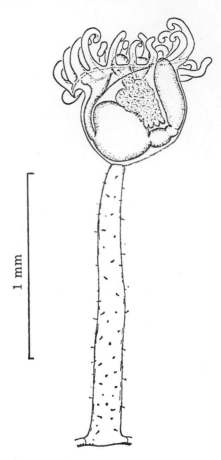

1 mm

Fig. 7. *Pedicellina cernua* adult in side view. Note the lack of a distinct muscular swelling at the base of the stalk where the polyp arises from the stolon (from Ryland, 1965).

In addition, the genera *Coriella, Pseudopedicellina, Chitaspis, Sangavella,* and *Loxosomatoides* have been described, but several of these genera may be referable to one of the above three (e.g., see Ryland, 1965, regarding the status of *Pseudopedicellina*). *Coriella* and *Pseudopedicellina* are both of the *Barentsia*-type with a basal muscular enlargement on the stalk. *Chitaspis, Sangavella,* and *Loxosomatoides* are all of the *Pedicellina*-type with no distinctive basal muscular enlargement on the stalk.

The freshwater, colonial family Urnatellidae contains only the single genus *Urnatella*, which possesses a beaded stalk and basal enlargement much like some members of the genus *Barentsia*. However, stalk budding, rather than stolon budding, is very highly developed in The beaded appearance of the stalk appears to result largely from c lar constrictions of a fairly thick stalk, rather than spaced muscular enlargements superimposed on a thin stalk, as is the case with most species of *Barentsia*.

Several recent works have reviewed what is known of entoproct biology (Cori, 1929, 1936; Hyman, 1951; Prenant and Bobin, 1956; Brien, 1959).

1.2 Asexual Reproduction

1.2.1 Occurrence and Types

Asexual reproduction by stolon, stalk, and calyx budding of the adults is very common among the various families of Entoprocta, and for some species may be the most common form of reproduction. In addition, a peculiar form of larval budding has been described for several loxosomatids (Jägersten, 1964; Franzén, 1967; Nielsen, 1966a).

Brien (1956, 1960) has discussed the budding of the three families of entoprocts from the view point of their evolution within the phylum. Regarding their adult morphology, Brien (1956) considers the solitary Loxosomatidae to have the most generalized and most primitive structure. The Pedicellinidae have a more complex stalk development, a junction between the calyx and stalk which allows for the loss and regeneration of the calyces and a ramifying stolon. The freshwater Urnatellidae, according to Brien, are the most highly evolved entoprocts and have carried even further what he feels are the significant evolutionary advances of the pedicellinids. He discusses the pattern of budding in each of the three families to further show these evolutionary trends, and each family will be considered in turn.

1.2.1.1 THE LOXOSOMATIDAE

In the Loxosomatidae, the buds arise either on the surface of or in depressions near the oral end of the calyx (e.g., Claparède, 1870; Atkins, 1932; Prenant and Bobin, 1956; Nielsen, 1964a) (Figs. 3 and 5). Generally one bud develops on each side, either simultaneously or alternately (Table I), but some loxosomatids such as *Loxosomella obesa*, *Loxosomella kefersteinii*, and *Loxosoma davenporti* may produce up to 6 per side for a total of 12 buds simultaneously (Nickerson, 1901; Atkins, 1932; Bobin and Prenant, 1954). *Loxosomella kefersteinii* is unusual for several other reasons as well: the buds are not attached at the base of the stalk, but by a small umbilicus located just below the calyx of the bud (Ryland and Austin, 1960). In addition, it is one of the few loxosomatids which may be found free-living and not associated symbiotically with a host (Nielsen, 1966b). Other loxosomatids which possess the distinctive umbilicus attaching the buds to the parent are *Loxosomella annulata* and *Loxosomella mepse*. Both of these species possess only two buds at most, with both restricted to only one side of the calyx in the case of *L. annulata* and one on each side in the case of *L. mepse*. The possession of an umbilicus led several authors to consider the possibility of establishing a new genus to contain these three species (Mortensen, 1911; Bobin and Prenant, 1954; Marcus, 1957). The record number of buds so far reported, however, is for *Loxosomella varians* with 15 (Nielsen, 1964a). In this case the buds are formed in a single group on the oral side of the calyx with new buds forming above the older ones, the total volume of which may far exceed that of the parent (Nielsen, 1964a).

In addition to slight differences in lateral positioning and arrangement, the buds may form at different levels of the calyx, which is often a species-specific characteristic (Bobin and Prenant, 1954; Brien, 1956).

In the case of loxosomatids, budding may be influenced by physiological conditions, such as the degree of nutrition, but does not appear to be strongly influenced by size, sex or age, since even immature buds may produce their own buds before being released from the parents. Although this may be an atypical situation (Atkins, 1932), there are many known examples of buds being produced on newly released juveniles which are not yet sexually mature (Brien, 1956).

The most extreme example of precocious budding is that exhibited by the larvae of some Florida Loxosomatidae, first described by Jägersten (1964). One or two small adults develop within pockets of the larvae of at least two species. The larva ruptures to release the miniature adults and then dies without ever undergoing the type of metamorphosis described by Harmer (1887) and Cori (1936) for *Pedicellina cernua*. The most striking phenomenon observed by Jägersten was that about half of these larval "buds" were sexually mature males with well-developed testes filled with sperms. One miniature adult, which was kept in a small bowl without food following release, was observed to produce its own small bud in 7 days.

Although Jägersten (1964) was unable to rear the adults released from his larvae and thus determine the species, Nielsen (1966a) working in the same area (Miami, Florida) collected a number of larvae which fit quite closely the description of Jägersten's "Type B" larvae. Nielsen was able to rear several of the larval-released adults and determined that they were a new species to which he gave the name *Loxosoma jaegersteni* (Fig. 3). With feeding, one specimen approximately doubled in size, grew three buds, and released its sperms all within 10 days (Nielsen, 1966a).

In addition, both Jägersten (1964) and Nielsen (1966a) have found that the larva of another loxosomatid species (corresponding to Jägersten's "Type A" larva) was capable of producing from two to four external buds. Neither Jägersten nor Nielsen was able to raise these successfully so the species remains unknown. Harmer (1885) apparently was the first to describe the external budding for the larva of what is presently known as *Loxocalyx leptoclini*, but it is not known whether Jägersten's and Nielsen's materials are referable to this species.

In addition to the above forms, Nielsen (1966a) has been able to follow the complete life cycle of another form with internal larval budding to which he has given the name *Loxosomella vivipara* (Fig. 6). In this species, however, the bud develops within the larva while the larva itself is developing in the ovary of its parent. By the time the larva is ready for release from its parent, the larval bud is essentially fully formed and nearly ready for release itself. The larva is then released and swims in the plankton for 1 to 3 days before releasing its fully formed bud through a rupture of its dorsal side, whereupon the larva dies (Fig. 6c). In contrast to the liberated larval buds of *Loxosoma jaegersteni*, those of *Loxosomella vivipara* have developing gonads, but are not sexually mature at the time of release. However, they may bear from one to two buds at the time of rupture from the planktonic larva (Nielsen, 1966a). Thus, up to four generations may be associated with a single parent: mother, larva developing in parent, bud developing in larva, and buds developing on bud contained within the larva. Since the life cycles of only a few of the some 90 species of loxosomatids are known, future work may reveal other equally dramatic types of reproduction.

Most of the individuals in a cluster of loxosomatids are of one sex (females, since males are rare in the case of *Loxocalyx neapolitanus*, *Loxosomella kefersteinii*, and *Loxosoma rhodnicola*, or nearly all males in the case of *Loxosomella crassicauda*, Franzén, 1962, Atkins, 1932), but few direct observations have been made to determine whether the buds are always of the same sex as the parents. Although such a distinction would perhaps be irrelevant in the case of protandric hermaphrodites, Atkins (1932) has reported that the buds of some individuals were not always the same sex as the parents, with some males producing female buds and some females producing male buds. However, since these individuals were from a cluster of individuals which Atkins (1932) considered to have retained their buds for an unusually long time, it is not clear what significance should be placed on these observations.

In the case of "normal" calyx budding of solitary loxosomatids, the buds tend to detach from the parent and become attached to the substrate in the vicinity of the parents by means of the pedal gland.

Just as there is no consistent pattern to bud development (i.e., they may form on the same side of the parental calyx or on alternate sides, either two at a time or with many in different stages of development), so there is no particular pattern to their detachment. Apparently, whenever development is complete, they either constrict or break away from the parent.

The development of buds and their detachment have been observed in Florida by Nielsen (1966b) for *Loxosomella kefersteinii* and in Great Britain by Ryland and Austin (1960). Nielsen (1966b) found that budding was quite rapid with buds being produced in only 5 to 6 days in laboratory aquariums. Ryland and Austin (1960), from field observations, similarly report that individuals of *L. kefersteinii* become sufficiently mature to produce buds in 2 weeks, although in some cases, buds may begin to develop on quite small individuals. When ready to detach, the large buds become very active and undergo extensive twisting and writhing movements which result in the breakage of the connection between bud and parent. This "detachment behavior" in aquariums was found to last up to an hour (Ryland and Austin, 1960). The newly detached buds of *L. kefersteinii* are unusual in that they are capable of swimming away from the parent in a fairly rapid and coordinated fashion. The locomotory power is apparently generated by the tentacle cilia directing the calyx anteriorly.

1.2.1.2 THE PEDICELLINIDAE

Pedicellinids bud primarily from the stolon but some species produce buds from the stalk as well (Fig. 1). No examples of calyx budding are known for the Pedicellinidae. As the stolons grow out over the substrate, new buds are produced at the growing tip. As the stolons and their developing buds proliferate, a colony develops, the members of which are all attached at their bases (Fig. 1). Buds are always perpendicular to the growing stolon and in some species a functional feeding individual may be developed within 36 hours (Mariscal, 1965). The buds appear in so-called fertile areas which are often rather evenly spaced along the stolon with infertile areas between (Brien, 1956). Members of the genus *Barentsia* are also capable of producing buds at the muscular enlargements on the stalk (Mariscal, 1965). In some cases, *B. benedeni* has been observed to initiate stolons from muscular nodes on the stalk, from which new feeding individuals are then produced in the usual manner (Fig. 1). Thus, if the lower portions of the stalks of a large colony become buried in silt, the colony may continue its growth above the surface of the original substrate.

1.2.1.3 THE URNATELLIDAE

Although *Urnatella* does not possess a well-developed stolon, a basal plate is present from which the stalks of several adults may arise to form what might be considered a small colony (Fig. 4). The basal plate appears to be homologous to the stolon of the pedicellinids but does not ramify out over the substrate as is common in the latter group. Stalk budding, which appears to be of somewhat minor significance in the Pedicellinidae, has been utilized as the major form of asexual reproduction in *Urnatella* (Davenport, 1893). New buds commonly proliferate from the more distal regions of the stalk so that a single large stalk may bear a number of secondary calyces in addition to the primary one. Davenport (1893) has discussed the several types of budding for *Urnatella*.

1.2.1.4 THE HISTOLOGY OF BUD FORMATION

The histological aspects of budding have been described by Hatschek (1877), Joliet (1877), Schmidt (1876, 1878), Salensky (1877), Seeliger (1890), Prouho (1891), Davenport (1893), Nickerson (1901), and Brien (1956, 1959), among others. Although Hatschek (1877) believed that all three embryonic layers were involved in the budding process, both Joliet (1877) and Schmidt (1876, 1878) consider budding to be primarily a mesodermal process. However, Seeliger (1889, 1890) appears to have given the most correct description of the process with both the mesoderm and ectoderm participating. At present, there appear to be no significant differences between budding in the Pedicellinidae, Loxosomatidae, and Urnatellidae; the following description thus applies to all three groups.

Budding begins by a thickening of the ectoderm. The increased rate of division causes the cells to push against each other to form an ectodermal outpocketing of the epidermis containing undifferentiated mesenchyme cells. A mass of the ectodermal cells then pinches off to form a ball just beneath the growing tip of the new bud. Two cavities separated by a thin, but incomplete wall of cells then begin to form within the ball of the cells. The distal-most cavity, still roofed over by the layer of cells which surrounds the ball, as well as the ectoderm of the growing tip of the bud, will go on to form the atrial cavity while the lower cavity will become the digestive tract. The small break in the wall of cells separating the two cavities will become the mouth of the adult and the anus will develop as a secondary opening into the atrial cavity. Many mesenchymal cells then move in beneath the rmal ball of cells, and proliferate to fill what will become the stalk of the adult. The tentacles, body wall, digestive tract, and subenteric ganglion all develop from the ectodermal precursors, while the muscles and gonads arise from the mesenchymal elements. A plug of mesodermal cells tends to mark the first indication of separation of the calyx from the stalk and this is shortly followed by an ingrowth of the cuticle and epidermal cells to form a definite constriction. The ectodermal cells roofing over the atrium divide to form fingerlike downgrowths which eventually become the tentacles. Finally the atrial cavity breaks through to free the ciliated tentacles and the young bud is ready to begin feeding. In the case of the loxosomatids, the pedal gland is another epidermal derivative.

1.2.1.5 REGENERATION

Another common form of what might be considered a type of asexual reproduction is regeneration of new calyces upon old stalks. This is very common among the pedicellinids (Nasonov, 1926a,b; Valkanov, 1951; Mariscal, 1965). In the case of *Barentsia benedeni* from San Francisco Bay, the onset of unfavorable environmental conditions (most noticeably high water temperatures) caused a rapid disintegration of the calyces in all the large reproductive colonies examined. However, within 1 month of placing these colonies in a running seawater system, about 10°C cooler, all stalks completely regenerated new feeding calyces. Many large active larvae were also present in the calyces, indicating that the colonies had become sexually active as well (Mariscal, 1965).

Although specialized asexual brooding stages are well known for the Ectoprocta, these are relatively rare in the case of the Entoprocta. However, Valkanov (1951), Toriumi (1951), and Emschermann (1961) have all described what are variously called resting buds, brood bodies, or hibernacula for certain pedicellinids. These consist of large, thickened, fingerlike outgrowths of the base of the stolon which, with a change in environmental conditions, are capable of producing new feeding individuals. Similarly, the freshwater *Urnatella* possesses a statoblastlike overwintering stage which regenerates a new individual in the spring (Davenport, 1893).

1.2.2 Factors Influencing Asexual Reproduction

Very little information is available regarding the specific factors, either environmental or physiological, responsible for controlling asexual reproduction. Bobin and Prenant (1954) suggest that the number and position of the buds of the solitary loxosomatids are strongly influenced by various physiological conditions, especially feeding, and that other factors such as the sex of the individuals, sexual maturity, and size do not appear significant. In general, one can say that with favorable environmental conditions and a steady supply of food, entoprocts, like other asexually reproducing animals, will continue to bud, often at a very rapid rate (Nielsen, 1966b; Ryland and Austin, 1960).

It is also clear that with the onset of unfavorable environmental conditions (e.g., high temperature), not only will budding cease, but the adult calyces degenerate and the active life of the colony or individual comes to a halt. When conditions again become favorable the stalks of colonial entoprocts will regenerate new calyces and budding will recommence (Mariscal, 1965).

One interesting problem among the colonial pedicellinids which might be mentioned here, is the determination of the sex of the newly formed buds. Here a distinction has to be made between the sex of the adult individuals of the colony and the sex of the colony as a whole. For example, among most of the Pedicellinidae the sexes of individuals and the colony appear to be separate (i.e., dioecious or unisexual) (Table I). In the case of the genus *Myosoma,* however, the adult individuals of both species so far described are dioecious, but the colony is monoecious, i.e., hermaphroditic, with both sexes represented. This has also been reported for *Barentsia macropus* by Ehlers (1890) and for *Pedicellina cernua* (= *americana*) by Dublin (1905a). However, other examples of supposedly the same species (*Pedicellina cernua*) have been variously reported as being either hermaphroditic or dioecious (both the individual zooids and the colony). One possible source of confusion among various authors in this regard is the distinct possibility that protogyny or protandry is involved in the life cycle of some species of entoprocts as discussed by Dublin (1905a). Since very few species have been checked at regular intervals over long periods of time, it would be easy to overlook the possibility of an alternation of male and female individuals within a single species.

Nothing is known of the physiological or environmental factors influencing the production of both male and female buds simultaneously along a single stolon or on a single loxosomatid (Atkins, 1932), nor of the factors which might be controlling protandry or protogyny within a single colony or individual. Nielsen (1966a) has been able to directly observe protandric hermaphroditism in *Loxosomella vivipara,* although the controlling factors remain unknown.

TABLE I

List of Reproductive Data for Nearly All Described Species of Entoprocts[a]

Species	Number of buds and budding time	Sex of adults and breeding time	Number of embryos and time in brood pouch	Locality	References
Family Loxosomatidae					
Genus Loxosoma					
1. L. annelidicola	2–4, NK	Di, NK	NK, NK	France, Wales	Saint Joseph, 1899; Prouho, 1891; Ryland and Austin, 1960
2. L. agile	1, Ma–Jn	NK, NK	2, Ma	Denmark	Nielsen, 1964a
3. L. breve	1, NK	Di, NK	NK, NK	Malaysia	Harmer, 1915
4. L. circulare	3, NK	Di, NK	2, NK	Malaysia	Harmer, 1915
5. L. cirriferum	3, NK	Di, NK	6–7, NK	Indonesia	Harmer, 1915
6. L. claparedi	2, A	Di, A	NK, NK	France	Bobin and Prenant, 1953f
7. L. cocciforme	1, NK	Di, NK	3, NK	Malaysia	Harmer, 1915
8. L. davenporti	12, NK	Mo, (Pg), Ju–Au	3, O	Massachusetts	Nickerson, 1898, 1899, 1900, 1901; Nielsen, 1966b
9. L. jaegersteni	3, M	Mo?(Pd?), NK	NK, NK	Florida	Nielsen, 1966a
10. L. lanchesteri	4, NK	Di, NK	NK, NK	Malaysia	Harmer, 1915
11. L. loricatum	1, NK	NK, NK	NK, NK	Malaysia	Harmer, 1915
12. L. loxalina	NK, NK	MO?, NK	NK, NK	England, France	Assheton, 1912; Prenant and Bobin, 1956
13. L. monensis	2, NK	NK, NK	NK, NK	Wales	Eggleston, 1965
14. L. pectinaricola	6, A–F	Di, NK	10, Jn–F	Baltic	Franzén, 1962; Nielsen, 1964a
15. L. pusillum	4, NK	Di, NK	1, NK	Malaysia	Harmer, 1915
16. L. rhodnicola	2, Jn–S	Di, NK	12, Jn–Au	Baltic	Franzén, 1962; Nielsen, 1964a
17. L. saltans	2, NK	Di, NK	NK, NK	England	Assheton, 1912
18. L. significans	2, Jn–Au	NK, NK	3, Jn–Au	Denmark	Nielsen, 1964a
19. L. singulare	2, Au	Di, M–Au	1, NK	France	Keferstein, 1862; Bobin and Prenant, 1953a
20. L. sluiteri	NK, NK	NK, NK	NK, NK	Malaysia	Harmer, 1915
21. L. spathula	2, N	NK, NK	O, N	North Carolina	Nielsen, 1966b

22. L. subsessile	2, NK	NK, NK	NK, NK	Malaysia	Harmer, 1915
23. L. troglodytes	NK, NK	NK, NK	NK, NK	Malaysia	Harmer, 1915
24. L. velatum	3, NK	Di, NK	NK, NK	Malaysia	Harmer, 1915
Genus Loxosomella					
1. L. annulata	2, NK	NK, NK	NK, NK	Malaysia	Harmer, 1915
2. L. antedonis	5, J–Ju	NK, NK	NK, NK	Greenland, Wales	Mortensen, 1911; Ryland and Austin, 1960
3. L. aripes	3, Au–S	NK, NK	3, S	Norway	Nielsen, 1964b
4. L. arynae	1, A	NK, NK	NK, NK	France	Bobin and Prenant, 1953b
5. L. atkinsae	4, A–S	NK, NK	1, Ma–Ju	Europe	Bobin and Prenant, 1953b; Nielsen, 1964a; Eggleston, 1965
6. L. bilocata	3, D, F–M	NK, NK	1, M	Florida	Nielsen, 1966b
7. L. bimaculata	2, NK	NK, NK	NK, NK	Puerto Rico	Rützler, 1968
8. L. bocki	2, NK	Di, NK	NK, NK	Gilbert Islands	Franzén, 1967
9. L. bouxini[b]	4, NK	Di, NK	NK, NK	France	Bobin and Prenant, 1953e
10. L. brucei	2, Ma–Au	NK, NK	7, NK	Britain	Eggleston, 1965
11. L. brumpti	NK, NK	NK, NK	NK, NK	Arctic	Nilus, 1909
12. L. claviformis	4, Au, N	Di, NK	6, Au–S, N, F	Europe	Atkins, 1932; Bobin and Prenant, 1953c; Eggleston, 1965
13. L. compressa	2–3, Ju–N	NK, O	NK, Ju, O	Norway	Nielsen and Ryland, 1961
14. L. constricta	NK, NK	NK, NK	NK, NK	South Africa	O'Donoghue, 1924
15. L. crassicauda	7–8, J–D	Mc?(Pd?), Ma	NK, NK	Europe	Salensky, 1877; Atkins, 1932
16. L. cricketae	3, N	NK, NK	NK, N	North Carolina	Nielsen, 1966b
17. L. cuenoti	2, NK	Di, NK	NK, NK	France	Bobin and Prenant, 1953e
18. L. discopoda	3, NK	NK, NK	NK, NK	Norway	Nielsen and Ryland, 1961
19. L. elegans	3, J–D	NK, NK	10, Jn–J	Denmark	Nielsen, 1964a
20. L. fagei	2, NK	NK, NK	NK, NK	France	Bobin and Prenant, 1953g, 1954
21. L. fauveli	3–6, M–N	Di, Ma	10, M–N	Europe	Atkins, 1932; Bobin and Prenant, 1953a; Nielsen, 1964a
22. L. follicola	3, F–M	NK, NK	NK, NK	Florida	Nielsen, 1966b
23. L. fungiformis	2, Au	Di, NK	NK, Au	France	Bobin and Prenant, 1953e
24. L. gauteri	NK, NK	Di, NK	NK, NK	Mediterranean	Bobin and Prenant, 1953c

(Continued)

TABLE I (*Continued*)

Species	Number of buds and budding time	Sex of adults and breeding time	Number of embryos and time in brood pouch	Locality	References
25. *L. glandulifera*	2, Ma–Ju	Di, NK	2, Ma–Ju, Au–O	Baltic	Nielsen, 1964a; Franzén, 1962
26. *L. globosa*	2, NK	Di, NK	NK, NK	Mediterranean	Bobin and Prenant, 1953c
27. *L. harmeri*	9, J–F	NK, NK	12, M–D	Denmark	Nielsen, 1964a
28. *L. illota*	3, F	NK, F	NK, NK	Florida	Nielsen, 1966b
29. *L. kefersteinii*	12, F, Ju–N	NK, NK	NK, NK	Mediterranean, Wales, Florida	Claparède, 1867; Nitsche, 1875; Ryland and Austin, 1960; Nielsen, 1966b
30. *L. macginitieorum*	2, NK	NK, NK	NK, NK	California	Soule and Soule, 1965
31. *L. marsypos*	1, O, J	NK, O, J	3, Ju, O	Baltic	Nielsen, 1964a; Nielsen and Ryland, 1961
32. *L. mepse*	2, NK	Di, NK	NK, NK	Brazil	Marcus, 1950, 1957
33. *L. minuta*	2, NK	NK, NK	NK, NK	Massachusetts	Schopf and Simon, 1965
34. *L. murmanica*[c]	2, A, Au	Mo, Au	NK, NK	Europe	Bobin and Prenant, 1953b; Eggleston, 1965; Nielsen, 1964a
35. *L. museriensis*	4, NK	Di, NK	4, NK	Red Sea	Bobin, 1968
36. *L. nitschei*	5, J–D	NK, NK	7, J–D	Europe	Bobin and Prenant, 1953b; Arvy and Prenant, 1952; Nielsen, 1964a; Ryland, 1961
37. *L. nordgaardi*	5, Ma–J	Di, Ma–Jn	5, Ma–Jn	Baltic	Ryland, 1961; Nielsen, 1964a
38. *L. obesa*	12, F–Jn	Di, NK	26, F–A, S	Britain	Atkins, 1932; Eggleston, 1965
39. *L. olei*	12, NK	Di, NK	NK, NK	Brazil	Marcus, 1957
40. *L. ornata*	4, J, M, Jn–N	NK, NK	4, Au, O–N	Denmark	Nielsen, 1964a
41. *L. parguerensis*	2, A	NK, NK	NK, NK	Puerto Rico	Rützler, 1968
42. *L. phascolosomata*	6, Jn–S	Di, Jn–S	12, Ju	Europe	Vogt, 1876; Atkins, 1927, 1932; Nielsen, 1964a
43. *L. polita*	2, J–F	NK, NK	2–3, M–J	Denmark	Nielsen, 1964a
44. *L. prenanti*	2, NK	NK, NK	NK, NK	California	Soule and Soule, 1965
45. *L. scaura*	2, Jn–Au	NK, NK	7, Jn–Au	Denmark	Nielsen, 1964a

Species					References
46. *L. shizugawaensis*	5, NK	NK, NK	NK, NK	Japan	Toriumi, 1949
47. *L. similis*	2, Ma	NK, NK	2, NK	Denmark	Nielsen, 1964a
48. *L. tedaniae*	3, D	NK, NK	NK, NK	Bermuda	Rützler, 1968
49. *L. triangularis*	3, M	NK, NK	1, M	Florida	Nielsen, 1966b
50. *L. varians*	15, J–O	NK, NK	5, J–N	Denmark, Massachusetts	Nielsen, 1964a; Schopf and Simon, 1965
51. *L. vivipara*	5, N–M	Mo?, N–J	2, J–M	Florida	Nielsen, 1966a
52. *L. worki*	3, D, F–M	NK, NK	NK, NK	Florida	Nielsen, 1966b
Genus *Lozocalyx*					
1. *L. alatus*	1, NK	NK, NK	NK, NK	France	Barrois, 1877; Prenant and Bobin, 1956
2. *L. cochlear*	2, D–M	Mo, NK	NK, NK	Italy, France	Schmidt, 1876; Prenant and Bobin, 1956
3. *L. leptoclini*	3, M–Jn	Di, NK	2–3, NK	Malaysia, Mediterranean	Harmer, 1885; Harmer, 1915; Prenant and Bobin, 1956
4. *L. lineatus*	2, NK	NK, NK	NK, NK	Malaysia	Harmer, 1915
5. *L. mortenseni*	NK, NK	NK, NK	NK, NK	Brazil	Marcus, 1950
6. *L. neapolitanus*	2, J–F	Di, J–F	2 NK	Mediterranean	Kowalewsky, 1866; Prenant and Bobin, 1956
7. *L. pes*	2, NK	Mo,(Pd), M	NK, M	Mediterranean	Schmidt, 1878; Prenant and Bobin, 1956
8. *L. raja*	2, D–M	Mo?	NK, NK	Mediterranean	Schmidt, 1876; Prenant and Bobin, 1956
9. *L. sawayai*	2, NK	NK, NK	NK, NK	Brazil	Marcus, 1939
10. *L. studiosorum*	2, NK	NK, NK	NK, NK	Japan	Toriumi, 1951
11. *L. teisseri*	4, A	Di?	NK, NK	France	Bobin and Prenant, 1953d; Prenant and Bobin, 1956
12. *L. tethyae*	6, N	Di, S	2, N	North Carolina, Mediterranean	Nielsen, 1966b; Bobin and Prenant, 1953d; Prenant and Bobin, 1956
Genus *Lozomespilon*					
1. *L. perezi*	4, A, Au–S	Di, Au–S	6, NK	France	Bobin and Prenant, 1953f; Gruijs-Faucher, 1959; Bobin, 1969; Prenant and Bobin, 1956

(Continued)

TABLE I (Continued)

Species	Number of buds and budding time	Sex of adults and breeding time	Number of embryos and time in brood pouch	Locality	References
Genus Loxostemma					
1. L. tetracheir	1–2, D, M	NK, NK	1, D, M	Florida	Nielsen, 1966b
Family Pedicellinidae					
Genus Pedicellina					
1. P. australis	NK, NK	NK, NK	NK, NK	Chile	Ridley, 1881
2. P. cernua[d]					
(= P. americana)	NK, Mo	Di(Pd?), Jn–Au	NK, NK	New York	Dublin, 1905a
(= P. echinata)	NK, Mo	Mo, Au–N	NK, NK	Mediterranean	Ehlers, 1890; Nitsche, 1869; Hatschek, 1877
(= P. echinata)	NK, Di	Di, NK	NK, NK	Belgium	Foettinger, 1887
(= P. glabra)	NK, Di	Di, Ma–S	NK, NK	France	Joliet, 1877; Cuenot, 1899
3. P. choanata	NK, NK	NK, NK	NK, NK	South Africa	O'Donoghue, 1924
4. P. compacta	NK, Di	Di, NK	NK, NK	Malaysia	Harmer, 1915
5. P. grandis	NK, NK	NK, NK	NK, NK	New Zealand	Ryland, 1965
6. P. hispida	NK, Di	Di, NK	NK, NK	New Zealand	Ryland, 1965
7. P. nannoda	NK, NK	NK, NK	NK, NK	Brazil	Marcus, 1937
8. P. nutans	NK, NK	NK, NK	NK, NK	France	Prenant and Bobin, 1956; Ryland, 1960
9. P. pernae	NK, NK	NK, NK	NK, NK	Wales	Ryland, 1965
10. P. pyriformis	NK, NK	NK, NK	NK, NK	New Zealand	Ryland, 1965
Genus Barentsia[e]					
1. B. benedeni[f]	J–Ma, Di	Di, F–Jn	8, M–Ju	California, Belgium	Mariscal, 1965; Foettinger, 1887
2. B. discreta[g]	NK, Di	Di, NK	NK, NK	Malaysia, Japan	Harmer, 1915; Oka, 1895
3. B. geniculata	NK, Di	Di, NK	1	Malaysia	Harmer, 1915
4. B. gorbunovi	NK, NK	NK, NK	NK, NK	Arctic	Kluge, 1946; Osburn, 1953
5. B. gracilis	NK, Di	Di, J–D	NK, S	Northern Europe, Malaysia, Brazil	Joliet, 1877; Foettinger, 1887; Harmer, 1915; Marcus, 1939

6. B. laxa	NK, Di	Di(?), NK	9, Ju–O	Massachusetts, Malaysia, Japan	Rogick, 1948; Harmer, 1915; Toriumi, 1951
7. B. macropus	NK, Mo	Di, M–A	NK, NK	Spain, France	Ehlers, 1890; Cuenot, 1899
8. B. matsushimana	NK, NK	NK, NK	NK, A	Japan	Toriumi, 1951
9. B. mutabilis	NK, NK	NK, NK	NK, NK	Japan, Wales	Toriumi, 1951; Ryland, 1960
10. B. robusta	NK, NK	NK, NK	NK, Au	Canada	Osburn, 1953
11. B. stiria	NK, NK	NK, NK	NK, NK	Spain	Prenant and Bobin, 1956
12. B. subrigida	NK, NK	NK, NK	NK, NK	California	Osburn, 1953
Genus Chitaspis	NK, NK	NK, NK	NK, NK	India	Annandale, 1916
Genus Coriella[h]					
1. C. stolonata	NK, NK	NK, NK	NK, NK	Arctic	Kluge, 1946; Osburn, 1953
Genus Loxosomatoides					
1. L. colonialis	NK, NK	NK, NK	NK, NK	India	Annandale, 1908, 1915
2. L. evelinae	NK, Mo	Di, NK	NK, NK	Brazil	Marcus, 1939
3. L. japonicum	NK, NK	NK, NK	NK, NK	Japan	Toriumi, 1951
Genus Myosoma					
1. M. hancocki	NK, Mo	Di, NK	NK, NK	Philippines	Soule, 1955
2. M. spinosa	NK, Mo	Di, NK	NK, NK	California	Robertson, 1900
Genus Sangavela					
1. S. vineta	NK, NK	Pd(?), NK	NK, NK	Brazil	Marcus, 1957
Family Urnatellidae					
Genus Urnatella					
1. U. gracilis	NK, Di	Di, NK	NK, NK	Pennsylvania, Europe	Leidy, 1883; Davenport, 1893; Sebesteyén, 1962; Emschermann, 1965
2. U. indica	NK, NK	Mo, NK	NK, NK	India	Seshaiya, 1949

[a] Abbreviations: Di, dioecious; Mo, monoecious; Pg, protogyny; Pd, protandry; NK, not known; J, January; F, February; M, March; A, April; Ma, May; Jn, June; Jl, July; Au, August; S, September; O, October; N, November; D, December.

[b] Loxosomella bouxini considered by Nielson, 1964a, to be a synonym of L. atkinsae.

[c] Nielsen (1964a), suggests that L. murmanica may be synonymous with L. nitschei.

[d] Pedicellina cernua is probably a protandric hermaphrodite, thus the reason for the variety of seemingly conflicting reports.

[e] Ryland (1965) considers the genus Pseudopedicellina Toriumi to be synonymous with Barentsia.

[f] Emschermann has been able to confirm that the San Francisco Bay species, originally identified as B. gracilis, is in fact, B. benedeni.

[g] Barentsia misakiensis synonymous with B. discreta.

[h] Because of the presence of a well-developed basal swelling on the stalks of individual Coriella zooids, this genus might be considered another variety of Barentsia.

1.3 Sexual Reproduction

1.3.1 Sexual Dimorphism

In general, male entoprocts cannot reliably be distinguished from females without sectioning unless embryos are visible in the brood pouch of females or the gonads are visible through the body wall. Some workers have noticed a size sexual dimorphism for some species, for example, females of *Loxosoma annelidicola* and *Loxosoma pectinaricola* are smaller than the males (Prenant and Bobin, 1956; Franzén, 1962). However, the males of *Loxosoma cirriferum* are reported to be smaller than the females (Harmer, 1915). In the case of some pedicellinids, the females often appeared to have larger calyces when the brood pouch was distended with developing larvae, but this may not have any direct relationship to the overall length of the adult animal which is usually used in size determinations.

1.3.2 Sex Determination and Hermaphroditism

Although the majority of entoproct species appear to be dioecious, some hermaphroditic species have been reported for both the loxosomatids and the pedicellinids (Table I). As previously mentioned (Section 1.2.2), one of the most interesting problems is the mechanism of sex determination of a colony in which both male and female individuals are produced along a single stolon. Another interesting problem involves the various factors controlling protandric hermaphroditism in some Entoprocta. Both problems are amenable to study and might provide information applicable to the general area of sex determination. Atkins (1932) discusses some of these problems for various loxosomatids.

1.3.3 Anatomy of the Reproductive System

The reproductive tract in dioecious entoprocts is relatively simple. In both males and females the two saclike gonads lie on either side of the body between the stomach and the wall of the atrial cavity (Becker, 1937; Mariscal, 1965). They connect to the exterior by a single straight duct which opens into the atrium (Figs. 2, 8, 9, and 10). The ovaries generally contain several primary oocytes, each in varying stages of development (Fig. 8). The testes may become quite extensive, often larger than the ovaries of the female, and occupy a large area

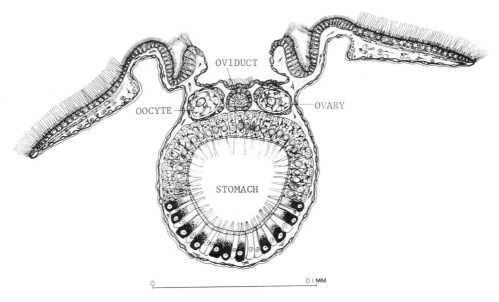

OVIDUCT

OOCYTE

OVARY

STOMACH

0.1 MM.

FIG. 8. Transverse section through the calyx of an adult female pedicellinid entoproct (*Barentsia benedeni*) showing the ovaries containing nearly full grown oocytes (from Mariscal, 1965).

on the sides of the calyx (Franzén, 1962; Mariscal, 1965) (Fig. 9). A large number of glandlike cells may lie along the gonoduct (Fig. 10) and apparently secrete the "vitelline" membrane which anchors the embryo in the brood pouch and appears to form the cuticle of the larva and presumably the adult (Mariscal, 1965; Atkins, 1932) (Figs. 11, 12, 13). The cuticle is capable of growth and regeneration because of the presence of long epidermal microvilli which penetrate it (Mariscal, 1965).

In hermaphroditic species, the overall anatomy is similar to that shown in Fig. 10, with the exception that both the testes and ovaries empty their contents into the common gonoduct with the ovaries lying anterior to the testes. It is not known for certain whether self-fertilization occurs in such cases.

1.3.4 Cytodifferentiation of the Gametes

Gametogenesis has been little studied in the Entoprocta. Dublin (1905a,b) working with *Pedicellina cernua* (= *americana*) completed the first and probably most extensive study of both oogenesis and sperma-

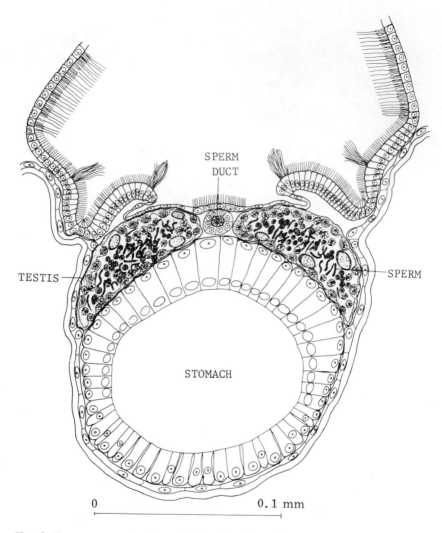

FIG. 9. Transverse section through the calyx of an adult male pedicellinid entoproct (*Barentsia benedeni*), showing the large, paired testes (from Mariscal, 1965).

togenesis (Fig. 7). Gametogenesis has also been discussed by Atkins, 1932; Franzén, 1956, 1962; and Gruijs-Faucher, 1959, among others.

According to Dublin (1905a), both oogenesis and spermatogenesis are quite similar. The primordial germ cells are located next to the oviduct, and give rise by means of repeated divisions to form several generations of oogonia and spermatogonia. The chromosome number of *P. cernua* appears to be 22, and with the exception of the last generation

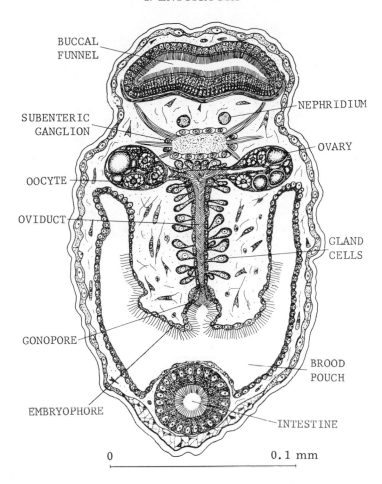

BUCCAL
FUNNEL

SUBENTERIC
GANGLION

OOCYTE

OVIDUCT

GONOPORE

EMBRYOPHORE

NEPHRIDIUM

OVARY

GLAND
CELLS

BROOD
POUCH

INTESTINE

0   0.1 mm

Fig. 10. Frontal section of the calyx of an adult female pedicellinid entoproct (*Barentsia benedeni*), showing the connection of the ovaries with the oviduct and its relationship to the brood pouch (from Mariscal, 1965).

of spermatogonia, the chromosomes are **V**-shaped and similar to those in the somatic cells.

In the case of *Loxosoma rhodnicola*, the mature sperm is filiform with a total length of about 70 μm and a diameter of about 1 μm (Franzén, 1956, 1962). The middle piece is composed of a mitochondrial layer surrounding the axial filament, the mitochondrial layer having developed from two distinct strands which migrate posteriorly along with the growing axial filament. The number of eggs and oocytes varies in *Loxosoma rhodnicola,* presumably depending on the age and physiologi-

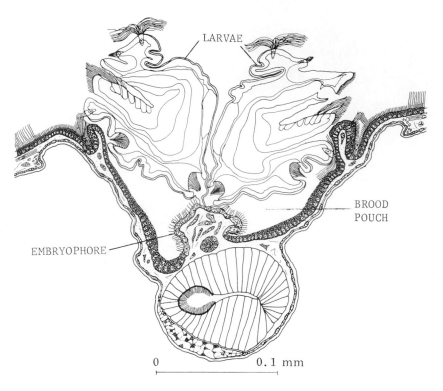

FIG. 11. Oblique transverse section through the brood pouch of an adult female pedicellinid entoproct (*Barentsia benedeni*), showing the position of two fully developed larvae and their connection with the embryophore in the brood pouch. Note that the left-hand larva was fixed in the act of lifting food from the food groove of the parent by means of its ciliary girdle cilia (from Mariscal, 1965).

cal condition of the adult, although the specific factors remain unknown. The young primary oocytes have a flattened, layered arrangement in the ovary, but the mature ova, which are very yolky, are nearly spherical with a diameter of about 80 μm. The growing primary oocytes tend to be surrounded by ovarian epithelial cells and are eventually expelled into the vestibule or atrial cavity. They are surrounded at this time by a thin membrane secreted by the gland cells of the oviduct (Fig. 13). The developing embryos remain attached to a projection (embryophore) in the brood chamber by this secreted membrane.

Nielsen (1966a) reports that initially the zygotes are very small (about 6 μm in diameter) in *Loxosomella vivipara* and that the embryos increase in size at the expense of nutrient from the surrounding ovarian cells which act as follicle cells.

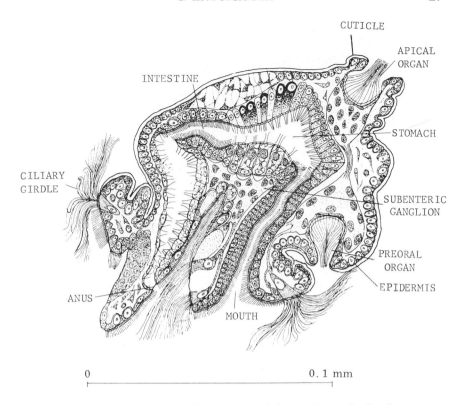

FIG. 12. Sagittal section of the fully developed larva of a pedicellinid entoproct (*Barentsia benedeni*) shown in swimming position (from Mariscal, 1965).

Emschermann (1965) studied sexual reproduction in *Urnatella gracilis* and found that only a single ovum develops in the ovary at any one time.

1.3.5 Gametogenic Cycles

Nothing known.

1.3.6 Factors Influencing Gametogenesis

Nothing known.

1.3.7 Reproductive Behavior

Not applicable.

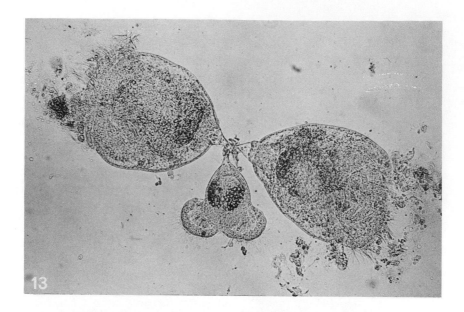

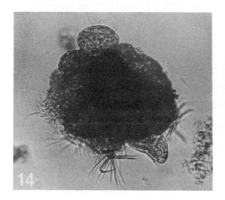

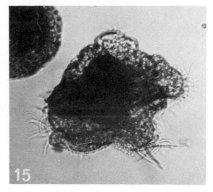

FIG. 13. Photomicrograph of three developing larvae removed from the brood pouch of a female pedicellinid entoproct (*Barentsia benedeni*) showing the common cuticular connection by which all three larvae were joined to the embryophore of the parent (photo by R. N. Mariscal).

FIG. 14. Photomicrograph of the left side of a free-swimming *Barentsia benedeni* larva. Note the cilia of the preoral organ extending from the anterior side of the larva (to left) and those of the apical organ extending from the top of the larva. This animal is about 100 μm long (photo by R. N. Mariscal).

FIG. 15. Photomicrograph of the right side of the same larva shown in Fig. 14. Note the extruded preoral organ (to the right). Periodically, while searching for a settling spot, the larva would stop swimming, extend its preoral organ momentarily and then continue on (scale approximately the same as in Fig. 14) (photo by R. N. Mariscal).

1.3.8 Spawning and Mating

1.3.8.1 MECHANISMS

Sperm release occurs when the adult contracts its tentacles and calyx sharply to emit a dense cloud of spermatozoa from the sperm duct into the surrounding medium by way of the vestibular cavity (Dublin, 1905a). In *Pedicellina cernua,* the calyx bends away from the area in which the sperms are released, decreasing the chance of intake of sperms in the feeding current by the male. Fertilization apparently occurs either in the ovaries or oviduct of both the dioecious and hermaphroditic species (Marcus, 1939; Nielsen, 1966a). Although sperms have been found in a "seminal vesicle" of the protandric hermaphrodite entoproct, *Loxosomella vivipara,* it is not clear whether self-fertilization occurs in this or in other hermaphroditic species (Nielsen, 1966a). The eggs are brooded by many, if not all species of entoprocts, until the feeding larvae are relatively large and capable of locomotion on their own.

Embryos are released by a rapid and repeated contraction of the tentacles and calyx which appears to tear loose the connection of the larva to the embryophore (Mariscal, 1965).

1.3.8.2 BREEDING PERIOD

Table I summarizes what little information is available on the breeding times of entoprocts. In most of these cases, however, the breeding times are indicative only of the dates of collection; very few entoprocts have been followed throughout the year to provide us with reliable seasonal data. In some, such as *Loxosomella nitschei* and *Loxosomella varians* from Denmark, embryos have been found in the brood pouch from January to November or December, suggesting that in some species sexual reproduction may occur throughout the year (Nielsen, 1964a). Nielsen (1964a) appears to be the only investigator who has attempted to tabulate the production of both buds and embryos of a large number of entoprocts from a single geographical area. His data indicate that many of the loxosomatids from Danish waters probably reproduce throughout the year with perhaps a peak in the fall. However, more data are necessary.

1.4 Development

1.4.1 Embryonic Development

Surprisingly, there have been a number of works on the embryology of entoprocts including those of Vogt, 1876; Barrois, 1877; Harmer, 1885;

Kowalewsky, 1866; and Nielsen, 1966a for the loxosomatids, and Barrois, 1875, 1877, 1882; Hatschek, 1877; Vogt, 1876, 1878; Uljanin, 1869; Balfour, 1885; Harmer, 1887; Lebedinsky, 1905; Seeliger, 1906; Cwiklitzer, 1909; and Marcus, 1939 for the pedicellinids. Most studied is the embryology of *Pedicellina cernua* upon which the following discussion is mainly based.

Following fertilization, the zygote of *P. cernua* is surrounded by the cuticular envelope secreted by the glands lining the oviduct. The yolky egg, about 40 to 60 μm in diameter, is discharged from the oviduct to be attached to the embryophore at the base of the atrium, where development is completed.

The egg of *P. cernua* displays spiral determinate cleavage which is total and unequal, similar, in general, to that of the polychaete annelids, molluscs, and others which possess a trochophore-type larva (Brien and Papyn, 1954). Although the third cleavage tends to be more or less equal, it is still possible to distinguish the four micromeres of the animal pole and the four macromeres of the vegetal pole. Intermediate between these two layers will develop the trochoblast cells which will go on to form the ciliated girdle of the mature larva. Altogether there are produced five quartets of micromeres and one quartet of macromeres. The first three of these quartets are presumptive ectoderm, the latter three, entoderm. The 4d cell appears to be the mesentoblast cell which gives rise to the mesoderm cells which will later form the ento-mesoderm. Although some workers have described, at the moment of gastrulation, a pinching off (to left and right) of two teloblast cells at the vegetal pole to form the mesoderm, Marcus (1939) considers these two cells to be ectodermal and not involved in mesoderm formation. A small hollow coeloblastula develops which becomes invaginated to form the gastrula.

At about the 90-cell stage, gastrulation takes place as the vegetal pole cells become less rounded and form a flattened antero-posterior axis. The vegetal pole cells in this region then invaginate initiating gastrulation, at which time the coeloblastic cavity is obscured. The ventral slitlike blastopore then closes and a new opening develops nearby which becomes the stomodeum. From the stomodeal invagination, cells are pinched off which contribute to the mesenchyme pool. The new stomodeum then joins the previously invaginated archenteron which in turn links up with a secondary proctodeal invagination to form the larval gut. The walls of the heavily ciliated larval gut then differentiate into stomach and intestine with the stomodeum forming the esophagus and the proctodeum forming the rectum. During this time a groove has formed on the ventral surface of the embryo between the mouth and anus which expands to form the vestibule or atrial cavity of the larva. This in turn later becomes the brood pouch of the adult female.

Shortly after the formation of the vestibule, the apical plate cells, located dorsally, invaginate to form the heavily ciliated and innervated apical organ (Fig. 12). In addition, another thickened area of ectoderm invaginates anteriorly to become the eversible preoral organ with the same configuration as the apical organ (Figs. 12, 14, and 15). From the ectodermal cells lining the vestibule, the subenteric ganglion presumably arises. The mesenchyme cells appear to give rise to the muscles and connective tissue, while the protonephridia appear to arise from the ectoderm. Finally a girdle of locomotory cilia develops from the trochoblast cells around the equator of the larva and the larva is ready to break free from the embryophore to take up a free-swimming existence. Prior to release from the parent, the larva has been described as feeding on nutritive cells derived from the walls of the parent's brood pouch (Cori, 1936). However, the larva has also been observed to filter-feed not only from the surrounding medium, but also from the food groove of the parent (Fig. 11) (Mariscal, 1965).

1.4.2 Larvae

1.4.2.1 TYPES

Reid (1845), Van Beneden (1845), Uljanin (1869), Hincks (1873), Barrois (1875, 1877), Joliet (1877), Salensky (1877), Seeliger (1906), Cwiklitzer (1909), Cori (1936), Marcus (1939), Toriumi (1944), Rogick (1948), Mariscal (1965), and Franzén (1970) have all described or considered various pedicellinid larvae. The loxosomatid larvae have been studied by Vogt (1876), Barrois (1877), Fewkes (1884), Atkins (1932), Cori (1936), Jägersten (1964), Nielsen (1966b, 1967a,b) and Franzén (1970), among others. The Urnatella larva has been described by Emschermann (1965).

The larval entoproct is often described as a "modified trochophore" but in the case of some of the pedicellinid larvae, at least, the larva is perhaps best described as a miniature swimming adult calyx (Figs. 12, 14, and 15). An outgrowth of the tentacles and stalk is essentially all that is necessary to derive the generalized adult (Mariscal, 1965). It seems that there are probably fewer gross anatomical modifications required in order to produce an adult pedicellinid entoproct from its free-swimming larva than for almost any other invertebrate group (cf. Figs. 2 and 12).

Although the gross anatomy of some loxosomatid larvae is similar to that of a pedicellinid larva, there are a number of rather bizarre modifications on this basic plan (e.g., Jägersten, 1964; Nielsen, 1966a,b) (Figs. 3 and 6). In addition to the presence of larval budding and the liberation of fully formed miniature adults from within the larval body previously mentioned, some of the larvae have the dorsal surface covered with up to seven pairs of glandular pouches of unknown function. The pouches are capable of being everted so that their lining of vesicular cells is exposed to the environment (Jägersten, 1964). Ciliated, paired preoral, and single apical organs are present and a ciliated prototroch or girdle of large active cilia surrounds the periphery and provides the main swimming force. Loxosomatid larvae generally possess a pair of eyespots anteriorly.

The surface of some entroproct larvae may be coated with a covering of fine particles (Jägersten, 1964; Mariscal, 1965). The adults also have a similar covering and particles of detritus adhere to them during secretion of new cuticle. Although possibly accidental, the coating enables both the larva and the adult to blend more closely with the substrate. The loxosomatid larvae described by Jägersten (1964) and Nielsen (1967b) have the dorsal surfaces covered with small stalked vesicles of unknown origin while the dorsal surface of the larva of *Loxosomella vivipara* is ornamented with a spiderweblike pattern of raised ridges (Nielsen, 1966a) (Fig. 5b). Although all previously described entoproct larvae have preoral or frontal organs and a footlike structure, Nielsen (1967b) has found that the larvae of *Loxosoma pectinaricola, Loxosoma rhodnicola,* and *Loxosomella elegans* all lack the preoral organ, eyes, and foot described for other loxosomatid larvae.

1.4.2.2 METAMORPHOSIS AND SETTLING

Following release, the entoproct larva becomes free-swimming except for brief periods of settling and creeping on the substrate (Figs. 16 and 17). While swimming, the larva may undergo many contractions and changes of shape along with extrusions of the preoral organ (Figs. 14 and 15) and the apical organ (Mariscal, 1965; Nielsen, 1967a). While creeping about, the larva appears to be testing the substrate in order to locate a suitable spot for metamorphosis (Figs. 16 and 17).

The elapsed time between release of the larva from the parent and metamorphosis varies from as little as 5 hours in *Barentsia* to as much as 14 days for some loxosomatids (Mariscal, 1965; Nielsen, 1967a). Although the metamorphosis of only a few species of entoprocts has been studied, what little information is available indicates that there are radically different types of metamorphosis in pedicellinids and loxosomatids.

The classical picture of entoproct metamorphosis has been until very recently that of Barrois (1881, 1882, 1886), Harmer (1887), and Cori (1936) for *Pedicellina*. The *Pedicellina* larva attaches by its creeping or oral surface to the substrate. The apical organ, the preoral organ, and the ciliary girdle all disappear as the edges of the girdle fuse together to form the attachment plate of the adult. The vestibule, or atrium, now sealed off from the exterior, sinks inward to be divided into three portions. The lower portion, which is made up of the ciliary girdle of the larva and the adjacent region of the vestibule, now contributes the components which make up the attachment disc or foot gland as well as the stalk. The middle portion of the vestibule degenerates into a mass of globules which Barrois (1881) considers to contribute to the nervous system of the adult and the colony. The third, or upper portion of the larval vestibule remains in association with the larval U-shaped gut which undergoes a remarkable 180° rotation so that the mouth and anus become directed upward to correspond to the adult configuration. The ectoderm of this free upper surface (formerly the aboral pole of the larval body) now becomes thickened and then invaginates to connect the rotated larval vestibular cavity to the outside. This cavity will thus form the vestibule of the adult and from its margin grow the tentacles of the adult. During this time the larval attachment plate and its adjacent region have elongated to become a definitive stalk which continues to grow to the length characteristic of the species. From this initial individual then, creeping stolons arise which spread out over the substrate to give rise eventually to other feeding individuals by asexual budding.

Although the above description applies specifically to the larva of *Pedicellina,* some workers, including Harmer (1887) have assumed that the metamorphosis of the loxosomatids follows much the same pattern. However, the investigations of Jägersten (1964) and Nielsen (1966a, 1967a) have demonstrated that the "metamorphosis" of some loxosomatids is radically different from that described for the pedicellinids.

For example, Jägersten (1964), working with planktonic larvae captured in the Bahama Islands, discovered that not only did some of these large pelagic loxosomatid larvae possess external budding, but also exhibited a remarkable type of internal budding, as discussed earlier. During the months of March through April, the majority of the two species of larvae collected by Jägersten contained one or two fully developed and sexually mature (male) miniature adults. These miniature adults "hatched" apparently by rupturing through the body wall of the larva, which then died (Fig. 6c). Nielsen (1966a) was able to verify and extend Jägersten's (1964) original observations and confirmed that the young adults were liberated through the dorsal side of the larvae

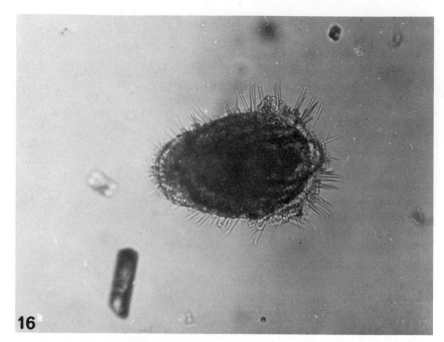

16

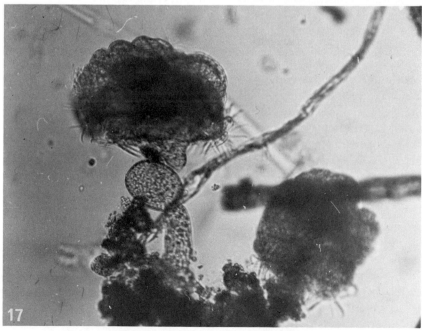

17

during the night. He also confirmed that sexually mature males with ripe sperms were among those individuals released from the larvae. Jägersten (1964) suggests that this may be an example of protandric hermaphroditism.

Nielsen (1966a) also has described a second loxosomatid from South Florida, *Loxosomella vivipara*, the larva of which is capable of liberating a single, large internal bud through a rupture in its dorsal surface (Fig. 6). This bud or miniature adult is released through the dorsal side of the free-swimming larva and already bears one to two buds of its own at the time of liberation from the larva. Following this, the larval body detaches and eventually disintegrates. Although the newly liberated individuals have developing reproductive organs, none were in an advanced enough state to determine the sex.

Nielsen (1967a) has more recently described a third type of entoproct metamorphosis, that of *Loxosomella murmanica*. At the start of metamorphosis, the larva of *L. murmanica* attaches to the substrate by its frontal or preoral organ. The ciliary girdle margin then contracts and seals over the larval vestibule while a stalk forms from the region of the attached preoral organ. After a week, the vestibule reopens and rudimentary tentacles may be seen within. Thus the gut of the larva becomes the gut of the adult. The metamorphosis of *L. murmanica* tends to resemble that of *Pedicellina cernua*; however, the *Pedicellina* larva attaches to the substrate by its contracted ciliary girdle while *L. murmanica* attaches by its preoral organ. However, Nielsen (1966a) does not clarify if there is also some rotation of the larval gut in the case of *Loxosomella* as has been reported for *Pedicellina*, or whether the gut becomes oriented in the proper position by differential growth of the stalk and calyx. Nielsen (1967a) also reports that the metamorphosis of *Loxosomella harmeri* is identical to that of *Loxosomella murmanica*. Nielsen (personal communication) considers the metamorphosis of *L. murmanica* to be the most primitive (and probably most common) type among the Loxosomatidae.

Thus, it seems clear that pedicellinid metamorphosis is not at all characteristic of all entoprocts and, in fact, may not hold for any loxosomatid. A situation in which the adult (in some cases, sexually mature) develops

FIG. 16. Photomicrograph of the dorsal side of a *Barentsia benedeni* larva which has settled to creep about on the substrate just before metamorphosis. The anterior end of the larva is to the left (scale approximately the same as in Fig. 14) (photo by R. N. Mariscal).

FIG. 17. Photomicrograph of the left side of *Barentsia benedeni* larva which is creeping about on a young bud of the same species just before metamorphosis (scale approximately the same as in Fig. 14) (photo by R. N. Mariscal).

essentially as an internal parasite of the larva and upon release kills the larva, would seem to be far removed from the settling and gradual metamorphosis of a larva into the adult form. It is hoped that, with further studies of entoproct metamorphosis, we may be in a better position to generalize regarding the selective advantages and evolutionary significance of this remarkably variable process among the Entoprocta.

The influence of various environmental factors on growth, metamorphosis, and settling has not been investigated. However, Ryland and Austin (1960), using settling plates, conclude that temperature may be one of the major factors affecting the reproduction of the two entoproct species they studied, *Loxosomella antedonis* and *Loxosomella kefersteinii*. Although a number of studies have analyzed the factors affecting settling and metamorphosis of ectoproct larvae (e.g., Maturo, 1959 and the numerous papers of Lynch, summarized in Lynch, 1961), no comparable studies have been carried out on entoprocts.

Acknowledgment

I would like to thank Dr. Peter Emschermann for suggestions concerning the preparation of this review. The literature review was completed in December, 1970.

1.5 References

Annandale, N. (1908). I. The fauna of brackish ponds at Port Canning, Lower Bengal. Part VII. Further observations on the Polyzoa, with the description of a new genus of Entoprocta. *Rec. Ind. Mus.* **2**, 11–19.

Annandale, N. (1915). Fauna of the Chilka Lake. The Polyzoa of the Lake and of brackish water in the Gangetic delta. *Mem. Ind. Mus.* **5**, 127–132.

Annandale, N. (1916). Zoological results of a tour in the Far East. Polyzoa, Entoprocta and Ctenostomata. *Mem. Asiatic Soc. Bengal* **6**, 13–37.

Arvy, L. and Prenant, M. (1952). *Loxosoma nitschei* Vigelius sur les Phascolions de Dinard. *Bull. Lab. Mar. Dinard* **36**, 2–5.

Assheton, R. (1912). *Loxosoma loxalina* and *Loxosoma saltans*—two new species. *Quart. J. Microsc. Sci.* (Ser. 2), **58**, 117–143.

Atkins, D. (1927). A new habitat for *Loxosoma phascolosomatum* Vogt. *J. Mar. Biol. Ass. U.K.* **14** [N.S.], 749–752.

Atkins, D. (1932). The Loxosomatidae of the Plymouth area including *L. obesum* sp. nov. *Quart. J. Microsc. Sci.* **75**, 321–391.

Balfour, F. M. (1885). "Comparative Embryology," Vol. 1. Macmillan, London.

Barrois, J. (1875). Sur les formes larvales des Bryozoaires. *C. R. Acad. Sci.* (*Paris*) **81**, 443.

Barrois, J. (1877). "Recherches sur l'embryologie des Bryozoaires." Librairie de Six-Horemans, Lille.

Barrois, J. (1881). Métamorphose de la Pédicelline. *C. R. Acad. Sci.* (*Paris*) **92**, 1527–1528.

Barrois, J. (1882). Embryogénie des Bryozoaires. *J. Anat. Phys.* **18**, 1–34.

Barrois, J. (1886). Mémoire sur la métamorphose de quelques Bryozoaires. *Ann. Sci. Nat. Zool.* (Ser. 7), **1**, 1–94.

Becker, G. (1937). Untersuchungen über den Darm und die Verdauung von Kamptozoen, Bryozoen, und Phoroniden. *Z. Morphol. Oekol. Tiere* **33**, 72–127.

Bobin, G. (1968). *Loxosomella museriensis* N. Sp. Entoprocte Loxosomatidae de Mer Rouge. *Israel J. Zool.* **17**, 175–189.

Bobin, G. (1969). Genèse et valeur du système d'attache à l'hôte chez *Loxosomespilon perezi* Bobin et Prenant (Entoprocte, Loxosomatidae). *Arch. Zool. Exp. Gen.* **110**(2), 225–265.

Bobin, G. and Prenant, M. (1953a). La classification des Loxosomes selon Mortensen et le *Loxosoma singulare* de Keferstein et Claparède. *Bull. Soc. Zool. Fr.* **78**(1), 84–96.

Bobin, G. and Prenant, M. (1953b). Sur les Loxosomes du *Phascolion strombi* (Montagu) et sur la spécificité de l'inquilinisme des *Loxosomes. Arch. Zool. Exp. Gén.* **90**, 18–41.

Bobin, G. and Prenant, M. (1953c). Sur trois Loxosomes méditerranéens. *Bull. Inst. Océanogs.* (*Monaco*) no. **1030**, 1–9.

Bobin, G. and Prenant, M. (1953d). Sur deux *Loxocalyx* des côtes françaises. *Bull. Soc. Zool. Fr.* **78**, 381–387.

Bobin, G. and Prenant, M. (1953e). Les populations de Loxosomes du *Phascolion strombi* (Montagu). *Arch. Zool. Exp. Gén.* **90**, 93–104.

Bobin, G. and Prenant, M. (1953f). Deux Loxosomes nouveaux de Roscoff. *Arch. Zool. Exp. Gen.* **91**, 25–35.

Bobin, G. and Prenant, M.(1953g). Sur les populations de Loxosomes des Aphrodites et des Hermiones. *Bull. Soc. Zool. Fr.* **78** (2–3), 122–132.

Bobin, G. and Prenant, M. (1954). Étude critique des principaux caractères utilisable dans la classification des Loxosomatidae. *Ann. Sci. Nat. Zool.* (Ser. 11), **16**(3), 7–33.

Brien, P. (1956). Le bourgeonnement des Endoproctes et leur Phylogenèse. A propos du bourgeonnement chez *Pedicellina cernua* (Pallas). *Ann. Soc. Roy. Zool. Belg.* **87**(1), 27–43.

Brien, P. (1959). Classe des Endoproctes ou Kamptozoaires. *In* "Traité de Zoologie" (P. Grassé, ed.), Vol. 5, pp. 927–1007. Masson, Paris.

Brien, P. (1960). Le bourgeonnement et la phylogenèse des Endoproctes et des Ectoproctes. Reflexions sur les processus de l'Evolution animale. *Bull. Acad. Roy. Belg.* Cl. Sci. (Ser. 5), **46**, 748–766.

Brien, P. and Papyn, L. (1954). Les endoproctes et la Classe des Bryozoaires. *Ann. Soc. Roy. Zool. Belg.* **85**, 59–87.

Claparède, M. E. (1867). Miscellannées zoologiques. Sur le *Loxosoma kefersteinii. Ann. Sci. Nat.* (Ser. 5), **8**, 28–30.

Claparède, M. E. (1870). Beitrage zur Anatomie und Entwicklungsgeschichte der Seebryozoen. *Z. Wiss. Zool.* **21**, 137–174.

Cori, C. J. (1929). Kamptozoa. *In* "Handbuch der Zoologie" (W. Kükenthal and T. Krumbach, eds.), Vol. 2(5), pp. 1–64.

Cori, C. J. (1936). Kamptozoa. *In* "Klassen und Ordnungen des Tierreichs" (H. G. Bronn, ed.), Vol. 4, pp. 1–119.

Cuenot, L. (1899). Sur la détermination du sexe chez les animaux. *Bull. Sci. Fr. Belg.* **32**, 462–535.

Cwiklitzer, R. (1909). Die Anatomie der Larve von *Pedicellina echinata*. *Arb. Zool. Inst. Univ. Wien* **17**, 157–186.

Davenport, C. B. (1893). On *Urnatella gracilis*. *Bull. Mus. Harvard* **24**, 1–44.

Dublin, L. I. (1905a). The history of the germ cells in *Pedicellina americana*. *Ann. N.Y. Acad. Sci.* **16**(1), 1–64.

Dublin, L. I. (1905b). On nucleoli in the somatic and germ-cells of *Pedicellina americana*. *Biol. Bull.* **8**(6), 347–364.

Eggleston, D. (1965). The Loxosomatidae of the Isle of Man. *Proc. Zool. Soc. London* **145**(4), 529–547.

Ehlers, E. (1890). Zur Kenntnis der Pedicellinen. *Abh. König. Ges. Wiss. Göttingen* **36**, 1–200.

Emschermann, P. (1961). Über Brutkörper bei dem Kamptozoon *Barentsia gracilis* Sars. *Zool. Jahrb. Physiol.* **69**(3), 333–338.

Emschermann, P. (1965). Über die sexuelle Fortpflanzung und die Larve von *Urnatella gracilis* Leidy (Kamptozoa). *Z. Morphol. Ökol. Tiere* **55**, 100–114.

Fewkes, J. W. (1884). A new pelagic larva. *Amer. Natur.* **18**, 305.

Foettinger, A. (1887). Sur l'anatomie des Pédicellines de la côte d'Ostende. *Arch. Biol.* **7**, 299–329.

Franzén, Å. (1956). On spermiogenesis, morphology of the spermatozoon and biology of fertilization among invertebrates. *Zool. Bidr. Uppsala* **31**, 355–482.

Franzén, Å. (1962). Studies on Entoprocta from the West Coast of Sweden. *Zool. Bidr. Uppsala* **33**, 311–326.

Franzén, Å. (1967). A new loxosomatid from the Pacific (Gilbert Islands) with a note on internal budding in Entoprocta. *Ark. Zool.* (Ser. 2), **19**(4), 381–390.

Franzén, Å. (1970). Morfologi och larvutveckling hos Entoprocta. *Svensk. Natur.* pp. 131–141.

Gruijs-Faucher, C. (1959). Monographie histologique de *Loxomespilon perezi* Bobin and Prenant (Entoprocte marin, Loxosomatidae). *Bull. Lab. Mar. Dinard* **45**, 3–41.

Harmer, S. F. (1885). On the structure and development of *Loxosoma*. *Quart. J. Microsc. Sci.* (Ser. 2), **25**, 261–337.

Harmer, S. F. (1887). On the life history of *Pedicellina*. *Quart. J. Microsc. Sci.* (Ser. 2), **27**, 239–263.

Harmer, S. F. (1915). The Polyzoa of the Siboga Expedition. Part I. Entoprocta, Ctenostomata and Cyclostomata. *Siboga Exped. Rep.* **28a**, 1–180.

Hatschek, B. (1877). Embryonalentwicklung und Knospung der *Pedicellina echinata*. *Z. Wiss. Zool.* **29**, 502–549.

Hincks, T. (1873). Contributions to the history of Polyzoa (*Pedicellina*). *Quart. J. Microsc. Sci.* **13**, 32–34.

Hyman, L. H. (1951). "The Invertebrates: Acanthocephala, Aschelminthes and Entoprocta," Vol. III. McGraw-Hill, New York.

Jägersten, G. (1964). On the morphology and reproduction of entoproct larvae. *Zool. Bidr. Uppsala* **36**(3), 295–314.

Joliet, L. (1877). Contribution à l'histoire naturelle des Bryozoaires des Côtes de France. *Arch. Zool. Exp. Gén.* **6**, 193–304.

Keferstein, W. (1862). Über *Loxosoma singulare* gen. et sp. n. den Schmarotzer einer Annelide. Untersuchungen über niedere Seetiere. *Z. Wiss. Zool.* **12**, 131–132.

Kluge, H. A. (1946). Kamptozoa from the Arctic Ocean. *In* "Trudy Dreifuiushchei èkspeditsii glavsevmorputi na ledokol' nom parokhode 'G. Sedov' 1937–1940" (V. Kh. Buinitski, ed.), Vol. 3, 149–155. Moscow.

Kowalewsky, A. (1866). Beiträge zur Anatomie und Entwicklungsgeschichte des *Loxosoma neapolitanum* sp. n. *Mém. Acad. Imp. Sci. St. Petersbourg.* (Ser. 7), **10**, 1–10.

Lebedinsky, J. (1905). Die Embryonalentwicklung der *Pedicellina echinata* Sars. *Biol. Zentralbl.* **25**, 536–548.

Leidy, J. (1883). *Urnatella gracilis,* a fresh water Polyzoan. *J. Acad. Nat. Sci. Philadelphia* (Ser. 2), **9**(1), 1–16.

Lynch, W. F. (1961). Extrinsic factors influencing metamorphosis in bryozoan and ascidian larvae. *Amer. Zool.* **1**, 59–66.

Marcus, E. (1937). Bryozoarios marinhos Brazileiros I. *Bol. Fac. Phil. Sci. Letr. Univ. São Paulo I, Zool. No.* **1**, 1–224.

Marcus, E. (1939). Bryozoarios marinhos Brazileiros III. *Bol. Fac. Phil. Sci. Letr. Univ. São Paulo XIII, Zool. No.* **3**, 111–353.

Marcus, E. du B.-R. (1950). A new loxosomatid from Brazil. *Bol. Fac. Fil. Cienc. Letras Univ. São Paulo Zool.* **15**, 193–202.

Marcus, E. du B.-R. (1957). Neue Entoprocten aus der Gegend von Santos. *Zool. Anz.* **159**, 68–75.

Mariscal, R. N. (1965). The adult and larval morphology and life history of the entoproct *Barentsia gracilis* (M. Sars, 1835). *J. Morphol.* **116**, (3), 311–338.

Maturo, F. J. S., Jr. (1959). Seasonal distribution and settling rates of estuarine Bryozoa. *Ecology* **40**(1), 116–127.

Mortensen, T. H. (1911). A new species of Entoprocta, *Loxosomella antedonis* from N. E. Greenland. *Medd. Greenland* **45**, 399–406.

Nasonov, N. (1926a). L'*Arthropodaria kovalewskii* n. sp. (Entoprocta) et la régénération de ses organes. *Trav. Lab. Zool. Sta. Biol. Sébastopol. Acad. Sci. U.R.S.S.* (Ser. 2), **5**, 1–38.

Nasonov, N. (1926b). Sur l'hivernage de l'*Arthropodaria kovalewskii* Nasonov et du *Balanus improvisus* Darwin dans l'eau douce. *C. R. Acad. Sci. U.R.S.S. 1926*, 51–52.

Nickerson, W. S. (1898). Preliminary notice of a new species of endoproct, *Loxosoma davenporti,* from the Massachusetts Coast. *Science* **7**, 220–221.

Nickerson, W. S. (1899). Notes on *Loxosoma davenporti. Science* **9**, 366–367.

Nickerson, W. S. (1900). Doubled Loxosomae. *Amer. Natur.* **34**, 891–895.

Nickerson, W. S. (1901). On *Loxosoma davenporti* (sp. nov.). An endoproct from the New England Coast. *J. Morphol.* **17**, 351–380.

Nielsen, C. (1964a). Studies on Danish Entoprocta. *Ophelia* **1**(1), 1–76.

Nielsen, C. (1964b). Entoprocta from the Bergen area. *Sarsia* **17**, 1–6.

Nielsen, C. (1966a). On the life cycle of some Loxosomatidae (Entoprocta). *Ophelia* **3**, 221–247.

Nielsen, C. (1966b). Some Loxosomatidae (Entoprocta) from the Atlantic Coast of the United States. *Ophelia* **3**, 249–275.

Nielsen, C. (1967a). Metamorphosis of the larva of *Loxosomella murmanica* (Nilus) (Entoprocta). *Ophelia* **4**(1), 85–89.

Nielsen, C. (1967b). The larvae of *Loxosoma pectinaricola* and *Loxosomella elegans* (Entoprocta). *Ophelia* **4**(2), 203–206.

Nielsen, C. and Ryland, J. S. (1961). Three new species of Entoprocta from Norway. *Sarsia* **1**, 39–45.

Nilus, G. (1909). Notiz über *Loxosoma murmanica* und *Loxosoma brumpti* sp. n. *Trav. Soc. Imp. Nat. St. Petersbourg* **40**(1), 157–169.

Nitsche, H. (1869). Beiträge zur Kenntnis der Bryozoen. II. Über die Anatomie von *Pedicellina echinata* Sars. *Z. Wiss. Zool.* **20**, 1–36.

Nitsche, H. (1875). Über den Bau und die Knospung von *Loxosoma kefersteinii* Claparède. *Z. Wiss. Zool.* **25**, 451–456.

O'Donoghue, C. H. (1924). The Bryozoa (Polyzoa) collected by the S. S. "Pickle." *Union S. Afr. Fish. Mar. Biol. Surv. Pretoria* **3**(10), 1–63.

Oka, A. (1895). Sur la *Barentsia misakiensis. Zool. Mag.* **7**, 76–86.

Osburn, R. C. (1953). Bryozoa of the Pacific Coast of America. Part 3, Cyclostomata, Ctenostomata, Entoprocta and Addenda. *Allan Hancock Pac. Exped.* **14**(3), 613–841.

Prenant, M. and Bobin, G. (1956). Bryozoaires. Première Partie. Entoproctes, Phylactolèmes, Cténostomes. *Faune Fr.* **60**, 1–398.

Prouho, H. (1891). Contribution à l'histoire des Loxosomes. Étude sur le Loxosome annelidicole *Cyclatella annelidicola* (Van Beneden et Hesse). *Arch. Zool. Exp. Gén.* (Ser. 2), **9**, 91–116.

Reid, J. (1845). Anatomical and physiological observations on some zoophytes. *Ann. Mag. Natur. Hist. Ser.* 1. **16**, 385–400.

Ridley, S. O. (1881). Account of the zoological collections made during the survey of H.M.S. "Alert" in the Straits of Magellan and on the coast of Patagonia. V. Polyzoa. *Proc. Zool. Soc. London*, pp. 44–61.

Robertson, A. (1900). Studies on Pacific Coast Entoprocta. *Proc. Calif. Acad. Sci.* **2**, 323–348.

Rogick, M. D. (1948). Studies on Marine Bryozoa II. *Barentsia laxa* Kirkpatrick 1890. *Biol. Bull. Woods Hole* **94**, 128–142.

Rützler, K. (1968). *Loxosomella* from *Tedania ignis*, the Caribbean Fire Sponge. *Proc. U.S. Nat. Mus.* **124**, 1–11.

Ryland, J. S. (1960). The occurrence of *Pseudopedicellina mutabilis* Toriumi (Entoprocta) in British waters with a note on *Pedicellina nutans* Dalyell. *Ann. Mag. Natur. Hist.* **3**(30), 371–383.

Ryland, J. S. (1961). Two species of *Loxosomella* (Entoprocta) from W. Norway. *Sarsia* **1**, 31–38.

Ryland, J. S. (1965). Some New Zealand Pedicellinidae (Entoprocta), and a species new to Europe. *Trans. Roy. Soc. N. Z. Zool.* **6**(19), 189–205.

Ryland, J. S. and Austin, A. P. (1960). Three species of Kamptozoa new to Britain. *Proc. Zool. Soc. London* **133**(3), 423–433.

Saint-Joseph, B. de. (1899). Annélides Polychètes de la rade de Brest et de Paimpol. *Ann. Sci. Natur.* (Ser. 8), **10**, 161–194.

Salensky, M. (1877). Études sur les Bryozoaires entoproctes. *Ann. Sci. Natur.* (Ser. 6), **5**, 1–60.

Schmidt, O. (1876). Die Gattung *Loxosoma. Arch. Mikrosk. Anat.* **12**, 1–14.

Schmidt, O. (1878). Bemerkungen zur den Arbeiten über *Loxosoma. Z. Wiss. Zool.* **31**, 68–80.

Schopf, T. J. M. and Simon, J. L. (1965). Entoproct records from northeastern United States: *Loxosomella varians* Nielsen and *L. minuta* (Osburn). *Biol. Bull.* **129**(2), 422.

Sebesteyén, O. (1962). On *Urnatella gracilis* Leidy (Kamptozoa Cori) and its occurrence in an industrial water-works fed by Danube water in Hungary. *Acta Zool.* **8**(3–4), 435–448.

Seeliger, O. (1889). Die ungeschlechtliche Vermehrung der endoprokten Bryozoen. Z. Wiss. Zool. **49**, 168–208.

Seeliger, O. (1890). Bemerkungen zur Knospenentwicklung der Bryozoen. Z. Wiss. Zool. **50**, 560–599.

Seeliger, O. (1906). Über die Larven und Verwandtschaftsbeziehungen der Bryozoen. Z. Wiss. Zool. **84**, 1–78.

Seshaiya, R. V. (1949). On *Urnatella indica* Seshaiya. A freshwater entoproctan from South India. *Rec. Ind. Mus.* **45**, 283–289.

Soule, D. F. and Soule, J. D. (1965). Two new species of *Loxosomella,* Entoprocta, epizoic on Crustacea. *Allan Hancock Found. Publ. Occas. Paper No.* **29**, 1–19.

Soule, J. D. (1955). A new species of *Myosoma* from the Pacific (Entoprocta). "Essays in the Natural Sciences in Honor of Captain Allan Hancock," pp. 173–177. Univ. of Southern California Press, Los Angeles.

Toriumi, M. (1944). *Zool. Mag.* **56**(1,2,3), 20–25. (In Japanese).

Toriumi, M. (1949). On some Entoprocta from Japan. *Sci. Rep. Tôhoku Univ. Ser. 4 Biol.* **18**(2), 223–225.

Toriumi, M. (1951). Some entoprocts found in Matsushima Bay. *Sci. Rep. Tôhoku Univ. Ser. 4 Biol.* **19**(1), 17–22.

Uljanin, B. (1869). Zur anatomie und Entwicklungsgeschichte der *Pedicellina. Bull. Soc. Imp. Natur. Moscou* **42**(1), 425–440.

Valkanov, A. (1951). Eigentumlichkeiten in dem Bau und der Organisation von *Arthropodaria kovalevskii* Nasonov in Zusammenhang mit ihrer Überwinterung. *Trav. Stat. Biol. Mar. Bulg.* **16**, 47–64.

Van Beneden, P. J. (1845). Recherches sur l'anatomie, la physiologie et le développement des Bryozoaires qui habitent la Côte d'Ostende. Histoire naturelle du genre *Pedicellina. Nouv. Mem. Acad. Roy. Sci. Bell. Lett. Bruxelles* **19**, 1–31.

Vogt, C. (1876). Sur le Loxosome des Phascolosomes (*Loxosoma phascolosomatum*). *Arch. Zool. Exp. Gén.* **5**, 305–356.

Vogt, C. (1878). Bemerkungen über Dr. Hatschek's Aufsatz über Embryonalentwicklung und Knospung von *Pedicellina echinata. Z. Wiss. Zool.* **30**, 374–378.

Chapter 2

TARDIGRADA

Leland W. Pollock

2.1 Introduction

Members of the phylum Tardigrada are microscopic and pseudocoelous as adults, possess a nonchitinous cuticle, and display limited metamerism. Although they are common, they usually are overlooked since none has direct economic importance and the habitats they frequent are not often studied microscopically. Tardigrades occur in leaf litter, among aquatic plants, in water films surrounding certain terrestrial plants (e.g., mosses and lichens), in capillary spaces between grains of sand in both freshwater and marine beaches, in association with intertidal or floating algae, and rarely as ectoparasites or commensals. Marine species are limited to the primitive order Heterotardigrada. Other heterotardigrades and members of the orders Mesotardigrada and Eutardigrada are found in freshwater or semiterrestrial habitats described above.

Ecological and anatomical studies of marine tardigrades are scarce.

Information regarding reproduction in marine forms is limited to the three most abundant genera, *Batillipes, Echiniscoides,* and *Stygarctus.* Additional information on this topic is based on nonmarine, eutardigrade genera, *Macrobiotus* and *Hypsibius.*

2.2 Asexual Reproduction

While asexual reproduction has not been reported for marine species, males are unknown in several species of Eutardigrada and in two common genera of moss-dwelling Heterotardigrada (Pennak, 1953; Rudescu, 1964). Ammermann (1967) described meiotic parthenogenetic reproduction in the eutardigrade *Hypsibius dujardini.* A meiotic division distributes the diploid chromosomes ($2N = 10$) between the oocyte and a polar body. Duplication of oocyte chromosomes, restoring the diploid number, precedes a second complete mitotic division which produces a diploid polar body and the diploid egg nucleus. The nucleus shifts toward the center of the egg and cleavage begins (see Section 2.4.1). A different parthenogenetic process occurs in some populations of the eutardigrades *Macrobiotus richtersi* and *Hypsibius oberhaeuseri* (Bertolani, 1971a,b,c, 1972a). In these cases, all cells are triploid ($3N = 18$) and ameiotic oocyte maturation occurs by mitotic divisions, producing triploid eggs.

2.3 Sexual Reproduction

2.3.1 Sexual Dimorphism

Tardigrades are dioecious. Traditionally, sexual determinations are limited to individuals bearing recognizable gametes. In addition, two vasa deferentia are found in males, while a single oviduct occurs in females. However, these internal characters are often difficult to discern.

Useful secondary characteristics have been described from five genera of marine tardigrades. In two species of *Halechiniscus,* sexes can be distinguished by relative lengths of their cephalic appendages, characteristic sensory structures (Richters, 1908; Schulz, 1955). In males, clavae are longer than the adjacent lateral cirri. Clavae of females are shorter than the lateral cirri. Dimorphism exists in the shape and location of gonopores in *Archechiniscus, Batillipes, Parastygarctus,* and *Stygarctus* (Renaud-Debyser, 1965; Renaud-Mornant, 1967; Pollock, 1970b). In fe-

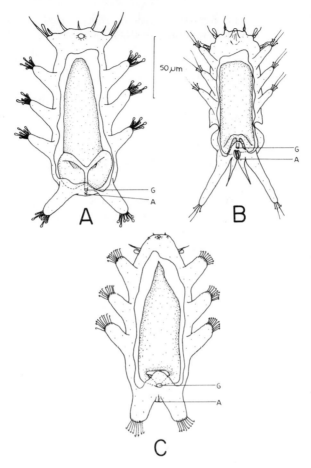

FIG. 1. Diagrammatic illustration of male reproductive systems of marine Tardigrada. Ventral views. (A) *Batillipes* sp.; (B) *Stygarctus* sp.; (C) *Echiniscoides* sp. A, anus; G, gonopore.

males (Fig. 2), a rosette of plates surrounds a gonopore located considerably anterior to the anus, while in males (Fig. 1), the gonopore is rounded and is just anterior to the anus. I have observed the occurrence of similar dimorphism in the genus *Echiniscoides*.

Other secondary characteristics are found in species of eutardigrades, including differences in shape of claws and the fourth pair of legs (see Rudescu, 1964 for review). Distinctions of this sort have not been reported for marine species. Although males are often smaller than females in Eutardigrada (Pennak, 1953), size discrepancy does not occur in the marine heterotardigrades studied morphometrically (Renaud-Mornant and Anselme-Moizan, 1969; Pollock, 1970b).

2.3.2 Sex Determination and Hermaphroditism

There are no known hermaphroditic tardigrades. Parthenogenesis, however, seems to occur in some moss-water genera where males are rare or have never been observed (e.g., *Echiniscus* and *Pseudechiniscus*, Pennak, 1953). In marine tardigrades, males occur regularly. The cumulative sex ratio in a population of *Stygarctus granulatus* was nearly equal (1.0 males:1.1 females), although males were slightly dominant in fall (1.1:1.0) and females were decidedly dominant in spring (1.0:1.4) (Pollock, 1970c). The ratio of males to females in *Batillipes* is also approximately equal (Schmidt, 1969; Pollock, 1970b). The mechanisms of sex determination remain unknown.

2.3.3 Anatomy of the Reproductive System

The gonad in both sexes is a sac suspended dorsally in the pseudocoelom by ligaments from its anterior end to insertions on the cuticle above the pharynx. Additional ligaments may be present near its posterior end. Gonads are narrow and flattened when animals are reproductively inactive, but swell to occupy most of the dorsum of the spacious pseudocoel as gametogenesis advances. Gonoducts terminate at a gonopore in Heterotardigrada, while in Eutardigrada, they join the hind gut, forming a cloaca. Although a seminal receptacle occurs in some female eutardigrades, none has been found in heterotardigrades.

In male *Batillipes* (Fig. 1A), the testis is widest posterolaterally where it curls forward ventrally to form lateral bulges. From the anterior tips of these bulges, vasa deferentia pass diagonally posterior, join midventrally, and continue as a common duct posteriorly to exit via the gonopore (Pollock, 1970b).

Lateral bulges are absent in male *Stygarctus* (Fig. 1B). Vasa deferentia pass anteroventrally from the posterolateral corners of the testis. These ducts join midventrally and continue posterior as a common duct to exit through the gonopore (Pollock, 1970b).

Lateral bulges are also absent in *Echiniscoides* (Fig. 1C). Vasa deferentia pass laterally around the gut and join at the midventral gonopore.

From the posterolateral extreme of the ovary in female *Batillipes* (Fig. 2A), a single oviduct passes dextrally around the gut. It exits midventrally through a rosette gonopore (Pollock, 1970b). The anatomy of female *Echiniscoides* is similar (Fig. 2C).

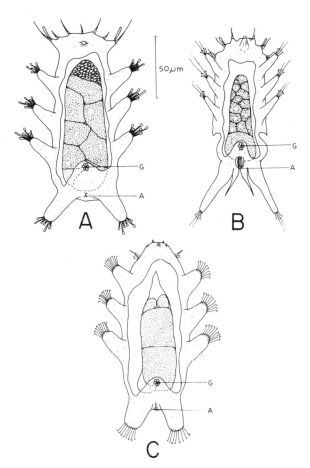

Fig. 2. Diagrammatic illustration of female reproductive systems of marine Tardigrada, ventral views. (A–C) Same as in Fig. 1.

"Annex glands" were described from female *Parastygarctus* (Renaud-Debyser, 1965) and *Stygarctus* (Pollock, 1970a). These consist of a pair of bulbs located in anterolateral extremes of the caudal body section. The bulbs connect by curved or convoluted tubules to separate pores located slightly posterolaterally to the gonopore. While their function remains undiscovered, they probably participate in reproduction or oviposition.

2.3.4 Origin of Germ Cells and Gonads

Gonads develop from the fusion of one of the four pairs of somatic coelomic pouches produced during organogenesis (see Section 2.4.1).

Apparently, the gonocoel is the only eucoelom retained in adults (Marcus, 1928, 1929). Nothing is known about the actual origin of the primordial germ cells.

2.3.5 Gametogenesis

Little information is available on this aspect of tardigrade reproduction. Marcus (1929) described oogenesis in *Hypsibius,* in which abortive oocytes functioned as "nurse cells" to nourish the developing oocytes. Later as oocytes expanded and filled the ovary, nurse cells were forced into the anterior end.

In tardigrades, egg coverings are either thin and transparent or thick and ornamented. They are secreted within the ovary by endothelial cells which have separated from the ovarian wall (Marcus, 1929). The diploid number of chromosomes ranges from 10 to 18 among species (Bertolani, 1972b).

In *Batillipes,* differences occur between oocytes produced during the spring and summer and those of the fall and winter (Pollock, 1970b). Sexually mature females collected in spring carry as many as eight pliable, irregularly distorted oocytes of approximately equal size and appearance, containing a clear nuclear area within a finely granular matrix (Fig. 2A). They are clustered within the distended ovary. In the fall, however, females rarely possess more than four oocytes, arranged linearly in the ovary. The posterior-most oocyte is larger than the others, but all maintain a more rigid appearance than do spring oocytes. Circumstances producing these distinctions have not been studied and differences in development have not been detected. Two types of oocytes also occur in some Eutardigrada (Pennak, 1953).

The spermatozoa of tardigrades range in length from 80 to 90 μm in some eutardigrade species to 35 μm in smaller marine forms (Ramazzotti, 1962). They possess a rounded or spiral-shaped anterior section and a long flagellum, constituting the only occurrence of a ciliumlike structure in the phylum. Baumann (1970) indicated that spermatids of *Macrobiotus hufelandii* have a rounded portion from which one or two flagella extend. On the other hand, Baccetti, *et al.,* (1971) found the mature spermatozoa of the same species possess a cylindrical acrosome (3.2 μm long), a helicoid nuclear section (6.5 μm long) and either a single long flagellum (80 μm) or a short flagellum (6-7 μm long) terminating in a tuft of 10 stringy fibers.

In reproductively active *Batillipes* and *Stygarctus,* motile spermatozoa pass posteriorly along the peripheral extremes of the testis and collect in the lateral bulges and vasa deferentia prior to their release (Pollock, 1970b). Such sperms have rounded head pieces in *Batillipes* but a spiral anterior section in *Stygarctus*.

2.3.6 Gametogenic Cycles and Factors Influencing Them

Reproductively mature adults and newly hatched young occurred in nearly every collection through a year-round study of the intertidal beach tardigrades at Woods Hole, Massachusetts (Pollock, 1970c). Year-round reproduction is typical of marine interstitial meiofauna, generally (Swedmark, 1959), and marine tardigrades, in particular (Schmidt, 1969; Pollock, 1970c).

Nevertheless, periodic peaks in abundance occurred in populations of *Batillipes* and *Stygarctus*. Thus, while successful reproduction is possible at all times, certain circumstances are more favorable. Some species show spring and fall maxima in population density (Renaud-Debyser, 1956; DeZio and Grimaldi, 1966; McGinty and Higgins, 1968; Pollock, 1970c) while others are abundant through the summer and fall (Pollock, 1970c).

2.3.7 Reproductive Behavior and Mating

Functions of tardigrades apparently are influenced by the pressure of body fluids in the pseudocoelom against the rigid cuticle. Tardigrades defecate and lay eggs only during molting, when the volume of the pseudocoelomic fluid is reduced. Presumably, internal fertilization is also limited to this period. Although mating has not been described for marine heterotardigrades, two patterns have been observed in eutardigrades. Females of some species deposit eggs in the exuvium out of which they crawl during a molt. Van Wenck (1914) observed males of one species to cluster around the cloaca of such an exuvium and ejaculate spermatozoa into the old cuticle where protected but external fertilization occurred. On the other hand, Marcus (1929) described the fertilization in *Hypsibius nodosus* to be internal, again as the female molted. Spermatozoa introduced through the cloaca in the cuticle of a partially shed female passed through her cloaca, up the oviduct, and into the ovary where fertilization occurred.

In any case, mating requires the proximity of females and males during the female's molting period. This is accomplished readily where tardigrades form dense clusters. However, mechanisms of attraction and sex identification under less aggregated circumstances remain unclear.

2.4 Development

2.4.1 Embryonic Development

Although embryonic development has not been studied in marine tardigrades, von Erlanger (1895) and Van Wenck (1914) described the embryology of the eutardigrade genus *Macrobiotus* and Marcus (1928, 1929) studied the postgastrulation development of several eutardigrades in the genus *Hypsibius*. Unfortunately, aspects of this work are confusing and would benefit from further study. Riggin (1962) outlined several difficulties in interpreting this information, particularly as past workers have applied it to phylogenetic considerations.

The first division spindle forms about 30° from one end of the ovaloid homolecithal zygote. Cleavages are total and blastomeres are produced which appear to be equal in size (Fig. 3A and B). The nuclei of blastomeres lie close to the surface of the embryo. The first few divisions are synchronous, but this synchrony is apparently lost by about the 16-cell stage. Eventually, a coeloblastula with a small blastocoel is produced (Fig. 3C). A stereogastrula is formed through gastrulation by unipolar proliferation (Fig. 3D), wherein a few cells from the vegetal pole are released into the blastocoel and subsequently divide to form 50–60 primary endodermal cells. In their midst, the archenteron develops, following the long axis of the gastrula (Fig. 3E, a cross section, and Fig. 3F). At this time, the embryo curls ventrally and an ectodermal invagination appears, extends, and fuses with the archenteron to form the proctodeum (Fig. 4A). The internal complement of endodermal cells then separates from the outer layer of ectodermal cells, leaving a space between them (Fig. 4B). Since no mesodermal formation or cell division is associated with this process, the term "schizocoel" should not be applied in referring to the cavity formed (Marcus, 1928, 1929). As the proctodeum joins the archenteron, the anlage of the stomodeum appears as an anterior ectodermal invagination.

Marcus (1928, 1929) described the formation of six enterocoelous outpocketings from the archenteron, including two cephalic pouches (Fig. 4C, 4D—CP_1 and CP_2) and four paired, somatic coelomic pouches (Fig. 4C and D—SP_1–SP_4). Unlike enterocoels in other invertebrate groups, enterocoels in tardigrades do not form a secondary body cavity or deuterocoel. The cephalic pouches contribute to the formation of the buccal apparatus, including the salivary glands and pharyngeal bulb (Fig. 4E,F, and G), and the anterior-most three pairs of somatic pouches disaggregate to form isolated endomesodermal structures, such

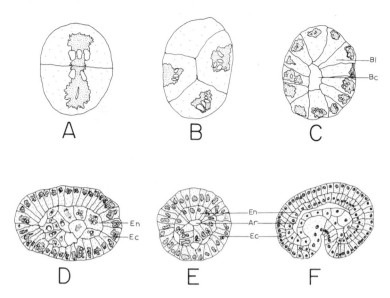

Fɪɢ. 3. Early embryology of Eutardigrada. (Redrawn from Van Wenck, 1914.) (A) Two-blastomere stage; (B) four-blastomere stage; (C) blastula, cross section; (D) gastrula, sagittal section; (E) gastrula, cross section; (F) late gastrula, sagittal section. Ar, archenteron; Bc, blastocoel; Bl, blastomere; Ec, ectoderm; En, endoderm.

as coelomocytes and muscle bands (Fig. 4E,F, and G). The fourth pair of pouches remains attached to the archenteron, fuses dorsally, and forms the single dorsal gonad (Fig. 4E,F, and G). Thus, the enterocoelous gonocoel is the only eucoelom retained in the adult.

Baumann (1961, 1964, 1966, 1970) has cultured several species of eutardigrades in the laboratory. He found that the time required for embryonic development is widely variable among and within species, depending upon surrounding conditions, such as moisture and temperature.

2.4.2 Postembryonic Development

Tardigrades undergo direct development and hatching is effected by penetration of the egg covering by stylets and claws (Marcus, 1929). Bertolani (1970a, 1970b) recently found that, with the possible exception of nervous tissue, eutely long attributed to the tardigrades (e.g., Marcus, 1929), apparently does not occur. Nevertheless, postembryonic growth results primarily from increases in cell size rather than in cell number. Changes in size and external morphology are accomplished during periodic molting and reformation of all cuticular parts.

In most cases, newly hatched young resemble adults but in miniature.

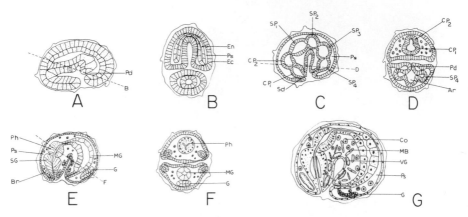

Fig. 4. Post gastrula embryology of Eutardigrada. (Redrawn from Marcus, 1929.) (A) Embryo during proctodaeal formation; dashed line "B" shows approximate location of next figure. (B) Cross section through embryo in (A). (C) Embryo during formation of cephalic and somatic pouches, sagittal section. (D) Embryo from (C), cross section. (E) Embryo during organ development. (F) Cross section through embryo in (E). (G) Well-developed embryo approaching hatching, sagittal section. Ar, archenteron; Br, brain; Co, coelomocyte; CP_{1-2}, coelomic pouches; Ec, ectoderm; En, endoderm; G, gonad; MB, muscle band; MG, midgut; Pd, proctodeum; Ph, pharynx; Ps, pseudocoelom; Sd, stomodeum; SG, salivary gland; SP_{1-4}, somatic pouches; VG, ventral ganglia.

McGinty and Higgins (1968) found that ontogenetic development in *Batillipes mirus* included a progressive decrease in the proportion of the total body length occupied by appendages, the development of gonadal tissue, and an increase in the number of toes on each leg. It has also been noted in other tardigrade genera that juveniles have fewer claws per leg than adults (Marcus, 1927; Ramazzotti, 1962; Renaud-Mornant and Anselme-Moizan, 1969; Pollock, 1970a).

Renaud-Mornant and Anselme-Moizan (1969) described the occurrence of four morphologically distinctive stages in the postembryonic development of *Stygarctus bradypus*. These stages are characterized by the ways in which they differ from adults. Stage I possessed only the middle two claws of the four occurring on each leg in the adult, a cuticle without the superficial segmentation of adults, no interplate space, no dorsal body spines, no caudal spikes, no gonopore, and no cuticular elaborations around the anus. Stage II showed the beginnings of dorsal segmentation of the cuticle, of the two dorsal body spines, and of papillae on the fourth pair of legs. Caudal spikes, additional claws, and gonopore were still absent. In Stage III, the dorsal cuticular plates and the dorsal body spines were nearly complete, lateral cuticular projections were transformed into their adult morphology, and caudal spines and two additional claws per leg were developed. Cuticular elaborations were evident around the anus. However, gonads and gonopores were absent. In Stage IV, the adult, the dorsal plates were completed and gonads and gonopores were apparent.

Estimations of natural life spans and growth rates of marine tardigrades are speculative at present. From sequential changes in size-class frequencies of populations of marine tardigrades, Pollock (1970c) suggested that a life span of 3–4 months seems most likely. McGinty and Higgins (1968) observed that gonadal tissue appeared only in animals larger than one-half the maximal adult size. If these are both reasonable generalizations, it would require 1–2 months for a newly hatched individual to reach reproductive maturity.

2.5 References

Ammermann, D. (1967). Die Cytologie der Parthenogenese bei dem Tardigraden *Hypsibius dujardini. Chromosoma* **23**, 203–213.

Baccetti, B., Rosati, F., and Selmi, G. (1971). Electron microscopy of tardigrades 4. The spermatozoon. *Monitore Zool. Ital.* **5**, 231–240.

Baumann, H. (1961). Der Lebensablauf von *Hypsibius. (H.) convergens* Urbanowicz (Tardigrada). *Zool. Anz.* **167**, 362–381.

Baumann, H. (1964). Uber den Lebenslauf und die Lebensweise von *Milnesium tardigradum* Doyère (Tardigrada). *Veroeff. Ueberseemus. Bremen* **3**, 161–171.

Baumann, H. (1966). Lebenslauf und Lebensweise von *Hypsibitus (H.) oberhauseri* Doyère (Tardigrada). *Veroeff. Ueberseemus. Bremen* **3**, 245–258.

Baumann, H. (1970). Lebenslauf und Lebensweise von *Macrobiotus hufelandii* Schultze (Tardigrada). *Veroeff. Ueberseemus. Bremen* **4**, 29–43.

Bertolani, R. (1970a). Mitosi somatiche e costanza cellulare numerica nei Tardigradi. *Atti. Accad. Naz. Lincei* **48**, 739–742.

Bertolani, R. (1970b). Variabilità numerica cellulare in alcuni tessuti di Tardigradi. *Atti Accad. Naz. Lincei* **49**, 442–445.

Bertolani, R. (1071a). Rapporto-sessi e dimorfismo sessuale in *Macrobiotus* (Tardigrada). *Atti. Accad. Naz. Lincei* **50**, 377–382.

Bertolani, R. (1971b). Partenogenesi geografica triploide in un Tardigrado (*Macrobiotus richtersi*). *Atti Accad. Naz. Lincei* **50**, 487–489.

Bertolani, R. (1971c). Osservazioni cariologiche su biotipi bisessuati e partenogenetici in *Hypsibius oberhaeuseri* (Tardigrada). *Atti Accad. Naz. Lincei* **51**, 411–413.

Bertolani, R. (1972a). Sex ratio and geographic parthenogenesis in *Macrobiotus* (Tardigrada). *Experientia* **28**, 94.

Bertolani, R. (1972b). Osservazioni cariologiche su alcuni *Macrobiotus* (Tardigrada). *Atti Accad. Naz. Lincei* **52**, 220–224.

DeZio, S. and Grimaldi, P. (1966). Ecological aspects of Tardigrada distribution in south Adriatic beaches. *Veroeff. Inst. Meeresforsch. Bremerhaven Sonderb.* **2**, 87–94.

McGinty, M. M. and Higgins, R. P. (1968). Ontogenetic variation of taxonomic characters of two marine tardigrades with the description of *Batillipes bullacaudatus* n. sp. *Trans. Amer. Microsc. Soc.* 87, 252–262.

Marcus, E. (1927). Zur Anatomie und Ökologie mariner Tardigraden. *Zool. Jahrb. Abt. Syst. Oekol. Geogr. Tiere* 53, 487–558.

Marcus, E. (1928). Zur Embryologie der Tardigraden. *Verh. Deut. Zool. Ges.* 32, 134–146.

Marcus, E. (1929). Zur Embryologie der Tardigraden. *Zool. Jahrb. Abt. Anat. Ontog. Tiere* 50, 333–384.

Pennak, R. W. (1953). "Fresh-Water Invertebrates of the United States." Ronald Press, New York, 769 pp.

Pollock, L. W. (1970a). *Batillipes dicrocercus* n. sp., *Stygarctus granulatus* n. sp., and other marine Tardigrada from Woods Hole, Massachusetts, U.S.A. *Trans. Amer. Microsc. Soc.* 89, 38–52.

Pollock, L. W. (1970b). Reproductive anatomy of some marine Heterotardigrada. *Trans. Amer. Microsc. Soc.* 89, 308–316.

Pollock, L. W. (1970c). Distribution and dynamics of interstitial Tardigrada at Woods Hole, Massachusetts, U.S.A. *Ophelia* 7, 145–165.

Ramazzotti, G. (1962). II Phylum Tardigrada. *Mem. Ist. Ital. Idrobiol.* 14, 1–595.

Renaud-Debyser, J. (1965). Répartition de deux tardigrades *Batillipes mirus* Richters et *Stygarctus bradipus* Schulz dans un segment de plage du bassin d'Arcachon. *C. R. Acad. Sci.* (*Paris*) 243, 1365–1367.

Renaud-Debyser, J. (1965). Étude sur un Stygarctidé (Tardigrada) nouveau de Madagascar. *Bull. Soc. Zool. Fr.* 90, 31–38.

Renaud-Mornant, J. (1967). *Parastygarctus higginsi* Renaud-Debyser, 1965, sur la côte orientale de Malaisie. Description de la femelle (Tardigrada). *Bull. Mus. Nat. Hist. Natur.* 39, 205–208.

Renaud-Mornant, J. and Anselme-Moizan, M. N. (1969). Stades larvaires du tardigrade marin *Stygarctus bradypus* Schulz et position systématique des Stygarctidae. *Bull. Mus. Nat. Hist. Natur.* 41, 883–893.

Richters, F. (1908). Marine Tardigraden. *Zool. Anz.* 33, 77–85.

Riggin, G. T. (1962). Tardigrada of Southwest Virginia: with the addition of a description of a new marine species from Florida. *Va. Agr. Exp. Sta. Bull.* 152, 1–145.

Rudescu, L. (1964). Tardigrada. *In* "Fauna Republicii Populare Romîne, Arthropoda," Vol. 55, Fasc. 7, pp. 1–400.

Schmidt, P. (1969). Die quantitative Verteilung und Populationsdynamik des Mesopsammons am Gezeiten-Sandstrand der Nordsee-Insel Sylt II. Quantitative Verteilung und Populationsdynamik Einzelner Arten. *Int. Rev. Gesamten Hydrobiol.* 54, 95–174.

Schulz, E. (1955). Studien an marinen Tardigraden. *Kiel. Meeresforsch.* 11, 74–79.

Swedmark, B. (1959). On the biology of sexual reproduction of the interstitial fauna of marine sands. *Proc. 15th Intern. Congr. Zool. London,* pp. 327–329.

Van Wenck, W. (1914). Entwicklungsgeschichtliche Untersuchungen an Tardigraden (*Macrobiotus lacustris* Duj.). *Zool. Jahrb. Abt. Anat. Ontog. Tiere* 37, 465–514 (plates 35–38).

von Erlanger, R. (1895). Beiträge zur Morphologie der Tardigraden. 1. Zur Embryologie eines Tardigraden: *Macrobiotus macronyx* Dujardin. *Morphol. Jahrb.* 22, 491–505 (plates XX and XXI).

Chapter 3

PRIAPULIDA

Jacob van der Land

3.1 Introduction

The Priapulida are a very small group of marine benthic worms, with only nine known species. Their relationship with other metazoan phyla is unclear, but they seem nearest to the kinorhynchs. Although often considered as "pseudocoelomates," sometimes in the phylum Aschelminthes, they do have a true coelomic cavity, lined with mesoderm (van der Land, 1970).

The well-known order Priapuloidea includes two families representing two very different types. The Priapulidae, with six species in four genera, are macrobenthic, predatory animals inhabiting more or less muddy sediments in cold water. *Tubiluchus corallicola,* the only known species of the Tubiluchidae, is a detritus feeder, adapted to an interstitial life in sandy sediments in warm water. I have published a systematic survey of this order (van der Land, 1970), with references to the literature on all aspects of its natural history. Recently the order Seticoronaria was erected for the tube-dwelling, tentaculate species *Maccabeus ten-*

taculatus, which was discovered in the Mediterranean, in muddy bottoms (Salvini-Plawen, 1974).

3.2 Asexual Reproduction

The Priapulida reproduce only sexually; asexual reproduction has not been observed. Their power of regeneration is limited and probably restricted to the tail, which regenerates rapidly after removal (Lang, 1948a); injuries to the body proper are fatal because of the loss of coelomic fluid.

3.3 Sexual Reproduction

All species are normally dioecious. Hermaphroditic individuals of *Priapulus* have been found (Molčanov, 1908; Lang, 1939), but they are apparently rare and should be considered abnormalities (Lang found one individual among hundreds of normal specimens). Parthenogenesis possibly occurs in *Maccabeus tentaculatus,* because all adult specimens that have been collected were females.

There are always two gonads (combined with the excretory organs) in the coelom of the posterior half of the abdomen. Separate urogenital pores open close to the anus at the posterior extremity of the abdomen.

3.3.1 Sexual Dimorphism

Priapulidae do not show external sexual dimorphism, but in *Tubiluchus* the sexes are markedly different (Figs. 1 and 2); the ventral surface of the abdomen in the male is covered with numerous cuticular specializations. Tubuli are sparsely strewn over the abdomen in both male and female, but, in the male, concentrations of these structures are present in the preanal region. Moreover, the male displays a number of organs altogether lacking in the female. There is a tiny club-shaped structure (clavula) just anterior to each of the two genital pores. Very small, rounded elevations of the cuticle (bullulae) are present on the midventral preanal bulge. Setae extend over the whole length of the abdomen, ending in two dense groups of tall setae anterior to the genital pores. In addition, there are many perigenital setae with a different morphology only occurring on the posterior part of the abdomen. A

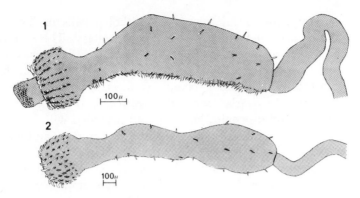

FIGS. 1 and 2. The tropical, meiobenthic *Tubiluchus corallicola* is the only priapulid showing a marked sexual dimorphism; the ventral surface of the abdomen is covered with numerous setae in the male. Figure 1, male (modified from van der Land, 1970); Fig. 2, female (modified from van der Land, 1968).

detailed description of all these structures can be found in my mono-graph (van der Land, 1970). Their function is as yet unknown but they probably play a role during copulation; it would be interesting to have observations on this point.

3.3.2 Development und Anatomy of the Reproductive System

The intricate structure of the gonads can best be explained by describing their development (a detailed description was given by Lüling, 1940). In larvae and juveniles there are only the two excretory organs, each consisting of a straight primary duct, which is attached to the lateral body wall by mesentery, and one or more solenocyte trees (bundles of solenocytes in the shape of a tree; Fig. 5). Their early ontogeny is still unknown.

In adults of *Tubiluchus* there is still only one solenocyte tree in each excretory organ, but in Priapulidae the number of solenocyte trees gradually increases from a few in the larvae to a considerable number in large adults. In *Halicryptus*, the solenocyte trees are distributed along the medial side of the primary duct, but in *Priapulus* the efferent canals of the solenocyte trees enter into one or more saclike expansions of the posterior part of the duct (nephridial sacs). In *Tubiluchus* the single solenocyte tree is also situated posterior to the gonads. Hence, in *Halicryptus* the whole primary duct becomes an urogenital duct, whereas in *Priapulus* and *Tubiluchus* the anterior (and by far the longest) portion serves solely as a genital duct.

The gonad originates from groups of epithelial cells of the anterior part of the primary duct, adjacent to the mesentery. The components of the gonad first develop as streaks of solid tissue, and later as secondary ducts, penetrating into the mesentery. In the male, testicular follicles develop from the walls of these secondary ducts, whereas in the female these ducts grow into ovarian sacs. The wall cells of the ovarian sacs and of the genital and urogenital ducts are all provided with long flagella (Fig. 6), undoubtedly aiding in the transport of secretions and gametes.

Apparently the gonads grow in size throughout the life of the animal (its life span is unknown, but probably many years); the primary duct and the mesentery grow in an anterior direction and new secondary ducts arise continuously anterior to the old ones. In large adults the gonads are intricate organs with thousands of testicular follicles or hundreds of ovarian sacs, so that their basically simple structure can no longer be easily recognized (Figs. 3–5, and 7).

The above description applies to *Priapulus* and *Halicryptus*, and probably also to *Priapulopsis* and *Acanthopriapulus*. The development of *Tubiluchus* has not yet been studied; it is probably basically the same as in the larger species, but the result is somewhat different. In this small animal the gonads have a tree-shaped structure, with the branches ending in testicular or ovarian follicles. The number of follicles is very small, in the order of ten to fifteen. The production of zygotes is also very small, as usual in animals adapted to an interstitial life. One might expect other differentiations of the internal reproductive organs for internal fertilization, but structures like seminal vesicles or receptacles have not yet been found.

A description of the ovary of *Priapulus* on the ultrastructural level was given by Nørrevang (1965).

3.3.3 Gametogenesis

According to Lang (1939) gametogenesis in *Priapulus* takes place throughout the whole year in Sweden. He could nearly always obtain viable spermatozoa and seemingly full-grown oocytes. However, experiments showed that fertilizable oocytes were only present in winter. Nørrevang (1965) observed all stages of oocyte growth in Denmark from early April until late October (winter conditions were not studied).

3.3.3.1 SPERMATOGENESIS

A short description of the spermatogenesis in *Halicryptus* and *Priapulus* was given by Lüling (1940). Young testicular follicles are nearly

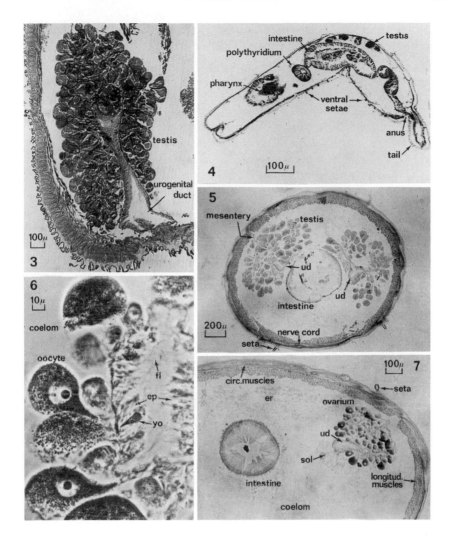

FIGS. 3–7. Figure 3 *Halicryptus spinulosus*, horizontal section through testis; Fig. 4. *Tubiluchus corallicola*, longitudinal section of male specimen; Fig. 5. *Halicryptus spinulosus*, cross section through testis region; Fig. 6. *Halicryptus spinulosus*, section through ovarian sac; Fig. 7. *Halicryptus spinulosus*, cross section through ovarian region. (Abbreviations: ep, epithelium of wall of ovarian sac; er, erythrocytes; fl, flagella; sol, solenocyte tree; ud, urogenital duct; yo, young oocyte).

completely filled with spermatogonia, originating from the epithelium of the secondary ducts. In mature follicles there is only a narrow outer layer of spermatogonia, the interior being filled with developing spermatocytes, spermatids, and spermatozoa. Franzén (1956) studied the

transformation of the spermatid into the spermatozoan. The young spermatids are still arranged in clusters. They contain one cup-shaped dictyosome, and many mitochondria. During development the mitochondria
accumulate on the peripheral side of the nucleus; they aggregate into
increasingly larger granules until only four large mitochondrial spheres
are left. These are situated around the insertion of the axial filament,
which has meanwhile grown out from a centriole at the nuclear membrane at the posterior pole of the nucleus. The dictyosome forms a
minute acrosome. The mature spermatozoan consists of a spherical head
with a small caplike acrosome, a short middle piece, containing the
four mitochondrial spheres, and a long, thin, undifferentiated tail. Hence,
the sperm is of the primitive type.

The spermatozoan of *Tubiluchus* has a strikingly different morphology
(Kirsteuer and van der Land, 1970). The acrosome is a slender, long
cone, with a spirally structured surface, and the head is ellipsoid, four
times as long as wide. These differences are probably correlated with
internal fertilization.

3.3.3.2 Oogenesis

A short description of the oogenesis in *Halicryptus* and *Priapulus*
was given by Lüling (1940); Nørrevang (1965) made a detailed electron
microscopical study of the development of the oocyte. The wall cells
of the ovarian sac are arranged in a one-layered epithelium. There seems
to be only one cell type and possibly all wall cells are prospective
oogonia. Each is provided with a long flagellum and a number of microvilli. During the early stages of the oogenetic process the flagellum and
the microvilli are lost, the nucleus enlarges, and a conspicuous nucleolus
becomes visible. At the same time neighboring wall cells surround the
oocyte and separate it from the lumen of the ovarian sac. During further
development the volume of the oocyte increases up to more than a
thousand times bulging into the surrounding coelom. In *Halicryptus*,
vitellogenesis starts although the oocyte is still in the epithelium (Fig.
6); in *Priapulus* this happens only after extrusion. Meanwhile the nucleus
and nucleolus grow to a considerable size and the full-grown oocyte
contains an enormous number of mitochondria, lipid droplets, and yolk
platelets. The eggs are shed as full-grown oocytes, and meiosis occurs
after fertilization.

The nutrients necessary for the growth of the oocyte come directly
from the surrounding coelomic fluid. The oocyte remains connected to
the ovarian epithelium by one or a few epithelial cells (acting as attachment cells), but these do not play a role in the transport of nutrients.
The oocyte is only separated from the coelom by the glycocalyx, a thin

membrane. It is still unknown how nutrients and waste products pass through the glycocalyx and the cell membrane [special modifications such as microvilli were not found by Nørrevang (1965)]. Because the oocytes hang freely in the fluid from which they receive their nutrients, *Priapulus* would be an ideal object for a biochemical and physiological study of the metabolism of oocytes, *in vivo* and possibly also *in vitro*.

3.3.4 Spawning and Mating

Priapulidae discharge both eggs and sperms into the water. Spawning has only been observed in the northern species *Priapulus caudatus* by Lang (1939, 1948a). He kept six pairs in separate small aquariums. The different pairs spawned on different days, but all at about the same time in the early evening. The male ejected its sperms first, giving the surrounding water a somewhat milky appearance. At first the spermatozoa were immobile, but after 10 to 15 minutes they started swimming. The female then showed signs of unrest, alternately contracting and expanding the body, and after a few minutes discharged the oocytes into the water. Occasionally the female spawns first, as was shown in later experiments; in these cases the oocytes are lost, because the male is apparently not stimulated by the presence of oocytes in the water.

Spawning in the male is likely to be stimulated by the presence of a female carrying full-grown oocytes. On the other hand, spawning of the female is apparently stimulated by the spawning of the male.

In Scandinavia, the breeding season of *Priapulus caudatus* seems to last from the second half of November to the first half of January. The only accurate observations on this point are those by Lang (1948a). On December 5th part of the smallest mature specimens of a population had finished spawning, whereas none of the medium-sized and large specimens had as yet spawned. By December 21st nearly all younger and some of the older specimens had spawned. On January 3rd only a few of the large specimens had not yet spawned and on January 8th only spent specimens were secured. Apparently younger specimens spawn earlier than the older ones or (in case the animals spawn more than once) they release all of their sexual products earlier.

Bull (1966) observed that the gonads of *P. caudatus* on the English coast were empty in March and "ripe" in October, and that sperms were liberated on November 29th.

Tubiluchus shows a conspicuous sexual dimorphism. It produces small numbers of gametes, and its sperms differ markedly from those of the normal primitive type. I once observed bodies that could possibly be sperm heads in the urogenital ducts of a female. These facts suggest

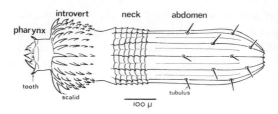

FIG. 8. *Tubiluchus corallicola*, morphology of larva, diagrammatic representation.

internal fertilization, but this has not yet been proved by direct observation of mating behavior. Probably mating is not a primitive character in the Priapulida; it is possibly one of the adaptations to an interstitial life.

3.4 Development

3.4.1 Embryonic Development

The embryology of the Priapulida is poorly known, perhaps because it is difficult to achieve normal development of the embryo in the laboratory situation. There are only a few observations on the early development of *Priapulus* by Lang (1953) and Žinkin (1949) and of *Halicryptus* by Žinkin and Korsakova (1953); their results are partly contradictory and need confirmation.

A fertilization membrane is formed immediately after fertilization; the zygote is slightly sticky and sinks to the bottom. Cleavage is total and equal (only occasionally are the cells of a markedly different size), of the bilateral type, occurring very slowly, the first cleavage only taking place after 20 hours and the fourth cleavage not before the 34th hour (temperature was not recorded). The polar body starts separating from the embryo at the time of the first cleavage. Usually it is completely detached during the 2-cell stage, but may remain attached until the 8-cell stage.

The presence of coeloblastula and gastrula stages is also well established, but further development is virtually unknown. According to Lang (1953) the embryo of *Priapulus* leaves the shell to become a contractile nonciliated larva, not much larger in size than the original zygote. This first larval stage has a quite simple morphology, consisting of an ectoderm enclosing a syncytial cell mass. A delicate cuticle is formed within a few days, but its further development has never been observed. Consequently, we do not yet know the genesis of the coelom, the fate of the blastopore, and the origin of the mouth and the anus, to mention but a few points of interest.

3.4.2 Larvae

Descriptions of the larvae are given in some detail by Hammarsten (1915), van der Land (1970), Lang (1948b), and Purasjoki (1944). The body consists of three regions (Fig. 8): introvert, neck, and abdomen; a tail is always lacking. The epidermis of the abdomen has a relatively strong cuticle, forming a lorica in which the animal can wholly withdraw (Fig. 11). Three types of larvae can be recognized at present (the larvae of *Acanthopriapulus* are still unknown): (1) in *Tubiluchus* (Fig. 9) the lorica is round in cross section and supplied with 20 equally spaced longitudinal ridges and up to about 20 tubuli (tactile organs); (2) in *Priapulus* (Fig. 10), *Priapulopsis,* and *Maccabeus* the lorica is dorsoventrally flattened (at least when the introvert

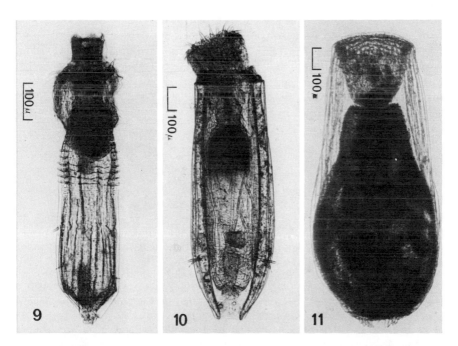

Figs. 9–11. The three types of larva occurring among the Priapulida. Figure 9, *Tubiluchus corallicola,* introvert and pharynx protruded; Fig. 10, *Priapulus caudatus,* introvert nearly completely protruded; Fig. 11, *Halicryptus spinulosus,* introvert withdrawn in the lorica.

is protruded) and provided with eight longitudinal ridges along its lateral sides and 4 or 8 (*Maccabeus*) tubuli; (3) in *Halicryptus* (Fig. 11) there are also eight lateral ridges but tubuli are lacking and the dorsal and ventral fields are more heavily cuticularized and provided with an areolate sculpture (so that one can speak of dorsal and ventral plates in this case).

Contrary to the peculiar external morphology, the internal anatomy is basically the same as in the adults. There is a partially protrusible pharynx (Figs. 8 and 9), a straight intestine, and a small excretory system. Typical larval organs are the adhesive glands in the introvert of *Halicryptus*.

The larvae of *Priapulus* dig in the mud and eat the superficial layer of mud, so they are considered detritus feeders (Lang, 1939). Their movements are very slow and inefficient because of the rigid lorica (Hammond, 1970).

The cuticle is molted between the successive larval stages. The number of stages has only been determined in *Halicryptus*, in which there are four stages (Purasjoki, 1944). The stages differ in size, but morphological differences are slight. It is also not known how long their development lasts, but it seems to be a considerable time. According to Lang (1939) development may take as long as 2 years in *Priapulus*. Schulz (1934) kept larvae of *Halicryptus* (probably not the youngest stages) for nearly 1 year in aquariums, but he did not observe metamorphosis.

During metamorphosis, which takes place in the cuticle of the last larval stage, the animal is contracted and motionless (Lang, 1939). After having left the larval exuvium the postlarva soon molts again. Some aspects of postlarval development have been described by van der Land (1970).

3.5 References

Bull, H. O. (1966). The marine fauna of the Cullercoats district. 3b. Priapulida, Echiurida, Sipunculida. *Rep. Dove Mar. Lab.* **15**(3), 13–16.

Franzén, Å. (1956). On spermiogenesis, morphology of the spermatozoon, and biology of fertilization among invertebrates. *Zool. Bidr. Uppsala* **31**, 355–480.

Hammarsten, O. (1915) Zur Entwicklungsgeschichte von *Halicryptus spinulosus* (von Siebold). *Z. Wiss. Zool.* **112**, 527–571.

Hammond, R. A. (1970). The burrowing of *Priapulus caudatus*. *J. Zool.* (*London*) **162**, 469–480.

Kirsteuer, E., and van der Land, J. (1970). Some notes on *Tubiluchus corallicola* van der Land (Priapulida) from Barbados, W. I. *Mar. Biol.* **7**, 230–238.

Lang, K. (1939). Über die Entwicklung von *Priapulus caudatus* Lam. *Kungl. Fysiografiska Saellsk. Lund Foerhandl.* **9**, 80–87.

Lang, K. (1948a). Contribution to the ecology of *Priapulus caudatus* Lam. *Ark. Zool.* **41A**(5), 1–12.

Lang, K. (1948b). On the morphology of the larva of *Priapulus caudatus* Lam. *Ark. Zool.* **41A**(9), 1–12.

Lang, K. (1953). Die Entwicklung des Eies von *Priapulus caudatus* Lam. und die Systematische Stellung der Priapuliden. *Ark. Zool.* **5**(2), 321–348.

Lüling, K. H. (1940). Über die Entwicklung des Urogenitalsystems der Priapuliden (Ein Beitrag zur Anatomie und Histologie dieser Tiere). *Z. Wiss. Zool.* **A153**, 136–180.

Molčanov, L. A. (1908). Beitrag zur Morphologie und Physiologie der Priapuliden. *Izv. Imp. Akad. Nauk* **2**(6), 957–967.

Nørrevang, A. (1965). Oogenesis in *Priapulus caudatus* Lamarck. An electron microscopical study correlated with light microscopical and histochemical findings. *Vidensk. Medd. Dansk. Natur. Foren.* **128**, 1–84.

Purasjoki, K. J. (1944). Beiträge zur Kenntnis der Entwicklung und Ökologie der *Halicryptus spinulosus*-Larvae (Priapulida). *Ann. Zool. Soc. Zool. Bot. Fenn.* **9**(6), 1–14.

Salvini-Plawen, L. v. (1974). Zur Morphologie und Systematik der Priapulida: *Chaetostephanus praeposteriens*, der Vertreter einer neuen Ordnung Seticoronaria. *Z. Zool. Syst. Evolut.-Forsch.* **12**, 31–54.

Schulz, E. (1934). Problematisches über das Larvenstadium von *Halicryptus spinulosus* v. Sieb. (Sipunculoidea). *Zool. Anz.* **105**, 59–61.

van der Land, J. (1968). A new aschelminth, probably related to the Priapulida. *Zool. Mededel.* **42**, 237–250.

van der Land, J. (1970). Systematics, zoogeography, and ecology of the Priapulida. *Zool. Verh.* **112**, 1–118.

Žinkin, L. (1949). Rannie stadii razvitija *Priapulus caudatus*. *Dokl. Akad. Nauk SSSR* **65**, 409–412.

Žinkin, L. and Korsakova, G. (1953). Rannie stadii razvitija *Halicryptus spinulosus*. *Dokl. Akad. Nauk SSSR* **88**, 571–573.

Chapter 4

SIPUNCULA

Mary E. Rice

4.1 Introduction

The phylum Sipuncula is a small group of unsegmented coelomate marine worms noted for its lack of morphological diversity. Comprised of 17 recognized genera and approximately 320 species, the group is distributed throughout the polar, temperate, and tropical oceans (Stephen and Edmonds, 1972). The worms dwell in benthic habitats, ranging from intertidal shores to abyssal depths. Commonly they burrow into sand, mud, or gravel and, in the tropics, are frequently found in burrows within dead coral or other calcareous rock. Some species inhabit discarded mollusc shells and others live under rocks or wedge themselves into crevices.

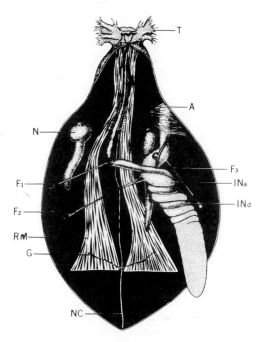

FIG. 1. Dissected specimen of *Themiste lissum*. A, anus; F₁, F₂, F₃, fixing muscles of the digestive tract; G, gonad; N, nephridium; NC, nerve cord; RM, retractor muscle; T, tentacles; INd, descending intestine; INa, ascending intestine. (Redrawn from Fisher, 1952.)

The adult body consists of an elongated cylindrical trunk with slender anterior introvert which may be withdrawn into the trunk by the contraction of one or more pairs of retractor muscles. The introvert is usually terminated by tentacles surrounding or dorsal to the mouth. The essential features of sipunculan anatomy are illustrated in Fig. 1, a dissected specimen of *Themiste lissum*. A long narrow esophagus, extending the length of the introvert, is continuous with the intestinal spiral in the trunk. Recurved, with descending and ascending limbs typically coiled about one another, the intestine opens to the exterior in a dorsal anus commonly located on the anterior trunk. Two nephridia, reduced in some species to one, open through ventrolateral pores usually at the approximate level of the anus. The nervous system includes a supraeso-phageal ganglion, circumesophageal connectives, and a ventral, un-paired and unsegmented median nerve cord. The gonad is commonly located at the base of the ventral retractor muscles.

The Sipuncula are classified as coelomate protostomes. Cleavage is spiral, the mouth forms in the position of the blastopore, and the coelom is formed by schizocoely. Characteristically there is a trochophore larva which in some species may be followed by a second larval type, the pelagosphera. A protonephridium, found in developmental stages of many protostomes, is absent in sipunculans. The complete lack of segmentation either during development or in the adult, distinguishes the Sipuncula from the Annelida and, in the opinion of most contemporary zoologists, justifies the recognition of the group as a separate phylum.

4.2 Asexual Reproduction

Asexual reproduction is known to occur in two species of sipunculans. A small, rock-boring species from the Caribbean Sea, *Aspidosiphon brocki,* reproduces asexually, dividing into two unequal parts (Rice, 1970). A constriction separates the smaller posterior daughter portion from the larger parent at a distance approximately one-fifth of the total length of the trunk from the posterior extremity (Fig. 2a). Before the completion of fission, the parent portion and the daughter portion each regenerates the structures essential to the formation of a new individual. In the case of the daughter the regenerated structures include the entire anterior body and introvert, anterior gut, retractor muscles, and nephridia. Adult structures contributed by the parent and incorporated into the daughter are one or two coils of the posterior intestinal spiral, posterior parts of the spindle muscle and ventral nerve cord, and coelomocytes (Fig. 3). The formation of the juvenile is completed at the time of detachment from the adult or shortly thereafter when the newly regenerated introvert is everted (Fig. 2c). The parent regenerates only the posterior body wall which is formed as an invagination anterior to the constriction and is everted at the time the daughter individual is detached. Asexual reproduction is a naturally occurring phenomenon in *Aspidosiphon brocki* and has been found in 15% of the specimens collected in a population in Key Largo, Florida. Sexual reproduction has not been observed in this species, although gonadal tissue is present in most individuals at the base of the retractor muscles.

A second asexually reproducing species, *Sipunculus robustus,* has been reported to reproduce both by transverse fission and lateral budding (Rajulu and Krishnan, 1969; Rajulu, 1975). The posterior third of the animal may constrict off by transverse fission to form a new individual or the posterior one-half or two-thirds may give rise to as many as 5 individuals simultaneously. The newly formed organs and structures of the daughter develop from a blastema composed of coelomocytes and pro-

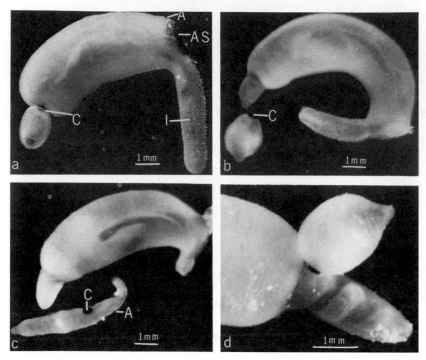

FIG. 2. Asexual reproduction in *Aspidosiphon brocki*. (a) Animal showing posterior constriction. (b) Recently separated parent and daughter. Note everted, newly regenerated posterior end of parent and black anterior cap of daughter. (c) Parent and juvenile, 2 days after separation. Regenerated anterior end of juvenile has been everted. (d) Posterior end of parent and daughter individual in process of separation. Regenerated posterior of parent has been everted and daughter remains attached only by fragments of the black material of the collar. A, anus; AS, anterior shield; C, collar on constricted animal or cap on daughter individual and juvenile; I, introvert. (From Rice, 1970.)

liferations of epidermal and gut cells of the parent. Most of the internal organs are formed before fission, but tentacles, introvert, and anus develop after detachment. Lateral buds may be formed in the region posterior to the anus, and at the time of separation from the parent the daughter individual is completely formed. Asexual reproduction in *Sipunculus robustus* has been observed only under conditions of stress such as maintenance of animals in stale sea water in the laboratory.

Although asexual reproduction has been only recently reported in sipunculans, their regenerative capabilities have been recognized for some time. In experimentally induced regeneration of the introvert of several species, formation of anterior gut, retractor muscles, and brain have been

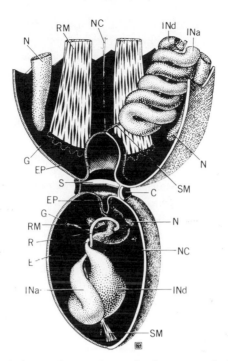

Fig. 3. Diagram of dissected posterior end of constricted *Aspidosiphon brocki* illustrating internal structures of daughter and posterior portion of parent. C, collar; E, esophagus; EP, epidermal invagination; G, gonad; INa, ascending intestine; INd, descending intestine; N, nephridium; NC, nerve cord; R, rectum; RM, retractor muscle; S, internal noncellular sheet of constriction; SM, spindle muscle. (From Rice, 1970.)

observed (Schleip, 1934; Wegener, 1938). Ectodermal elements have been presumed to form from a strand of regenerative cells in the nerve cord and mesodermal elements from coelomocytes. The literature on regeneration in sipunculans is reviewed by Hyman (1959).

4.3 Sexual Reproduction

4.3.1 Sexual Dimorphism

As a rule sipunculans are dioecious; however, external signs of sexual dimorphism are entirely lacking. Only in those species in which the body wall is translucent and the oocytes are of a distinctive pigmentation is it possible to ascertain by gross examination whether an individual is male or female. For example, in mature specimens of *Phascolosoma perlucens* the female will appear bright red in coloration due to the pigmentation of the oocytes within the coelom, whereas the males, because of concentrations of coelomic sperms, will be pale yellow or white. In many sipunculans it is possible through most of the year to determine sex without permanent injury to the animal by extracting a small sample of coelomic gametes with hypodermic syringe and needle and examining the sample microscopically for developing oocytes or spermatocytes.

4.3.2 Hermaphroditism

Hermaphroditism has been documented in only one species of sipunculan, *Golfingia minuta* (Åkesson, 1958). In this species, reported to be a protandrous hermaphrodite, coelomic oocytes and spermatocytes may occur simultaneously or only oocytes may be present. Coelomic oocytes of those animals with gametes of both sexes are immature. As the breeding season progresses, the percentage of animals with both male and female coelomic gametes decreases, and the percentage with oocytes only increases. It has been assumed, therefore, that the spermatocytes require a shorter time than the oocytes to mature and that an animal functions as a male before the definitive maturation of the oocytes is completed. The gonad is divided into specialized regions, the more median section near the ventral nerve cord producing only oocytes and the more lateral parts giving rise to both oocytes and spermatocytes.

An unexplained prevalence of females has been recorded in populations of some species. In a collection of 200 specimens of three different species of *Golfingia* (*G. elongata*, *G. vulgaris*, and *G. minuta*), Keferstein (1863) found only females. Claparède (1863) examined hundreds of specimens of *Golfingia elongata* and found only one or two males. In an Indian population of *Themiste signifer*, Awati and Pradhan (1936) reported a ratio of 1 male to 60 females. Cole (1952), in a study of the morphology of *Golfingia pugettensis*, found only 2 males in 100 specimens. In a later study on the development of this same population, the male to female ratio was 50:37 with 13% of undetermined sex (Rice, 1966).

4.3.3 Anatomy of the Reproductive System

The gonad of most sipunculans extends as a narrow digitate band of tissue along the base of the two ventral retractor muscles, extending from or near the lateral edge of one muscle, under the ventral nerve cord to the lateral edge of the other muscle (Fig. 1). In species such as *Phascolion strombi*, in which the number of ventral retractors has been reduced to one, the gonad may be asymmetric, extending from the ridge on the base of the muscle anteriorly for a short distance along one side of the ventral nerve cord. Male and female gonads are similar in form, although in the male the gonadal digitations are frequently more numerous and thinner. The length of the digitations may vary from 0.1 to 0.5 mm and in those species with a definite breeding season, the gonad may disappear or be considerably reduced in size during part of the year.

Enclosed by a peritoneal sheath and suspended by a peritoneal mesentery, the gonad has been presumed to originate from peritoneal cells which are transformed into gonia in the proximal region of the organ (Andrews, 1889; Hérubel, 1908). Within the gonad the gonocytes are found in a gradient of progressively advanced stages from proximal to distal ends of the digitations. At the distal border of the gonad the gonocytes are released into the coelom where, as freely floating cells bathed in coelomic fluid, they continue to grow and differentiate until the time of spawning when they are accumulated from the coelom into the nephridia.

The nephridium of a sipunculan is an elongated tubular structure with two openings, one external, the nephridiopore, and one internal, often referred to as the nephrostome (Fig. 4). Both openings are located

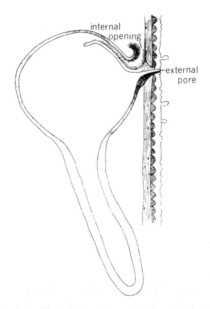

Fig. 4. Diagram of longitudinal section through nephridium and body wall of *Phascolosoma varians*. (From Shipley, 1890.)

at or near the anterior point of attachment of the nephridium to the
body wall. The internal opening is ciliated and funnel-shaped and frequently bordered by a crescentic lip. It joins the nephridial and coelomic
cavities by way of a narrow ciliated canal, and it is through this canal
that coelomic gametes are directed into the nephridium before spawning
by some still unknown mechanism. After a short period of storage in
the nephridium the gametes are discharged to the exterior through the
nephridiopore.

4.3.4 Origin of the Germ Cells

There is no evidence concerning the origin of the germ cells in sipunculans. The similarity in cytology of peritoneal cells and oogonia has led
to the assumption in the past that germ cells arise from the peritoneum
surrounding the base of the gonads. Andrews (1889) proposed that
". . . the nuclei of the peritoneum multiplied rapidly to form a mass
of germ nuclei." Later he amended his proposal to suggest that germ
cells might be retroperitoneal rather than peritoneal in origin (Andrews,
1890). Hérubel (1908), in an extensive treatise on the biology of sipunculans, stated that the oocytes ". . . sont certaines cellules péritonéales
qui se différencient."[*] Although Gonse (1956a) did not refer to the
source of the oogonia, he noted that the cells in the proximal region
of the gonad resembled those of the peritoneum. A similarity between
peritoneal and oogonial cells as well as an apparent elaboration or thickening of the peritoneum during the period of gonadal growth has been
reported (Rice, 1973).

The assumption that germ cells are derived from peritoneum does
not take into account the possibility of segregation of the germ line
in early developmental stages. The only observation of germ cells during
development was made by Gerould (1907). He found what he termed
"reproductive cells" at the base of the ventral retractor muscles in larvae
of *Golfingia vulgaris* and *Phascolopsis gouldi* at 2 to 3 weeks of age,
but he did not follow the embryological derivation of these cells. Until
more information is available any consideration of the origin of the
germ cells in sipunculans remains speculative.

4.3.5 Cytodifferentiation of the Gametes

4.3.5.1 Differentiation of Ovarian Oocytes

In the ovary of sipunculans, cytodifferentiation occurs in sequential
stages from proximal to distal ends, beginning with the oogonia or mitoti-

---

[*] Oocytes "are certain peritoneal cells which are differentiated."

cally dividing cells at the proximal end and progressing distally through the successive nuclear states of the meiotic prophase of the primary oocyte. At the distal border the cells have progressed to the diplotene stage with notable increases in nuclear and cytoplasmic volumes. It is at this stage that the oocytes are liberated into the coelom to undergo the remainder of their growth. Based on the successive nuclear changes, Gonse (1956a) has recognized seven regions in the ovary of *Golfingia vulgaris* (Fig. 5). In the first and most proximal region, oogonia are characterized by small nuclei similar to peritoneal cells and the divisions are mitotic. Cells of the second region are transitional from oogonia to the leptotene stage of the first meiotic prophase. The chromosomes are first condensed into balls in the telophase of the last mitotic division, then despiralized and grouped at the thin end of the nucleus. A nucleolus appears in this stage and the nucleus is increased in size. Region 3 consists of the leptotene stage. The chromosomes are strongly despiralized and the zone is narrow, indicating rapid change. The fourth and fifth regions are, respectively, the zygotene stage of chromosomal synapsis and the pachytene stage with thickened and elongated chromosomes in the bouquet form. The diplotene stage marks the sixth region; chromosomes are spiralized and tetrads are formed. In the seventh and most distal region the prophase is completed. Chromosomes are despiralized, nucleoli appear, and the nuclear and cytoplasmic volumes increase. Other authors have mentioned similar nuclear gradients in ovaries of *Phascolion strombi*, *Golfingia minuta* (Åkesson, 1958), *Phascolosoma agassizi*, *Golfingia pugettensis* (Rice, 1974), and *Phascolosoma arcuatum* (Green, 1975).

Gonse (1956a) noted, in addition to the oocytes in the ovary of *G. vulgaris*, small cells which he designated as "annex cells" and assumed to be abortive oocytes. These cells, later to become the follicle cells of the coelomic oocytes, are first distinguishable in region 2 and undergo the same meiotic changes as the developing oocytes. In the ovaries of other species, similar small cells, assumed to become the follicle cells of later stages, have been found among the oocytes; however, meiotic changes have not been observed and they have been interpreted as infolded peritoneal cells rather than as abortive oocytes (Åkesson, 1958; Rice, 1974).

4.3.5.2 Differentiation of Coelomic Oocytes

After liberation from the ovary, oocytes of sipunculans undergo vitellogenesis and the major portion of their growth as freely floating cells suspended in coelomic fluid and surrounded by coelomocytes. Commonly the oocytes break off from the ovary in clumps of 10–20 cells, interspersed

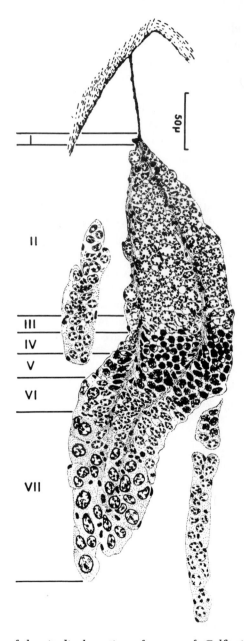

Fig. 5. Diagram of longitudinal section of ovary of *Golfingia vulgaris*. Ovary is suspended by peritoneal mesentery from base of retractor muscle. Seven regions of successive nuclear changes are designated. I, oogonia; II, transitional from oogonia to first meiotic prophase; III, leptotene stage; IV, zygotene stage; V, pachytene stage; VI, diplotene stage; VII, completion of first meiotic prophase; chromosomes despiralized, nuclear and cytoplasmic volumes increased. (From Gonse, 1956a.)

with smaller "annex" or peritoneal cells. Once in the coelom the clumps soon disperse into single oocytes, the smaller cells arranging themselves around the periphery of the oocytes to become the follicle cells. An exception is found in oocytes of *Phascolosoma,* which are usually detached from the ovary as single cells and lack a covering of follicle cells. During the period of coelomic growth the volume of an oocyte increases as much as 200 or more times.

Coelomic oogenesis has been studied in four species: *Golfingia vulgaris* (Gonse, 1956a,b, 1957a,b), *Golfingia ikedai* (Sawada *et al.,* 1968), *Golfingia pugettensis,* and *Phascolosoma agassizi* (Rice, 1974). Incidental observations on oogenesis in additional species are found in various studies on development or reproductive cycles (Gerould, 1907; Åkesson, 1958; Green, 1975). In the most comprehensive of the studies on oogenesis, Gonse has investigated cytological, cytochemical, and physiological properties of coelomic oocytes of *Golfingia vulgaris.* He distinguished six stages of coelomic oocytes, which he referred to as 0, 1, T, 2, 3, and M. Stage 0 is represented by clumps of oocytes and "annex" cells recently detached from the ovary and stage M is the final or mature stage which precedes nephridial accumulation and spawning. Stages 0, T, and M are considered to be transitory. Each stage is defined cytologically by size, appearance of nucleus, and kind and localization of cytoplasmic inclusions. Characteristics of the stages are summarized in Fig. 6. Similar stages of growth have been demonstrated in coelomic oocytes of other species, although specific variations occur in the persistence of follicle cells, size and shape of oocytes, structure of egg envelope, and localization and relative time of appearance of cell inclusions.

Follicle cells, when present, are detached during coelomic oogenesis, but the relative time of detachment varies in different species. Detachment occurs in *Golfingia vulgaris* (stage T) and *G. ikedai* when the oocytes reach a diameter of 60 μm. In *G. pugettensis* it is much later; in this species at the time the oocytes attain a diameter of 90 μm the follicle cells are raised up from the surface of the oocyte, standing out as blebs over the egg's surface, but they are not detached until the final stage of coelomic oogenesis when the oocytes are approximately 150 μm in diameter (Fig. 7a–d). In coelomic oocytes of *Themiste pyroides,* ranging in diameter from 30 to 190 μm, follicle cells are elevated from the surface of oocytes at a diameter of 100 μm and are retained to a diameter of 186 μm at which stage they are replaced by a jelly layer 50 μm in thickness.

While in the coelom as freely suspended cells, oocytes may manifest their first indication of polarity. Oocytes of species of *Phascolosoma,* released from the ovary as spherical cells usually 20 μm in diameter,

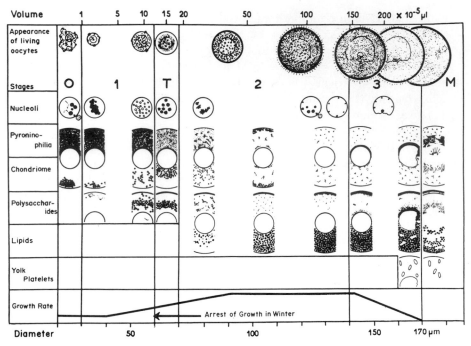

FIG. 6. Pictorial summary of differentiation of coelomic oocytes of *Golfingia vulgaris*. Six stages of coelomic differentiation are designated: O, oocytes in clumps, recently detached from ovary along with annex cells; 1, clumps dispersed as single oocytes with surrounding follicle cells, growth begins; T, transitory stage, follicle cells lost; 2, main phase of growth; 3, growth decelerated, burst of RNA production, appearance of yolk platelets; M, final or "mature" stage preceding nephridial accumulation, dissolution of germinal vesicle. (From Gonse, 1956a, p. 222.)

are changed during coelomic oogenesis to slightly flattened ellipsoids (Fig. 7e–h) which in *P. agassizi* measure $140 \times 110 \times 90$ μm. The polarity and bilateral symmetry exhibited at this phase of oogenesis are retained in the developing embryo. Polarity in coelomic oocytes of *Golfingia vulgaris* first becomes apparent when the nucleus undergoes an asymmetric flattening in stage 3. Oocytes of most other species of *Golfingia* and *Themiste* are spherical throughout coelomic oogenesis with little or no manifestation of polarity.

The egg envelope of the fully differentiated coelomic oocyte of sipunculans is comprised of several layers and perforated by pore canals. Cytoplasmic extensions or microvilli extend through the pore canals and in a few species the entire egg is covered by an outer jelly coat. The egg envelope has been characterized cytochemically as a mucoprotein

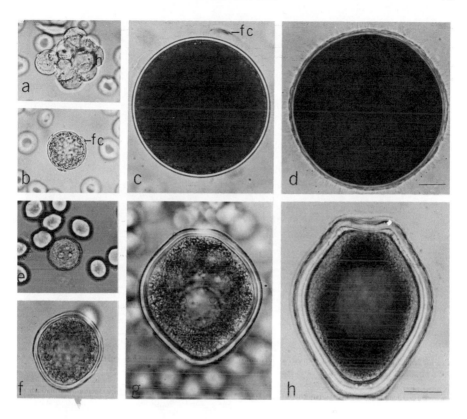

Fig. 7. Photographs of living coelomic oocytes. (a–d) *Golfingia pugettensis*; (c–h) *Phascolosoma agassizi*. Coelomocytes are seen out of focus in background. Scale, 25 μm. fc, Follicle cells. [From (in part) Rice, 1974.]

which is secreted by the oocyte during the period of coelomic growth (Gonse, 1956a; Rice, 1974).

The envelope of oocytes of *Phascolosoma agassizi* reaches a thickness of 10 μm and fully developed envelopes of *Golfingia pugettensis* and *G. ikedai* are 5 and 2 μm, respectively. In these latter species the number of layers may be as many as 14. In *P. agassizi*, three layers are distinguishable. As shown in electron micrographs, microvilli pass through pores in the inner and middle layers, branching into fan-shaped structures in the outer layer (Fig. 8). Tips of the microvilli are marked by an amorphous material or fuzz which consolidates in the latest oocytes as a homogeneous fringe around the egg and is expanded in the regions of the animal and vegetal poles. The inner layer gives a strongly positive reaction for mucoprotein and a weak response for protein whereas the

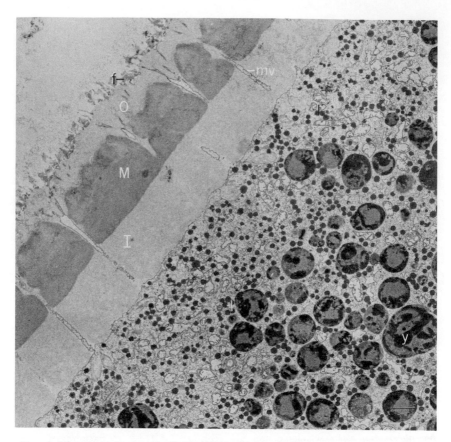

FIG. 8. Electron micrograph of a coelomic oocyte of *Phascolosoma agassizi,* showing egg envelope. O, M, I, Outer, middle, and inner layers of egg envelope; f, surrounding fuzz; 1, lipid; mv, microvillus; y, yolk granule. Scale, approximately 1 μm.

reverse is true of the middle layer. The outer layer has been identified as an acid mucopolysaccharide and the surrounding fuzz as a mucoprotein.

On the basis of their electron microscopic studies on coelomic oocytes of *Golfingia ikedai,* Sawada *et al.* (1968) have proposed a scheme of envelope formation (Fig. 9). The completed envelope of this species is composed of a fibrous outer layer, a multilamellate middle layer, and a diffuse inner layer. The earliest oocytes show no indication of an envelope, but projecting from the surface are numerous microvilli with pinocytotic vesicles at their bases. As the oocyte grows, the diffuse material of the inner layer, presumed to be precursor substances extruded from the cell surface, condenses to form in succession the 14 middle

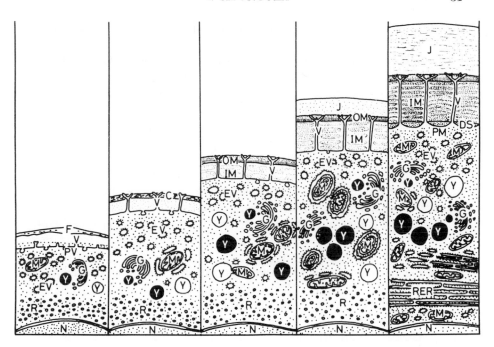

Fig. 9. Schematic representation of growth of coelomic oocytes of *Golfingia ikedai*, based on electron microscopic observations. C, chorion; DS, diffused substances; EV, endoplasmic vesicles; F, follicle; G, golgi body; IM, inner membrane; J, jelly layer; M, mitochondria; N, nucleus; OM, outer membrane; PM, plasma membrane; PV, pinocytotic vacuole; R, ribosome cluster; RER, rough-surfaced endoplasmic reticulum; V, microvillus; Y, yolk granule. (From Sawada *et al.*, 1968, p. 37.)

layers. The origin of the fibrous outer layer is uncertain, but the microvillar tips have been implicated. A jelly layer surrounding the oocyte in the latest stages may also be a secretion of the microvilli.

The nucleus in early oocytes is always rounded, occupying a central position in the cell. Later, but at characteristic times in different species, the nucleus is marked by numerous peripheral infoldings. In *Golfingia vulgaris* and *Phascolosoma agassizi* this occurs in the final stage of oogenesis; in the former the folds occur on only one side of the nucleus. The nucleus of *G. pugettensis* becomes scalloped much earlier, when the diameter of the oocyte is only one-half of its mature dimension. The nucleoli appear as numerous fragments in the early oocytes of *G. pugettensis* and *G. vulgaris*, later diminishing in *G. pugettensis* to two or three discrete nucleoli with few fragments and in *G. vulgaris* to several discrete spherical structures arranged around the periphery, one of which in the final stage dominates the others in size. The oocytes of *P. agassizi* have 2–5 nucleoli throughout oogenesis, but in younger stages one is larger than the others.

Mitochondria, as observed in two species of *Golfingia* (Gonse, 1956a; Sawada *et al.*, 1968), are localized in early oocytes near the periphery of the cell, but as their numbers increase in later stages they move throughout the cytoplasm. In electron microscopic studies of *G. ikedai* the mitochondria are found to be fused with peripheral endoplasmic vesicles in early coelomic oocytes; in later stages they are associated with one of two kinds of yolk granules.

Pyroninophilia, interpreted as a probable indication of the presence of ribonucleic acid, has been demonstrated in the cytoplasm of coelomic oocytes of *Golfingia vulgaris*. Strong perinuclear concentrations occur in the earliest stages and again in the latest stage. Electron microscopy of the oocytes of *G. ikedai* has similarly revealed an early concentration of rough endoplasmic reticulum.

Vitellogenesis in *Golfingia vulgaris* is initiated soon after detachment of oocytes from the ovary when carbohydrate yolk granules, presumed to be galactogen, make their appearance at the periphery of the cells. As the oocytes develop the granules move toward the nucleus, then are dispersed toward the cortex where, in the final stage (stage M), they form a layer of cortical granules. A second polysaccharide, recognized histochemically as glycogen, appears as a diffuse substance in a perinuclear position late in coelomic oogenesis. Lipid does not appear until stage 2 when it is found in association with a mitochondrial mass in the perinuclear zone; later it disperses through the cytoplasm. At the time of germinal vesicle breakdown (stage M) the lipid is concentrated in the peripheral cytoplasm. Prominent elongate bodies or yolk platelets, not defined histochemically, but presumed to be proteinaceous yolk, occur in well advanced oocytes of stage 3 (Fig. 6). From investigations of physiological properties of the oocytes of this species, two peaks of exogenous respiration have been demonstrated during coelomic oogenesis, the earlier corresponding to the beginning of carbohydrate synthesis and both corresponding to periods of high concentrations of ribonucleic acid. The peaks of respiration were found to be coupled with metabolism of hexoses and pentoses (Gonse, 1957a,b).

In the elaboration of yolk there is little consistency among different species in the sequence in which carbohydrate, protein, and lipid make their appearance in the developing oocytes, or in the form and localization of the yolk granules within the cells. The carbohydrate yolk of *Phascolosoma agassizi* differs from that of *Golfingia vulgaris* in that it is associated with protein as a carbohydrate–protein complex in the form of distinctive yolk granules of irregular staining pattern. The granules are visible first in the perinuclear zone, rather than at the periphery as in *G. vulgaris,* and after dispersal through the cytoplasm they move in later stages away from the periphery leaving a clear cortical area. The earliest oocytes of *G. pugettensis* that have been examined show carbohydrate yolk granules midway between nucleus and periphery; no tests for protein have been made. Lipid in this species is present as large droplets at the periphery in the early oocytes, appearing relatively sooner than in *G. vulgaris.* It is then dispersed and finally, in late oocytes, as in *G. vulgaris,* it is localized in the peripheral cytoplasm. The large and rather rare lipid droplets of *P. agassizi* have not been seen until late in oogenesis, but very small granules, suspected to be lipid, are distributed through the cytoplasm in the earliest stages of coelomic oocytes. Characteristic of lipid, these small granules appear after centrifugation in the centripetal half of the egg and in electron micrographs they are seen as homogenous spheres without surrounding membranes (Fig. 8).

Recently spawned, unfertilized eggs of many species of sipunculans have been described in studies on development. Sizes of the eggs of 17 species can be found in Table II (see Section 4.4.2). All sipunculan eggs are encompassed by a thick characteristic envelope which is composed of several layers and perforated by pores. There is considerable variation among species in the thickness of the envelope and the number of its layers as well as in the size, shape, pigmentation, and yolk content of the eggs (Fig. 10 a–e). The eggs of most sipunculans are spherical, but those of a few species, such as *Paraspidosiphon fischeri* and *Golfingia minuta* are oval and those of the genus *Phascolosoma* are typically flattened ellipsoids, frequently with depressed apexes. Sipunculan eggs are commonly various shades of red or yellow. Large eggs with high yolk content, such as eggs of *Themiste,* may be white or grayish and eggs low in yolk, as found in species of *Sipunculus* and *Siphonosoma,* are transparent. At the time of spawning, the germinal vesicle of the eggs has broken down and meiosis is arrested in the first meiotic metaphase (see Sections 4.3.8, 4.4.1).

4.3.5.3 DIFFERENTIATION OF MALE GAMETES

Little is known about cytodifferentiation of male gametes. The cytology of the male gonad of *Phascolosoma arcuatum* has been reported to resemble that of the ovary, as described by Gonse (1956a) for *Golfingia vulgaris,* with the exception that the cytoplasmic volume of the distal spermatocytes is less than that of the oocytes in that position (Green, 1975). Spermatocytes break off from the testis as loosely associated clumps of cells which undergo two meiotic divisions and differentiate into spermatids while floating as morulae in the coelomic fluid. The cells within

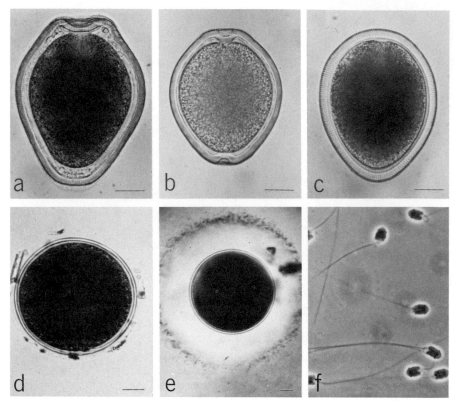

Fɪɢ. 10. Recently spawned gametes of sipunculans. Photographs of living eggs and sperms. Eggs are unfertilized. (a) Egg of *Phascolosoma agassizi.* (b) Egg of *Phascolosoma perlucens.* (c) Egg of *Phascolosoma antillarum.* (d) Egg of *Phascolion cryptus.* (e) Egg of *Themiste pyroides.* Note thick jelly layer. (f) Spermatozoa of *Themiste pyroides.* Scale a–e, 25 μm. (a and e from Rice, 1967. b, c, and d from Rice, 1975).

a clump increase in number as divisions occur and in each cell the cyto-plasmic volume decreases and a flagellum develops. The clumps of differentiated spermatids break up into free spermatozoa in the coeloms of *G. pugettensis* and *Themiste pyroides* where they may be seen through most of the year. In *P. agassizi* a few spermatozoa are free in the coelom at the time of the breeding season, but the majority remain in clusters. In *P. arcuatum,* spermatids and free spermatozoa occur in the coelomic fluid only a short time before spawning.

Spermatozoa of sipunculans have the typical morphology of the primi-tive sperm as defined by Franzén (1956). Similar to those of other species which discharge their gametes freely into the seawater, they

are comprised of low acrosomal caps, midpieces each with four mitochondrial spheres, and long filamentous tails. The sperms of *Themiste pyroides*
are somewhat modified in that they possess a highly developed acrosomal
cap with pointed tip and swollen basal rim (Fig. 10f). It is probable
that this increased complexity is associated with the thick jelly coat
of the egg through which the sperms must pass (Rice, 1974).

4.3.6 Gametogenic Cycles

Annual reproductive cycles are known for four species of sipunculans:
Golfingia vulgaris (Gonse, 1956b), *G. pugettensis* (Rice, 1966), and two
populations of *Phascolosoma agassizi* (Rice, 1966; Towle and Giese,
1967), all from temperate waters, and one tropical species, *Phascolosoma
arcuatum* (Green, 1975). Cycles have been defined in these studies by
estimations of breeding seasons from observations on spawning, cytological examination of gonads, and measurements of coelomic oocytes
throughout the year. Oocyte measurements have been expressed in relative frequencies and no absolute estimations are available of the various
sizes of cells.

Golfingia vulgaris has a limited breeding season lasting from June
to September in Roscoff, France. The gonad, however, is continuously
active, releasing small oocytes into the coelom throughout the year,
with reduced activity in the winter. Monthly measurements of coelomic
oocytes over a year show that although oocytes are always present in
the coelom their growth is arrested during part of the year (Figs. 6
and 11). In the winter all of the oocytes are small, i.e., less than 61
μm, and the size-frequency curve is unimodal, indicating arrested
growth. Growth commences in the spring and with the appearance
of a slightly larger group of cells the size-frequency curve becomes
bimodal, the second mode at 66 μm, close to the first. In the summer
the latter mode disappears and all sizes of oocytes are found: the
smallest oocytes persist as a prominent mode, intermediate oocytes
are present but in low frequencies, and a population of large oocytes
makes its appearance, forming a second mode at 154 μm. In the autumn, at the conclusion of the breeding season, growth of small oocytes
is again arrested and oocytes of intermediate and large size gradually
disappear. Low frequencies are interpreted as indicative of a rapid
rate of growth and high frequencies as slow growth. Thus, those populations that form prominent modes, that is, the small and large cells,
grow more slowly than the cells of intermediate size (Fig. 6).

In the annual cycle of *Golfingia pugettensis*, endemic to the Northwest
Pacific Coast of the state of Washington, the breeding season occurs

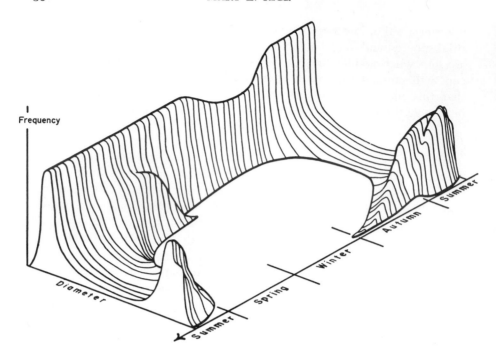

Frequency

Diameter

Summer Spring Winter Autumn Summer

Fig. 11. Three-dimensional diagram of size-frequency curves of coelomic oocytes of *Golfingia vulgaris* over a period of 1 year. The absolute value of the frequency of the small oocytes is arbitrarily kept constant except in the autumn when the ovary is less active and the production of coelomic oocytes is decreased. (From Gonse, 1956b, p. 232.)

in October, November, and December. During this period size-frequency curves show, as for *G. vulgaris*, two major populations of coelomic oocytes, one of small oocytes and one of large, with only a small proportion of oocytes of intermediate size (Fig. 12). In January, after the breeding season, large oocytes are no longer present but the population of small oocytes persists and a new population of slightly larger cells becomes evident. This exists through the spring but disappears in the summer at the same time several small populations of cells of intermediate size are formed.

Although the breeding periods of *Golfingia pugettensis* and *G. vulgaris* are at different seasons, the latter spawning during the summer months, there is, nevertheless, the same sequence of oocyte growth with similar phases of apparent arrest and acceleration. However, the relative time of appearance of the various phases differs. In *G. pugettensis* there is an apparent arrest of growth of small oocytes during the breeding season, and after the large eggs are discharged by spawning, the small

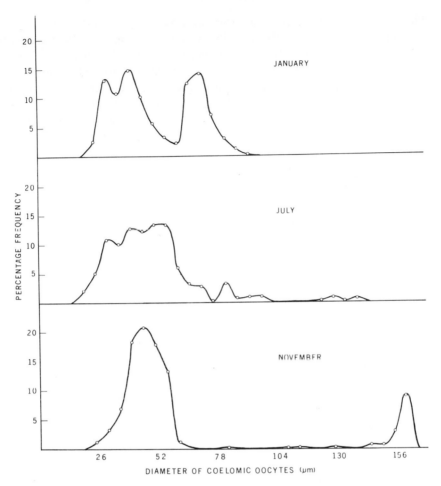

Fig. 12. Size-frequency curves of coelomic oocytes of *Golfingia pugettensis* (1962–63). In January 450 eggs were measured, 50 from each of 9 animals. In both July and November, 300 eggs were measured, 100 from each of 3 animals. (From Rioo, 1966.)

eggs begin to grow. In *G. vulgaris* the frequency of small eggs passing into the intermediate stage is much higher; growth of small oocytes continues throughout the breeding season and is arrested only after spawning, the period of arrest lasting for several months. Thus during the period of spawning in *G. vulgaris* intermediate oocytes are prevalent. Those remaining after spawning are presumably resorbed. In *G. puget-tensis* by the time of spawning most of the intermediate cells have already given rise to mature cells and the growth of the oocytes has ceased or been considerably reduced.

The oogenic cycle of a population of *Phascolosoma agassizi* in the San Juan Archipelago off the Northwest Pacific Coast of Washington, differs from that of both *Golfingia pugettensis* and *G. vulgaris* in that oocytes of all sizes, small, intermediate, and large, are present throughout the year (Rice, 1966). An examination of the frequency polygons reveals that there are always two principal size groups of cells, the largest oocytes usually predominating except immediately after spawning (Fig. 13). This population of *P. agassizi* spawns from early June through August.

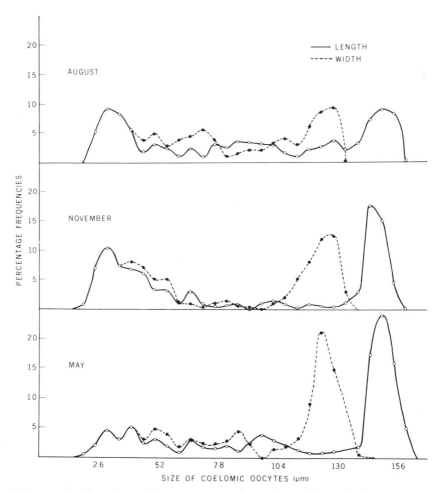

FIG. 13. Size-frequency curves of coelomic oocytes of *Phascolosoma agassizi* from the San Juan Archipelago, Washington (1964–1965). Measurements each month represent a total of 500 oocytes, 100 from each of 5 animals. (From Rice, 1966.)

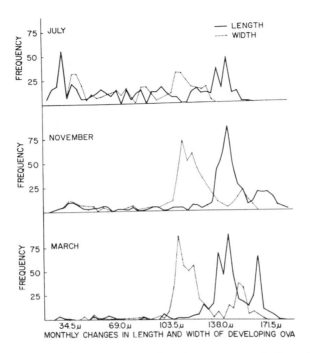

FIG. 14. Monthly changes in size of coelomic ocytes of *Phascolosoma agassizi* from Monterey Bay, California (1960–1961). Frequency distribution each month represents measurements of 500 oocytes, 20 from each of 25 females. [From Towle and Giese, 1967, fig. 1 (in part), p. 232.]

A population of the same species at Monterey, California spawns three months earlier, in March, and for a 3-month period following spawning both male and female gametes are absent in the majority of animals (Towle and Giese, 1967). By the fourth month, July, all sizes of coelomic oocytes are found, including high proportions of both small and large oocytes (Fig. 14). During the fall and winter the frequency of the small oocytes diminishes while the large oocytes proportionately increase until a month before the March spawning, when approximately 90% of the coelomic oocytes are large. By contrast, the month before the breeding season in the population from the San Juan Archipelago only 55% of the coelomic oocytes are in the large size group. An important difference in the oogenic cycles of the two populations seems to be the degree and duration of reduction in gonadal activity. The gonad of the Monterey animals is active for only a limited time during the year and after a temporary disappearance, again develops following the spawning season. In the animals from San Juan Island the gonad is continually active but partially reduced before and during the breeding season.

In summary, the growth of the coelomic oocytes of *Golfingia pugettensis* and *G. vulgaris* is not continuous, but is characterized by phases of arrest and acceleration. In contrast, the coelomic oocytes of *Phascolosoma agassizi*, including both the population from Monterey and San Juan Island, appear to undergo continual growth, although not necessarily at a continuous rate, until the definitive size is attained.

A tropical species, *Phascolosoma arcuatum* from Queensland, Australia, breeds from December through February and, like the population of *P. agassizi* from Monterey, California, but unlike other species from higher latitudes, it lacks coelomic oocytes entirely for a 2-month period following the breeding season (Green, 1975). Moreover, in *P. arcuatum*, spermatocytes are released from the testis only a short time before breeding begins in December and development into spermatozoa is rapid. This is in contrast to other species of sipunculans in which male gametes are present in the coelom much of the year (Section 4.3.5.3).

4.3.7 Factors Influencing Gametogenesis

Little is known of the factors controlling the growth of gametes in sipunculans. It has been suggested (Åkesson, 1961a) that neurosecretory products may play some role in the reproductive cycle since Carlisle (1959), in studies of *Sipunculus nudus*, found a greater amount of neurosecretory material in animals before the breeding season than following it. Neurosecretory material has been demonstrated in other sipunculans (Gabe, 1953; Åkesson, 1961a), but its significance in reproduction has not been investigated.

The possible influence of exogenous factors such as temperature on gametogenesis of sipunculans is suggested by the difference in the duration of breeding seasons of some tropical and temperate species. Observations on spawning in certain tropical or subtropical species, e.g., *Phascolosoma perlucens*, *Themiste alutacea*, and *T. lageniformis*, indicate that breeding may occur throughout the year (Section 4.3.8). However, no information is available on gametogenesis in these species, and in other tropical species, such as *P. arcuatum*, the breeding season is known to be restricted, as it is in all species from temperate waters. In species with well-defined breeding seasons differences may be found in the time of initiation of breeding in the same species at different latitudes. At Monterey, California, *Phascolosoma agassizi* breeds 3 months earlier than at 11° farther north in the San Juan Archipelago, Washington (Rice, 1966; Towle and Giese, 1967). However, in two populations of *P. arcua-*

tum in Australia, separated by 8° in latitude, correlation between temperature and spawning is the reverse, that is, the population at the higher latitude spawns first (Green, 1975).

4.3.8 Spawning and Breeding Periods

Gametes are spawned from the nephridia by forceful ejection through the nephridiopores into the seawater where fertilization occurs. When spawning is imminent the nephridiopores are often swollen and the animals may become quite active, frequently extending and retracting the introverts. *Phascolosoma agassizi,* when maintained in an aquarium with sand and gravel, has been observed to extend the anterior end above the surface of the substratum so that the nephridia are well exposed, and at the time gametes are released the body is extended and turgid (Rice, 1966). Although usually in sipunculans all of the gametes from both nephridia are spawned at once, variations of this process have been observed. A male specimen of *Paraspidosiphon fischeri* spawned 7 times in the laboratory over a period of 40 minutes, first several times from the right nephridium, then from the left. *Phascolion cryptus,* with only a single nephridium, was observed to spawn short intermittent spurts of sperm over a period of 15 minutes (Rice, 1975).

Since sipunculans, along with the majority of marine invertebrates, spawn their gametes into the seawater, synchronization of spawning is essential to assure fertilization and the survival of the species. Synchronization in sipunculans is dependent on the regulation of two separate events: (1) uptake of coelomic gametes by nephridia; (2) release of gametes from the nephridia into the seawater. Factors controlling these events have not been analyzed, but hypotheses have been proposed for nephridial uptake and data have been accumulated on such aspects of spawning as sequence of male and female activity, time of day that spawning occurs, and intervals between spawnings within a breeding season.

Before spawning, gametes are accumulated into the nephridium from the coelom by way of the internal funnels or nephrostomes. The process of nephridial selectivity by which the most mature gametes, either oocytes or sperm, are taken into the nephridium, while the immature gametes and coelomocytes are rejected, is poorly understood. Since oocytes are known to enter the nephridium soon after breakdown of the germinal vesicle (Gerould, 1907; Åkesson, 1958; Rice, 1966) it can be assumed that nephridial selectivity is related to the initiation of maturation of the gametes. (It should be noted, however, that maturation divisions do not occur until after sperm entry; see Section 4.4.1.) Gerould (1907) noted in *Golfingia vulgaris* that before spawning the nephridia became greatly distended with fluid, which he presumed to be seawater taken in through the nephridiopores. Assuming that the eggs selected by the nephridium were hydrotropic because of their lower specific gravity, Gerould proposed that these eggs were accumulated in the region of the nephrostome and directed by ciliary currents into the nephridia where they resorbed water. Åkesson (1958) further suggested that a chemical change occurring in the oocyte at breakdown of the germinal vesicle might alter the direction of nephridial cilia which would then move these eggs into the nephridium. In explaining nephridial uptake of mature sperm, Åkesson noted that seawater induced motility of coelomic sperm. He then hypothesized that as the nephridia filled with seawater, some of the water passed into the coelom through the nephrostomes and the mature sperm were activated to swim toward the increasing concentration of seawater to the nephridium.

Breakdown of the germinal vesicle in the coelom and dispersal of coelomic sperm clusters are apparent prerequisites for nephridial uptake. In many species oocytes of maximum size with intact nuclei are present in the coelom through much of the year (Section 4.3.6), but only during the breeding season shortly before spawning does dissolution of the germinal vesicle occur. Few studies have been made of the mechanisms regulating the maturation of gametes. Pasteels (1935) carried out *in vitro* studies on the breakdown of the germinal vesicle in coelomic oocytes of *Phascolion strombi*. He found that the addition of increasing quantities of calcium ions induced maturation of the eggs. In preliminary studies (Rice, 1966) it was found that crude extracts of coelomic sperm, coelomic oocytes, coelomocytes, brain, and muscle all induced germinal vesicle breakdown *in vitro* of coelomic oocytes of *Phascolosoma agassizi*. The majority of coelomic sperm in *P. agassizi* remain in the form of clusters until a short time before nephridial uptake when the clusters break down into free spermatozoa. Experimental dispersal of spermatid clusters was accomplished *in vitro* by exposure to hypertonic seawater. The resultant free spermatozoa became motile within 24 hours after transfer to seawater.

After nephridial uptake gametes remain in the nephridium only a brief period before spawning occurs. Gerould (1907) noted in *Golfingia vulgaris* and *Phascolopsis gouldi* that the nephridia filled with gametes a few hours before spawning. Åkesson (1958) reported that oocytes of *Phascolion strombi* underwent germinal vesicle breakdown 12 hours before spawning. In small transparent specimens of *Paraspidosiphon fischeri* in which nephridia were distended with gametes at the time of collection, spawning occurred within periods varying from 1 to 24 hours.

The sequence of male and female spawning is variable in different species. At least two species of sipunculans, *Golfingia vulgaris* and *Phascolion strombi* (Gerould, 1907, Åkesson, 1958), follow Thorson's rule of epidemic spawning in which the males spawn first, stimulating the spawning of the females (Thorson, 1946). In populations of *Phascolosoma agassizi* from Monterey, California (Towle and Giese, 1967) and from the San Juan Archipelago (Rice, 1967), spawning of females has been observed to precede that of males. In the latter population the female spawning may be followed by the male or either sex may spawn independently. Only a small proportion of the animals placed in one container spawned at the same time. It was found that spawning can often be triggered in the laboratory, if gametes are present in the nephridia, by a change to freshly aerated water or a sudden change in temperature. No observations have been made on the spawning of sipunculans under natural conditions in the field.

Gerould (1907), in his studies of *Golfingia vulgaris* and *Phascolopsis gouldi*, reported that spawning is confined to hours of darkness between 2000 and 0400 or 0500. He stated further that when animals were maintained continuously in a dark aquarium the rhythm was interrupted and spawning sometimes occurred during the day. His explanation for spawning at night was that the sipunculan's body relaxes in darkness, resulting in distension of the nephridia with consequent nephridial uptake of surrounding seawater and accumulation of mature hydrotropic eggs from the coelom. Spawning at night has also been reported for *Phascolion strombi* and *Golfingia elongata* (Åkesson, 1958, 1961a). However, in *Phascolosoma agassizi* spawning did not occur exclusively at night. Of a total of 89 spawnings, 9% spawned during daylight hours, 38% at night but in the artificial illumination of the laboratory, and 53% in darkness. Thus Gerould's explanation for night spawning would not be applicable in 47% of these cases. Again, in *Themiste pyroides*, of the spawnings of 20 animals recorded over a 3-year period, 55% occurred overnight, the remainder during daylight hours (Rice, 1967).

The only suggestion for periodicity in spawning in sipunculans is found in records for *Phascolosoma agassizi* (Rice, 1967). A total of 99 spawnings were recorded in the laboratory over a period of 3 years. The data, summarized in Fig. 15, reveal a peak of maximum spawning each year as well as a certain periodicity throughout the breeding season. In 1962 and 1963 the periodicity between spawning peaks ranged from 21 to 29 days. There was no obvious correlation of spawning periods with either tides or phases of the moon. Since these observations were made under uncontrolled conditions in the laboratory it cannot be assumed that the same periodicity exists in the natural habitat.

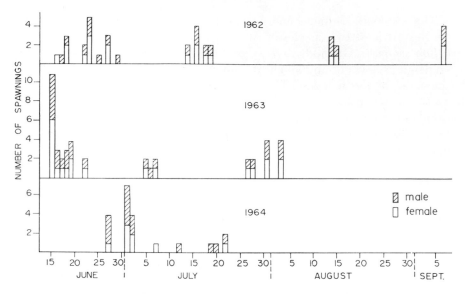

FIG. 15. Spawning dates and number of spawnings of *Phascolosoma agassizi* from the San Juan Archipelago, Washington. Spawnings were recorded in the laboratory from 1962 to 1964. (From Rice, 1967, p. 147.)

Although there may be a periodicity in spawning during the breeding season there is not a complete gametogenic cycle between each spawning in sipunculans. A single animal may spawn repeatedly during a breeding season as evidenced in the six recorded spawnings of a single female of *Themiste pyroides* during an 8-week period (Rice, 1967). It seems probable that each spawning is preceded by a maturation of only a certain percentage of the largest oocytes in the coelom which are then collected by the nephridium.

Estimations of breeding periods have been made for 10 species of sipunculans by one or more of the following methods of study: observations on spawning in the laboratory over a period of a year, annual studies of gametogenic cycles with periodic examination of coelomic oocytes and gonads, or from studies concerned primarily with development. Additional observations on spawning of 9 species have been made at sporadic intervals, but not carried out over a period of a year. It is important to note, therefore, that in these instances the times of recorded spawning are not indicative of the range or duration of breeding seasons, but show only the months in which spawning was observed and can be expected to occur. The information is included for its possible value to persons who may wish to carry out further investigations in the localities named. The data are summarized in Table I.

TABLE I

BREEDING PERIODS OF SIPUNCULANS

Species	Locality	Reference	Breeding period of months of observed spawning
Golfingia elongata	Roscoff, France	Åkesson, 1961a	July, Aug.
Golfingia minuta	Kristineberg, Sweden	Åkesson, 1958	Sept.–Nov.[a]
Golfingia pellucida	Fort Pierce, Florida	Rice, unpublished	Feb.–May, Aug., Sept. Nov.
Golfingia pugettensis	San Juan Islands, Washington	Rice, 1967	Oct.–Jan.[a]
Golfingia vulgaris	Roscoff, France	Gerould, 1907; Gonse, 1956b	June–Sept.[a]
Paraspidosiphon steenstrupi	British Honduras	Rice, unpublished	June
Paraspidosiphon fischeri	Isla Margarita, Venezuela	Rice, 1975	Oct.–Dec.
	Isla Perico, Panama	Rice, unpublished	July
	Galeta, Panama	Rice, unpublished	June, July
	Key Largo, Florida	Rice, unpublished	Oct.
Phascolion cryptus	Virginia Key, Florida	Rice, 1975; Rice, unpublished	July–Nov.
Phascolion strombi	Kristineberg, Sweden	Åkesson, 1958	Sept., Oct., Nov.[a]
Phascolopsis gouldi	Newport, Rhode Island	Gerould, 1907	June–Aug.
	Woods Hole, Massachusetts	Gerould, 1907	Aug.–Sept.
Phascolosoma agassizi	Monterey, California	Towle and Giese, 1967	Mar.–May[a]
	San Juan Islands, Washington	Rice, 1967	June–Sept.[a]
Phascolosoma antillarum	Key Biscayne, Florida	Rice, 1975; Rice, unpublished	July–Sept.
Phascolosoma arcuatum	Ross River, Queensland, Australia	Green, 1975	Dec.–Feb.[a]
Phascolosoma perlucens	La Parguera, Puerto Rico	Rice, 1975	April
	Barbados	Rice, 1975	Feb., Mar.
	Isla Margarita, Venezuela	Rice, 1975	Oct.–Feb.
Themiste alutacea	Key Biscayne, Florida	Rice, 1975, unpublished	Nov.–Apr., June–Sept.[a]
	Isla Margarita, Venezuela	Rice, 1975	Oct.–Dec.
Themiste lageniformis	Fort Pierce, Florida	Rice, unpublished	Mar.–Aug., Oct.[a]
Themiste pyroides	San Juan Islands, Washington	Rice, unpublished	Mar.–Aug., Oct., Dec.[a]
	Vancouver Island, British Columbia	Rice, 1967	March–August[a]
Siphonosoma cumanense	Tampa Bay, Florida	Rice, unpublished	June
Sipunculus nudus	Tampa Bay, Florida	Rice, unpublished	June
	Naples, Italy	Hatschek, 1883	July

[a] Indicates duration of breeding season; population observed during all seasons of the year.

Of the 10 species for which the breeding season has been determined with some certainty, 6 are temperate and 4 tropical or subtropical. All of the temperate species and one of the tropical species (*Phascolosoma arcuatum*) have well-defined breeding seasons. Three of the tropical or subtropical species, *Themiste lageniformis*, *T. alutacea*, and *Phascolosoma perlucens*, appear to spawn throughout much of the year.

4.4 Development

4.4.1 Fertilization

Sipunculan eggs when spawned into the seawater are arrested in the first meiotic metaphase. The process of fertilization, beginning with the entrance of the sperm into the egg, includes the extrusion of the first and second polar bodies and enlargement of the sperm nucleus to form the male pronucleus, and, finally, the fusion of the female and male pronuclei.

Gerould (1907) described fertilization in the eggs of *Golfingia vulgaris* and *Phascolopsis gouldi*, but he was unable to observe sperm penetration. Both Gerould (1907) and Åkesson (1958) supposed the sperm to enter through a pore of the egg membrane. However, the width of the sperm head exceeds that of the pore several times and, in more recent studies (Rice, 1966), it has been demonstrated that sperm penetration is effected by the formation of a hole in the egg envelope. Penetrating sperm or the resultant sperm entry holes have been observed in living and/or sectioned eggs of the following species: *Phascolosoma agassizi*, *P. perlucens*, *P. varians*, *Golfingia pugettensis*, *G. pellucida*, and *Themiste pyroides*. In addition to the sperm entry hole in the egg envelope, the penetrating sperm of *Themiste pyroides* also leaves behind it a clearly discernible track in the thick jelly coat of the egg. Before penetration this sperm extends an acrosomal filament 50 μm in length from the edge of the jelly to attach to the egg envelope. The sperm entry hole may persist for 2 to 3 days in the developing embryo.

Following sperm penetration, maturation of the egg is completed with the formation of the two polar bodies. In the elliptical eggs of *Phascolosoma* the cytoplasm rounds up, pulling away from the egg envelope at the two poles, and the polar bodies are formed at the animal pole. In spherical eggs such as those of most *Golfingia* and *Themiste*, a space between egg envelope and cytoplasm forms at the animal pole at the time or shortly before extrusion of the polar bodies.

Details of sperm migration and zygote formation have been studied in *Golfingia vulgaris* and *Phascolopsis gouldi* (Gerould, 1907). In these species, the sperm, after entrance into the egg, rotates until its long axis is parallel to the surface of the egg. An aster with centrosome appears at the base of the nucleus and within about 10 minutes the sperm, led by the aster, migrates to the center of the egg where it increases in size and the chromatin becomes separated into a loose network. At the same time the astrosphere enlarges and from it prominent fibers radiate throughout the cytoplasm except in the direction of the animal pole.

During this period of sperm enlargement the two polar bodies are given off by the egg. According to Gerould's account, the reduction or transverse division occurs at the first polar body division and the equational or longitudinal division takes place at the second division, the reverse of the usual process. During maturation divisions the number of chromosomes is reduced from 20 to 10, the haploid number. After extrusion of the second polar body the nucleus of the egg is broken up into 10 chromatic vesicles which soon fuse to form the female pronucleus. A centrosome, from which astral fibers radiate, lies in a fold of the pronucleus toward the center of the egg. As the two pronuclei move toward each other, the astral rays of the female pronucleus become less prominent and the centrosome moves to one side, about 90° from that point which will first contact the sperm nucleus. As the sperm nucleus moves toward the animal pole, its astrosphere moves to one side and decreases in size. At the time the two pronuclei unite, their astrospheres are of approximately the same size, but the astral fibers of the male pronucleus, are much more prominent. As the first cleavage approaches, the two asters are equal in size.

4.4.2 Embryonic Development

4.4.2.1 CLEAVAGE

Cleavage of the sipunculan egg is spiral, unequal, and holoblastic (Fig. 16a,b). Cell lineage of one species, *Golfingia vulgaris*, was studied through the 48-cell stage by Gerould in 1907. Starting with a dexiotropic cleavage at the third division, the cleavage planes alternate in direction to the 48-cell stage, after which the spiral pattern continues only in certain areas of the egg. A characteristic feature of the cleavage of *Golfingia vulgaris* is the relatively large size of the micromeres at the 8-cell stage. At the 16-cell stage, the 8 micromeres of the animal hemisphere, all of approximately equal volume, exceed in size all of the cells of

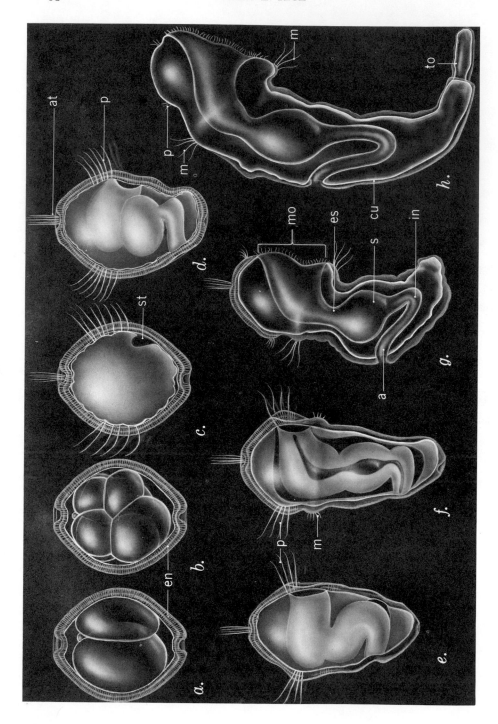

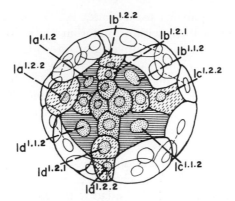

FIG. 17. Forty eight-cell stage of *Golfingia vulgaris*, anterior hemisphere, showing the molluscan cross and intermediate cells. Rosette cells are dotted, cross cells are dashed, and intermediate cells barred. Prototroch cells are around the periphery. (Redrawn from Gerould, 1907, fig. D, p. 99.)

the vegetal hemisphere except those of the D quadrant; the 2d cell or somatoblast is the largest cell at this stage and the second largest is the 2D. The relative size of the micromeres in early cleavage is reflected in the extraordinarily large size of the prototroch cells in the embryo. At the 48 cell stage the animal hemisphere consists of the 32 cells which make up the apical plate and the prototroch. The spiral pattern is interrupted at the 48-cell stage by the radial division of the $1q^{12}$, the result of which is the formation of the apical cross. The cross cells are in the position of the molluscan cross, that is, the arms extend out from the rosette cells along the future sagittal and frontal planes of the embryo in a radial direction rather than in the interradial direction characteristic of the annelidan cross (Fig. 17). This feature has been misunderstood by many authors probably because of Gerould's adoption of terminology

FIG. 16. Diagrammatic representation of developmental stages of *Phascolosoma perlucens*, illustrating events in the metamorphosis of the trochophore. (a) Two-cell stage. (b) Eight-cell stage. (c) Early trochophore. (d) Late trochophore. (e–f) Premetamorphosis stages. (g) Planktotrophic pelagosphera larva, immediately after trochophoral metamorphosis, 2½–3 days of age. (h) Pelagosphera larva, about 1 week old. As the trochophore elongates during metamorphosis (e,f) the egg envelope loses its porosity and lamellation and is transformed into the cuticle of the pelagosphera larva (g,h). Immediately before metamorphosis (f) the gut cavity and coelomic cavity are formed and metatrochal cilia appear. At the time of metamorphosis the anus and mouth are opened to the exterior, the prototroch reduced and terminal organ everted (g). a, anus; at, apical tuft; cu, cuticle; en, egg envelope; es, esophagus; in, intestine; m, metatroch; mo, mouth; p, prototroch; s, stomach; st, stomodeum; to, terminal organ.

previously used for polychaetes (Clark, 1969); however, when using these terms he places them within quotation marks. Thus he labels the cross cells "intermediate" cells, apparently referring to the intermediate cells of annelids, not of sipunculans. In his figure of the apical region of the 48-cell stage, reproduced here (Fig. 17), the prominent cross cell are clearly seen to be in the radial position. The 4 cells of the rosette and the 16 cells of the prototroch are notable for their large size. The prototroch, derived from the trochoblasts, $1q^2$ cells, consists of a girdle of 16 primary cells at the 48-cell stage; later three secondary cells, probably derived from intermediate cells, are added to complete the prototrochal girdle of the trochophore. A gap persists in the prototroch between cells of the C and D quadrants, joining apical and somatic plates, and has been designated as the dorsal cord. The division of 3D gives rise to 4d which immediately divides to form two equal cells which sink below the surface to form the mesodermal teloblasts.

Additional information available on early cleavage of other species indicates that the relatively large micromeres are a feature of yolky eggs and lecithotrophic development, whereas in eggs with relatively small amounts of yolk, micromeres are smaller than, or in some cases equal to, the macromeres. For example, in the 8-cell stages of *Phascolosoma agassizi* and *P. perlucens*, species with eggs low in yolk content, the micromeres and macromeres in the A, B, and C quadrants are all approximately the same size; 1d is slightly larger and 1D is by far the largest of the 8 cells. The largest cell of the 16-cell stage of *P. agassizi* is the somatoblast 2d which is nearly twice the size of 2D. The relative size of micromeres and macromeres at the 8-cell stage of 10 species of sipunculans is found in Table II.

4.4.2.2 GENERAL DEVELOPMENTAL PATTERNS

Several patterns of development are known to occur in the Sipuncula. One is direct development, defined here as nonpelagic development in which ciliated stages are absent and the embryo transforms directly into a small, crawling worm, usually hatching from one or more egg coats. The remaining patterns are classified as indirect development and include either one or both of the following swimming larval stages: (1) trochophore, characterized by ciliated prototroch and apical tuft, and (2) pelagosphera with prominent ciliated metatrochal band, lacking a prototroch or with considerably reduced prototroch.

Information is now available on the development of 17 species of sipunculans, including 5 species of *Golfingia*, 4 species of *Phascolosoma*, 3 of *Themiste*, 2 of *Phascolion*, and one each of *Phascolopsis*, *Paraspidosiphon*, and *Sipunculus*. Table II lists the species according to developmental patterns. Four common categories are designated: I, direct development; II, indirect development with one larval stage, the trochophore, which transforms into the vermiform stage; III, indirect development with two larval stages, the trochophore and lecithotrophic pelagosphera; and IV, indirect development with two larval stages, the trochophore and planktotrophic pelagosphera. The species listed range in distribution from the cold waters of the Northeastern Pacific and North Atlantic to the tropical waters of the Atlantic.

Three species are known to undergo direct development: *Golfingia minuta* from Kristineberg, Sweden (Åkesson, 1958), *Phascolion cryptus* from Virginia Key, Florida (Rice, 1975), and *Themiste pyroides* from San Juan Archipelago, Washington (Rice, 1967). The eggs are rich in yolk, blastulae are solid, and gastrulation occurs by epiboly. In two of the species in which 8-cell stages have been examined (*Themiste pyroides* and *Phascolion cryptus*), the macromeres in the A, B, and C quadrants are smaller than their respective micromeres. Derivatives of the large micromeres, the prototroch cells, form a prominent band of noncilated cells of extraordinary size. At this stage, comparable to the trochophore although lacking cilia, the cytoplasm is retracted from the egg envelope in the pretrochal area around the apex of the embryo, forming an apical groove which delimits the cells of the nonciliated rosette. Joining pretrochal and posttrochal hemispheres a dorsal groove interrupts the prototrochal band and immediately posterior to the prototroch a space between cytoplasm and egg envelope marks the position of the stomodeum.

At the age of 2 days (25°C), *Phascolion cryptus* hatches as a crawling worm from the anterior portion of the egg envelope and the surrounding jelly coat (Fig. 18). The embryo first elongates posttrochally, rupturing the jelly coat posteriorly and then the pretrochal region, followed by the prototrochal, separates from the egg envelope resulting in a large anterior cavity. As the retractor muscles become functional the head is extended and withdrawn and through this activity the anterior egg envelope is severed from its posttrochal connection to the embryo. Thus the jelly coat and anterior egg envelope are discarded, a new cuticle is formed to cover the pretrochal and prototrochal regions, and the posttrochal egg envelope is transformed into the posterior cuticle of the vermiform stage.

In *Themiste pyroides*, the entire egg envelope appears to be retained as the cuticle of the vermiform stage. The embryo develops within the envelope and surrounding jelly coat until the age of 8 to 9 days (12°–13°C) when, by a series of movements extending and contracting the body, it hatches out of the jelly coat as a small, crawling worm.

TABLE II

A Summary of Developmental Characteristics of the Sipuncula

Species[a]	Egg size diameter or length × width (μm)	8-Cell stage relative size of micro- and macromeres in quadrants A, B, C	Gastrulation	Trochophore	Pelagosphera	
					Lecitho-trophic	Plankto-trophic
Category I						
Golfingia minuta[1]	260–280 × 215–230	?	Epiboly	0	0	0
Themiste pyroides[5]	190	Micromeres > Macromeres	Epiboly	0	0	0
Phascolion cryptus[6]	136	Micromeres > Macromeres	Epiboly	0	0	0
Category II						
Phascolion strombi[1]	125	Micromeres > Macromeres	Epiboly	8 Days	0	0
Phascolopsis gouldi[3]	150–180	Micromeres > Macromeres	Epiboly	3 Days	0	0
Category III						
Golfingia vulgaris[3]	150–180	Micromeres > Macromeres	Epiboly	3 Days	2 Days	0
Golfingia elongata[2]	125	?	Epiboly + invagination	2 Days	4 Days	0
Golfingia pugettensis[5]	160	Micromeres = Macromeres	Epiboly	8 Days	13 Days	0
Themiste alutacea[6]	138	?	Epiboly	2 Days	6 Days	0
Themiste lageniformis[8]	145	Micromeres > Macromeres	Epiboly	0	8–12 Days	0

Category IV

Golfingia pellucida[7]	165	?	?	3 Days	0	1 Month
Paraspidosiphon fischeri[6]	103 × 94	?	?	2 Days	0	1 Month
Phascolosoma agassizi[5]	140 × 110	Micromeres = Macromeres	Epiboly + invagination	8–10 Days	0	1 Month
Phascolosoma antillarum[6]	127 × 97	?	?	3 Days	0	1 Month
Phascolosoma perlucens[6]	112 × 91	Micromeres = Macromeres	Epiboly + invagination	3 Days	0	1 Month
Phascolosoma varians[6]	104 × 90	?	?	3 Days	0	1 Month
Sipunculus nudus[4]	120	Micromeres < Macromeres	Invagination	3 Days	0	1 Month

[a] References: 1. Åkesson, 1958; 2. Åkesson, 1961a; 3. Gerould, 1907; 4. Hatschek, 1883; 5. Rice, 1967; 6. Rice, 1975; 7. Rice, unpublished; 8. Williams, 1972.

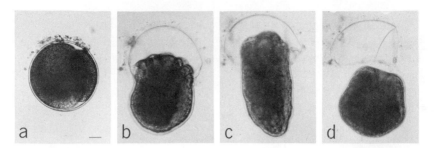

Fig. 18. Developmental stages of *Phascolion cryptus,* depicting direct development and partial hatching from egg envelope. Photographs of living embryos. Scale, 25 μm. (a) Two-cell stage. Presence of adhesive jelly layer is indicated by attached debris. (b–c) Egg envelope has detached in pretrochal and prototrochal regions. Anterior space develops between envelope and embryo. Anterior end can be retracted as in (b) or extended as in (c), and by such movements the envelope is eventually ruptured at the posttrochal junction. (d) Hatched vermiform stage. Head is retracted. Anterior egg envelope is completely detached and posterior egg envelope has been transformed into cuticle. [From Rice, 1975 (in part).]

The sticky jelly coats of the eggs of *Themiste pyroides* and *Phascolion cryptus* cause adhesion to any surface contacted; thus these embryos develop attached to the substratum in the area of spawning. The direct development of *Golfingia minuta,* exemplifying a form of primitive brood protection, takes place over a period of 3 to 4 weeks within the burrows occupied by the females. The embryos do not develop prototrochal cilia, but cytoplasmic blebs, similar to those noted in other species before formation of cilia, have been interpreted by Åkesson to indicate that lack of prototrochal cilia in this species is secondary rather than primitive. An extremely weak and temporary ciliation with no loco-motory function forms on marginal and rosette cells. The egg en-velope transforms into the cuticle of the juvenile, but beneath it a definitive cuticle is apparent. Development is slow in this species; young worms begin to migrate from the burrow at 1 month, but heads are not retractable until the age of 6 weeks.

The development of both *Phascolion strombi* from Kristineberg, Sweden (Akesson, 1958) and *Phascolopsis gouldi* from Newport, Rhode Island and Woods Hole, Massachusetts (Gerould, 1907) is marked by a single pelagic larval stage, a lecithotrophic trochophore (Category II, Table II). Eggs of the two species are macrolecithal with micromeres exceeding macromeres in size in the A, B, and C quadrants at the 8-cell stage, the blastulae are solid, and gastrulation occurs by epiboly. Trocho-phores are characterized by rosette cells with long cilia and a band of ciliated prototroch cells (Fig. 19a). Marginal cells on either side of

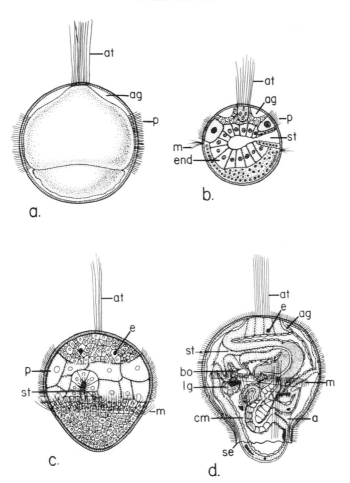

Fig. 19. Trochophores of sipunculans. (a) Trochophore of *Phascolopsis gouldi*. Surface view; 20 hours. (Redrawn from Gerould, 1907, plate 8, fig. 59.) (b) Trochophore of *Golfingia elongata*. Diagram of sagittal section. (Redrawn from Akesson, 1961a, p. 516.) (c) Trochophore of *Golfingia vulgaris*. Ventral view, indicating cell outlines; about 40 hours. (Redrawn from Gerould, 1907, plate 7, fig. 50.) (d) Trochophore of *Sipunculus nudus*. Lateral view. Egg envelope an underlying cell layer (serosa) have split at posterior end, preceding shedding of envelope. (Redrawn from Hatschek, 1883, plate 3, figure 45.) a, anus; ag, apical groove; at, apical tuft; bo, buccal organ; cm, coelom; e, eye; end, endoderm; lg, lip gland; m, metatroch; p, prototroch; se, serosa; st, stomodeum.

the prototroch of *Phascolion strombi* are weakly ciliated and anterior to the prototroch of *Phascolopsis gouldi* is a band of well-developed preoral cilia. At metamorphosis the prototroch cells degenerate and the trochophore is transformed into an elongate worm, capable of extending and retracting the anterior portion of the body. The egg envelope of *Phascolopsis gouldi* is shed at metamorphosis, but in *Phascolion strombi* it is retained as the cuticle.

A developmental pattern consisting of two pelagic larval stages, a lecithotrophic trochophore and a lecithotrophic pelagosphera, is found in the following species: *Golfingia elongata*, *G. pugettensis*, *G. vulgaris*, and *Themiste alutacea* (Category III, Table II). All are temperate or cold-water species except *T. alutacea* which ranges from the temperate water off North Carolina and Brazil to the tropical waters of the Caribbean Sea. In these species a pelagic lecithotrophic trochophore metamorphoses into a lecithotrophic pelagosphera which swims for a short time, often near the bottom, undergoes a gradual loss of cilia, and transforms into the veriform stage. The 4 species within this pattern show a gradient in development from least to most highly modified from that of the preceding pattern. The development of *Golfingia vulgaris* most nearly resembles that previously described, whereas that of *G. pugettensis* is probably the most highly modified. In *G. vulgaris* the egg is rich in yolk and gastrulation is epibolic (Fig. 20a). There is no abrupt formation of metatrochal cilia since these are developed in the trochophore stage (Fig. 19c). Metamorphosis of the trochophore occurs at 2 days when the egg envelope is ruptured and cast off and the anterior end of the body becomes retractable (Fig. 21c,d). Prototrochal cilia appear often to be destroyed, but metatrochal cilia slip through the envelope without injury. The larva continues to twirl, usually near the bottom, by means of the metatroch until the cilia are lost at 5 days of age and the young vermiform stage assumes an elongate shape and begins to creep along the bottom (Gerould, 1907).

Development of *Golfingia elongata* is similar, except that during gastrulation an archenteron is formed. The trochophore is spherical with a long apical tuft, short prototrochal cilia, and long metatrochal cilia (Fig. 19b). Metamorphosis consists of elongation, beginning at 48 hours (17°C), and disintegration of the prototrochal cells. The egg envelope is transformed into the cuticle of the vermiform stage.

The trochophore of *Themiste alutacea* changes from the spherical shape of the blastula to an oval shape with a prominent equatorial band of prototrochal cilia and a long apical tuft. Metamorphosis, beginning at 28 hours (25°C) with elongation of the trochophore, results at 32 hours in a lecithotrophic pelagosphera. During metamorphosis a well-developed circlet of metatrochal cilia and a postmetatrochal sphincter appear, the stomodeum opens through the egg envelope to form the mouth and ventral ciliated surface of the head, and the egg envelope

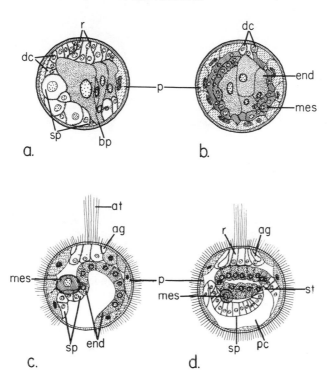

FIG. 20. Gastrulation and mesoderm formation. Prototroch cells are marked by dashes, other ectoderm cells are clear; endoderm is dotted; mesoderm is barred. (a) Sagittal section of embryo of *Golfingia vulgaris* showing blastopore. Gastrulation is epibolic; 14½ hours. (Redrawn from Gerould, 1907, plate 6, fig. 40b.) (b) Cross section of embryo of *Golfingia vulgaris*; 24 hours. Mesodermal bands have split into splanchnic and somatic layers. (Redrawn from Gerould, 1907, plate 6, fig. 45. Prototrochal cilia, although described in text, are not shown in Gerould's figures.) (c) Optical median section of embryo of *Sipunculus nudus* showing formation of ectodermal somatic plate and embolic gastrulation. (Redrawn from Hatschek, 1883, plate 2, fig. 15.) (d) Optical median section of embryo of *Sipunculus nudus*, later than (c). Prototroch cells have surrounded embryo. (Redrawn from Hatschek, 1883, plate 2, fig. 23.) ag, apical groove; at, apical tuft; bp, blastopore; dc, dorsal cord; end, endoderm; mes, mesoderm; p, prototroch; pc, posterior cavity; r, rosette cells; sp, somatic plate; st, stomodeum.

is transformed into the larval cuticle. Transformation into the juvenile may begin as early as 7 to 8 days with loss of metatrochal cilia, and by 2 weeks the coelomic yolk has been absorbed and the gut completed. Within 1 month it has assumed the shape of the juvenile with elongated introvert and 4 ciliated tentacular lobes.

The trochophore of *Golfingia pugettensis* resembles that of *Themiste alutacea,* but a characteristic feature in this species is the presence

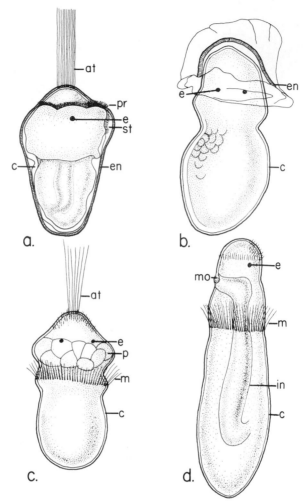

FIG. 21. Metamorphosis of *Phascolopsis gouldi* and *Golfingia vulgaris*. (a) Trochophore of *P. gouldi*, 48 hours, showing circlet of preoral cilia and developing cuticle
beneath egg envelope. Lateral view. (Redrawn from Gerould, 1907, plate 8, fig.
62.) (b) Metamorphosing trochophore of *P. gouldi* in process of shedding egg
envelope; 43 hours. Dorsal view. (Redrawn from Gerould, 1907, plate 9, fig. 68.)
(c) Metamorphosing trochophore of *G. vulgaris*. Egg envelope has been cast off,
prototrochal cilia lost, and prototrochal cells are in process of degeneration. Dorsal
view. (Redrawn from Gerould, 1907, plate 7, fig. 52.) (d) Lecithotrophic pelagosphera larva of *G. vulgaris;* 60 hours. Lateral view. Gut is incomplete. (Redrawn
from Gerould, 1907, plate 7, fig. 54.) at, apical tuft; c, cuticle; e, eye; en, egg
envelope; in, intestine; m, metatroch; mo, mouth; p, prototroch; pr, preoral cilia;
st, stomodeum.

of conspicuous lipid droplets in the prototroch cells. Metamorphosis of the trochophore occurs at 8 days (8°–9°C). Differing from others with this developmental pattern, the lecithotrophic pelagosphera of *Golfingia pugettensis* has a well-developed terminal organ by which it attaches to the substratum.

A species with an apparently highly modified development, similar to that described above but not readily classified in any of the developmental categories commonly found in sipunculans, is *Themiste lageniformis*. Its early development is essentially the same as that of directly developing species in that it lacks a pelagic trochophore stage. However, the nonpelagic stage, comparable to a trochophore, metamorphoses to a lecithotrophic pelagosphera with metatroch and terminal attachment organ (Williams, 1972).

The final developmental pattern known to occur in the phylum includes two larval stages: a pelagic lecithotrophic trochophore and a planktotrophic pelagosphera (Category IV, Table II). The development of 7 species falls within this classification. Four belong to the genus *Phascolosoma*: *P. agassizi*, *P. antillarum*, *P. perlucens*, and *P. varians*. Other species are *Golfingia pellucida*, *Paraspidosiphon fischeri*, and *Sipunculus nudus*. *Phascolosoma agassizi* is a cold- to temperate-water species; others are either tropical or range from temperate to tropical waters. With the exception of *Sipunculus nudus* in which the trochophore is uniquely modified, the development is very similar in all species.

Some of the characteristic features of this developmental pattern are summarized in Fig. 16, a diagrammatic representation of development through metamorphosis to the pelagosphera larva in *Phascolosoma perlucens*. The egg of this species is relatively low in yolk content, the blastula has a small blastocoel in the anterior hemisphere, and gastrulation, although mostly epibolic, is in part achieved by invagination with the formation of a small archenteron. During the trochophore stage the gut and mesodermal bands are differentiated and posttrochal elongation begins (Fig. 16c–e). Long prototrochal cilia, extending through the pores in the egg envelope function to propel the larva in a spiral swimming movement while the cilia of the apical tuft are directed forward. Metamorphosis of the trochophore into the planktotrophic pelagosphera takes place at $2\frac{1}{2}$ to 3 days of age (25°C) and lasts over a period of several hours. Metamorphosis is marked by formation of a new ciliary band, the metatroch, reduction of the prototroch, loss of the apical tuft, posttrochal elongation, expansion of coelom, rupture of the egg envelope in the region of the stomodeum to give rise to the ventral ciliated surface of the head and mouth, opening of the anus to complete the gut, and formation of the adhesive terminal organ (Fig. 16f–h). Pelagosphera larvae of *P. perlucens* have been maintained in the laboratory as long as 6 months, but a second metamorphosis into the juvenile form has not been observed.

Other species included within this pattern, except *Sipunculus nudus,* show essentially the same developmental features. The early development of *S. nudus* is unique, particularly in the elaboration of prototroch and embryonic cavities. Blastomeres are nearly equal in size, but slightly larger at the vegetal pole. In the blastula a nearly central blastocoel is displaced slightly toward the animal pole. Gastrulation occurs by invagination of the endoderm cells which pull away from the egg envelope at the vegetal pole, leaving a posterior cavity (Fig. 20c,d). At the same time the cells of the apical plate surrounding the rosette cells sink from the envelope forming an apical groove or cavity of the head around the central rosette. The posterior cavity and the apical groove are connected by a narrow dorsal canal which results from the sinking of a double row of small ectoderm cells. Ciliated cells, assumed to be homologs of trochoblasts (Gerould, 1903), then spread out to surround the embryo and form in conjunction with the egg envelope a structure which has been termed a "serosa" (Hatschek, 1883). The trochophore develops within the serosa and at the time of metamorphosis to the planktotrophic pelagosphera the entire serosa is shed (Figs. 19d and 22a,b). The pelagosphera has a ciliated metatrochal band and is similar to other such larvae in this developmental category. After 1 month in the laboratory a second metamorphosis occurs into the juvenile form (Fig. 22c).

4.4.2.3 GASTRULATION AND FORMATION OF MESODERM

In species with macrolecithal eggs, as in the first three developmental categories, gastrulation occurs entirely by epibolic movements with the exception of *Golfingia elongata* in which a small archenteron is formed. This mode of gastrulation, best described for *Golfingia vulgaris* and *Phascolopsis gouldi* (Gerould, 1907), occurs by an overgrowth of the cells of the dorsal somatic plate ventrally and laterally to enclose the solid mass of large endoderm cells. The blastopore is represented by the narrow ends of the club-shaped endoderm cells which maintain contact with the egg envelope until closed over by the somatic plate (Fig. 20a,b). Descendants of the stomatoblasts divide rapidly and invaginate in the region of the trochoblasts at the anterior end of the blastopore to form the stomodeum.

In species with less yolky eggs and planktotrophic larvae, as found in the fourth developmental category, gastrulation is accomplished in part by invagination, but the major role is usually played by epiboly.

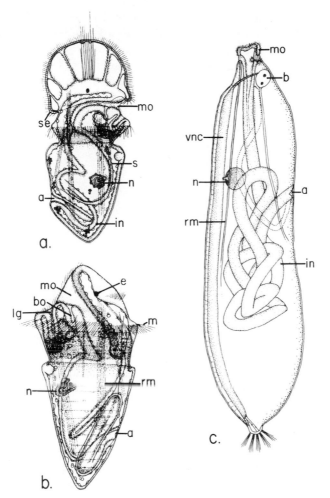

FIG. 22. Metamorphosis of *Sipunculus nudus*. (a) Trochophore hatching from "serosa," i.e., egg envelope plus prototroch cells. Serosa is still attached in head region. (Redrawn from Hatschek, 1883, plate 5, fig. 49.) (b) Pelagosphera larva, 2 days after shedding of serosa. (Redrawn from Hatschek, 1883, plate 5, fig. 51.) (c) Juvenile, after metamorphosis of pelagosphera. (Redrawn from Hatschek, 1883, plate 6, fig. 71.) a, anus; b, brain; bo, buccal organ; e, eye; in, intestine; lg, lip gland; m, metatroch; mo, mouth; n, nephridium; rm, retractor muscle; s, stomach; se, serosa; vnc, ventral nerve cord.

Endoderm cells in blastulae of species of *Phascolosoma*, such as *P. agassizi* and *P. perlucens*, are separated from the egg envelope by a large space around the vegetal pole. During gastrulation, cells of the somatic plate grow down over the entomeres filling the space in an epibolic gastrulation. A narrow slit in the region of the blastopore marks an invagination leading into a narrow archenteron. Endodermal invagination is the primary process of gastrulation in *Sipunculus nudus* (Fig. 20c,d). Closure of the blastopore in this species is achieved by a forward and ventral growth of ectoderm of the dorsal lip; this ectodermal growth at the same time gives rise to the median somatic plate.

Mesoderm is derived from a pair of mesodermal teloblasts located on either side of the endoderm cells. Proliferation of these cells, observed in *Golfingia vulgaris*, *Phascolopsis gouldi* (Gerould, 1907), *Sipunculus nudus* (Hatschek, 1883), and *Phascolosoma agassizi* (Rice, 1967), results in the formation of the lateral bands of mesoderm (Fig. 20b).

4.4.3 Larvae

4.4.3.1 TROCHOPHORE LARVA

A trochophore stage is a characteristic phase in the development of all sipunculans, although in some species it may be highly modified. Trochophores of several species are illustrated in Fig. 19. The apical plate of the pretrochal hemisphere consists of numerous small ectoderm cells which later give rise to the head ectoderm and brain and of a central circle of large rosette cells, usually bearing the long cilia of the apical tuft. The apical groove around the rosette cells may persist in the early trochophore, but later it is filled by a growth of ectoderm cells. Embedded in the apical plate in a dorsal position just anterior to the prototroch is a pair of small eyespots. The somatic plate covers the posterior hemisphere and is the source of the ectoderm of the trunk. The stomodeum is in a medioventral position, extending anteriorly into the region of the prototroch. The large cells of the prototroch (19 in *Golfingia vulgaris*) form a prominent equatorial band which spreads out over a large proportion of the surface of the larva. Prototroch cells are usually ciliated and always concentrated with yolk granules. The early trochophore retains the shape of the egg, but then an oval shape is assumed and in later stages the length increases as posttrochal elongation begins. Enclosed by the overlying egg envelope, trochophores of all species of sipunculans are lecithotrophic. Rudiments of most internal adult organs are present at this stage. Three regions of the gut (esophagus, stomach, and intestine) are differentiated and mesodermal bands, two cell layers in thickness, are present on either side of the gut.

The pattern and functional significance of ciliation varies among different species. Directly developing species, such as *Phascolion cryptus* and

Themiste pyroides, lack cilia entirely; *Golfingia minuta* shows a weak and temporary ciliation on pretrochal, posttrochal, and rosette cells, but not on prototrochal cells. All other species possess ciliated prototroch cells and an apical tuft of cilia. Prototrochal cilia are well developed and serve as the means of locomotion in species of *Phascolosoma* and in *Golfingia pugettensis, G. pellucida,* and *Paraspidosiphon fischeri.* In trochophores of *G. elongata, G. vulgaris,* and *Phascolopsis gouldi,* pretrochal and metatrochal bands of cilia are also present; metatrochal cilia are by far the longest in the two former species and most significant for larval locomotion, whereas in the latter the pretrochal cilia are the longest and the metatrochal vestigial. Metatrochal cilia are present in the trochophore of *Sipunculus nudus* but, enclosed by the serosa, are not functional (Fig. 19d).

4.4.3.2 Metamorphosis of the Trochophore

Metamorphosis of the trochophore may result in one of several developmental stages. In those species with only one pelagic stage, it may end the pelagic phase of development, the trochophore metamorphosing into a vermiform stage which gradually transforms into the juvenile form. It may result in a second pelagic larval stage, the lecithotrophic pelagosphera, which swims for a short time in the plankton or near the bottom before transforming into the vermiform stage. Finally, it may give rise to a planktotrophic pelagosphera which swims in the plankton for a prolonged period before a second metamorphosis into a juvenile form. Main events of metamorphosis into a planktotrophic pelagosphera are illustrated in the diagram of development of *Phascolosoma perlucens* (Fig. 16).

Metamorphosis is characterized by reduction or loss of the prototroch, formation or expansion of the coelom, transformation or shedding of the egg envelope, and, in those species with pelagosphera larvae, by the elaboration of the metatroch as the principal locomotory organ. Muscular activity is initiated either at metamorphosis or shortly thereafter with the introversion and extrusion of the introvert.

In species which undergo direct development without a pelagic trochophore stage, changes occur at the time of transformation into the vermiform stage which are comparable to those of the metamorphosis of the trochophore, e.g., elongation, formation of the coelom, dissolution of prototroch cells, and the beginning of muscular activity.

The coelom is formed by schizocoely or a splitting of the mesodermal bands into two layers, one of which forms the inner layer of the body wall and the other an outer covering of the gut. In species with relatively microlecithal eggs and planktotrophic pelagosphera the coelom is formed before metamorphosis of the trochophore, but undergoes a great expansion at this time. The coelom of species with yolky eggs forms simultaneously with other events of metamorphosis.

The egg envelope of most species appears to be transformed at the time of metamorphosis into the larval cuticle. Transformation is accompanied by a loss of porosity and lamellation, first immediately posterior to the prototroch in the region of initial elongation, and later throughout. At the same time, as the larva stretches and the muscles become functional, the previously rather rigid egg envelope becomes elastic and flexible. Two species, *Phascolopsis gouldi* and *Golfingia vulgaris,* have been reported to shed the egg envelope at metamorphosis and another species, *Phascolion cryptus,* loses the prototrochal and pretrochal portions (Figs. 18b–d and 21a,b). *Sipunculus nudus* sheds not only the egg envelope, but also the underlying ciliated cells (Figs. 19d, and 22a).

Prototroch cells, usually heavily laden with yolk, release their yolk granules during metamorphosis into the coelomic cavity, thus contributing a substantial source of nutrition for the developing larva. The significance of this contribution is greatest in species with macrolecithal eggs and lecithotrophic development. Entire cells of the prototroch of *Golfingia vulgaris* and *Phascolopsis gouldi,* including nuclei and cytoplasm, are cast into the coelom at metamorphosis and the region of the prototroch is overgrown by ectoderm cells. In *Golfingia elongata* and *Phascolion strombi* the prototroch cells degenerate, releasing lipid and yolk granules into the coelom, after which they are replaced with ectoderm. Prototroch cells of *G. pugettensis* are marked by characteristic large lipid globules which at the time of metamorphosis are released into the coelom where they are readily recognizable; the cells, although retained as a prototrochal band on the head of the pelagosphera, are much reduced in size. The large yolk-laden prototroch cells of *G. minuta* and *Themiste pyroides* begin to degenerate at the time of coelom formation releasing their granules into the coelom. In species of *Phascolosoma,* release of yolk from the prototroch cells begins at an early stage of development when granules are passed from the cells into cavities formed along the inner sides of the prototroch. As the prototrochal cells decrease in size, the size of the "prototrochal cavities" becomes larger. The granules appear to break down in the cavities where presumably they provide nutrition for the developing embryo. At metamorphosis the prototrochal cells, reduced in size, persist in the pelagosphera as a weakly developed dorsal band of cilia on the head. Before metamorphosis in *Sipunculus nudus,* the size of the cells of the serosa is diminished as they gradually liberate nutrients into the anterior and posterior cavities (Figs. 19 and 20). By the time of metamorphosis of *P. agassizi* and *S. nudus* all of the prototrochal yolk has been discharged into embryonic cavities.

At metamorphosis of the trochophore the metatroch assumes the role of the principal organ of locomotion. In *Golfingia vulgaris* and *G. elongata* the metatroch is present and well developed in the trochophore stage along with the prototroch, but at metamorphosis the prototroch diminishes and the metatroch remains as the sole locomotory organ. In *G. pugettensis* the metatroch is established late in the trochophore stage, but is elaborated at the time of metamorphosis when the prototroch regresses. In species of *Phascolosoma* metatrochal cilia are developed at the time of metamorphosis. In *Sipunculus nudus* metatrochal cilia appear before metamorphosis but become functional only when the serosa is shed.

4.4.3.3 PELAGOSPHERA LARVA

Following the stage of trochophore, the pelagosphera larva, whether lecithotrophic or planktotrophic, is characterized by a prominent band of metatrochal cilia and a regionalization of the body into head, metatrochal region or "thorax," and elongated trunk (Figs. 23–25). The head and thorax are retractable into the trunk and later become the introvert of the adult. Dorsally the head bears at least one pair of eyespots. Posterior to the eyespots and along either side of the head, a prototrochal ridge forms a **U**-shaped band of weakly developed cilia. The ventral head is ciliated and divided into two lobes by a median groove which continues into the mouth. Posterior to the head the metatrochal region or thorax is greatly distended when the larva is swimming and at its maximum extension is by far the widest portion of the larva (Fig. 25a). The posterior boundary of the thorax is marked by the metatrochal sphincter muscle, which is usually contracted when the larva is swimming and when the anterior end is completely retracted (Figs. 23–25). The elongated trunk may vary somewhat in shape depending on the degree of extension or contraction of the flexible larval body. At the posterior tip of most larvae there is a terminal attachment organ. Terminal organs are absent in a few planktotrophic larvae and in the lecithotrophic larvae of *Golfingia vulgaris* and *Themiste alutacea* (Figs. 21d and 24a), although the posterior end of the latter species does possess adhesive properties. Terminal organs of planktotrophic pelagospheras, in contrast to those of lecithotrophic larvae, are retractile, being withdrawn into the body by the contraction of a single pair of retractor muscles, originating from the dorsal body wall near the anus (Fig. 24b,c). In lecitho-

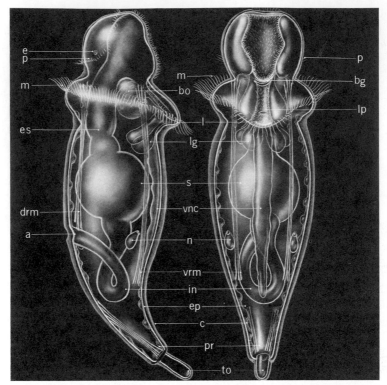

FIG. 23. Pelagosphera larva of *Phascolosoma perlucens*. Diagram showing internal structures at 1 week of age. Left, lateral view; right, ventral view; a, anus; bg, buccal groove; bo, buccal organ; c, cuticle; drm, dorsal retractor muscle; e, eye; ep, epidermis; es, esophagus; in, intestine; l, lip; lg, lip gland; lp, lip pore; m, metatroch; n, nephridium; p, prototroch; pr, posterior retractors; s, stomach; to, terminal organ; vnc, ventral nerve cord; vrm, ventral retractor muscle. (From Rice, 1975.)

trophic pelagospheras the gut cavity is incomplete and, although the mouth is usually open, the anus has not broken through the larval cuticle. The gut of planktotrophic larvae is complete, the anus opening dorsally on the middle or the anterior region of the trunk and the larvae are independent feeders. Associated with the feeding process are structures not present in lecithotrophic larvae: the lower lip, a projection forming the posterior boundary of the mouth, and two organs associated with the lip, the buccal organ and lip glands. The buccal organ is extrusible through a transverse slit or buccal groove at the base of the lip and the lip glands open through a pore on the lip.

In addition to the 7 species of planktotrophic larvae known from developmental studies (Table II), numerous larvae of uncertain species

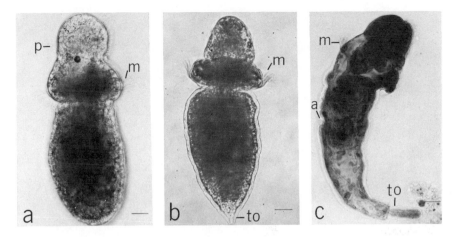

FIG. 24. Photographs of living pelagosphera larvae, reared from spawnings in the laboratory. (a) Lecithotrophic pelagosphera of *Themiste alutacea;* 3 days. Lateral view. Terminal organ is absent. (From Rice, 1975.) (b) Lecithotrophic pelagosphera of *Golfingia pugettensis;* 13 days. Dorsal view. Note presence of terminal organ. (From Rice, 1967.) (c) Planktotrophic pelagosphera of *Phascolosoma perlucens;* about 7 days. Lateral view. (From Rice, 1975.) Scale, 25 μm. a, Anus; m, metatroch; p, prototroch; to, terminal organ.

have been reported from the oceanic plankton. Such larvae were first recognized as belonging to the Sipuncula by Häcker (1898) from plankton collections in the North and South Atlantic Ocean. Comparing them with *Sipunculus nudus* known from Hatschek's studies, he described three larval types which he named "Baccaria oliva," "Baccaria citrinella," and "Baccaria pirum," distinguishing them by body shape and form of cuticular papillae. These larvae are less spectacular in size and form than the large, transparent planktonic larvae described by later workers.

Unaware of earlier work, Mingazinni (1905) described a larva from one preserved and contracted specimen collected in the plankton from a depth of 500 m between New Caledonia and New Zealand. He erroneously assumed the larva to be an adult, creating a new genus and species, *Pelagosphaera aloysii.* Senna (1906), working with material from the same expedition but from waters off India and Ceylon, suggested that structures identified by Mingazinni as gonads were instead the glandular appendages of the esophagus which are now referred to as lip glands (Jägersten, 1963). Although Mingazinni's error was soon realized, the name pelagosphera remains entrenched in the literature and is today used to designate the larval stage of sipunculans which succeeds the stage of trochophore (Rice, 1967).

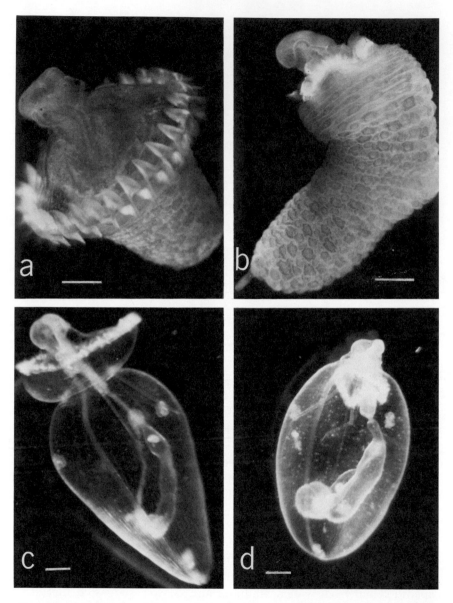

Fig. 25. Photographs of living planktotrophic pelagosphera larvae from oceanic plankton. (a) Larva of unknown adult. From plankton of Florida Current. Antero-dorsal view of swimming larva with extended metatroch. (b) Same specimen as (a) in quiescent position with terminal organ extended. Metatroch partially extended. (c) Larva of *Sipunculus polymyotus*. From plankton of Florida Current. Larva swimming with metatroch extended. (d) Same specimen as (c). Larva quiescent; metatroch withdrawn. Scale, 250 μm.

Other reports of oceanic planktotrophic larvae of sipunculans have been made by Heath (1910), Dawydoff (1930), Stephen (1941), Fisher (1947), Åkesson (1961b), Damas (1962), Jägersten (1963), Murina (1965), Scheltema and Hall (1965), and Hall and Scheltema (1975). Most of the earlier observations were made on preserved and contracted specimens, and only in more recent studies on living material (Jägersten, 1963; Murina, 1965; Hall and Scheltema, 1975) has the morphology of oceanic larvae been correctly interpreted. Jägersten gave the first accurate description of the head region, pointing out the relationships between lip, lip glands, and buccal organ. He examined several larval types which he divided into two groups, those with a "smooth" body surface and those with a "rough" or papillated body cuticle.

Characters generally used to distinguish pelagosphera larvae externally are texture of cuticle, presence and form of cuticular papillae, pigmentation, shape of head and lip, number and color of eyespots, general body form, and presence or shape of the terminal organ (Fig. 25). Some distinctive internal characters are number of lobes of lip gland, number of retractor muscles, shape and pigmentation of nephridia, and presence and number of longitudinal muscle fibers. Pigmentation in different larval forms may vary from brilliant orange to various shades of pink, yellow, green, and brown. The body may be white or transparent with pigmented nephridia and buccal groove. Eyespots are either red or black and usually two in number, but in some species there are accessory spots lateral or anterior to the larger ones. The retractable terminal organ may be a thick elongated cylinder or thin narrow rod, or in some instances no more than a rounded knob. The terminal organ in some species may be extended frequently, rarely in others, and in a few no terminal organ has been observed (Hall and Scheltema, 1975).

Specific affinities have been tentatively assigned to three of the oceanic planktotrophic larvae. Fisher (1947), basing his identification on the number of muscle bands, designated a larva collected off Cape Hatteras, North Carolina as *Sipunculus polymyotus*. Murina (1965) identified larvae from the Gulf of Aden as *S. aequabilis* and from the Northwest Pacific as *S. norvegicus*. Hall and Scheltema (1975) described external and internal morphology of 10 larval types from the North Atlantic. Of these, 7 were previously undescribed and the following 3 were redescribed: *Sipunculus polymyotus*, "Baccaria oliva," redescribed as Type C, and "Baccaria citrinella," redescribed as Type A. The latter was assigned to the genus *Aspidosiphon*, but no other adult affinities were determined.

The pelagosphera larva of *Sipunculus polymyotus* is large and transparent (Fig. 25c,d), often 4–5 mm in contracted length, commonly found in the Gulf Stream (Fisher, 1947; Hall and Scheltema, 1975). The head is simple with a single pair of black eyespots and accessory pigment spots. The lip is bifurcated by a groove marked by a yellowish-green pigment. Similarly pigmented are the duct of the lip gland, the metatrochal band at the base of the cilia, the entire recurved gut, and the small rounded nephridia. The cuticle is smooth and iridescent. Internally the body wall musculature is divided into approximately 50 longitudinal muscle bundles. There are three pairs of retractor muscles of the introvert, two dorsal and one ventral.

The most commonly occurring larva in the North Atlantic is "Baccaria citrinella" or Type A (Hall and Scheltema, 1975). The cuticle is "rough" with numerous cuticular elevations arranged in a characteristic pattern of regularly crossing rows on the trunk. Each elevation or cuticular papilla is rounded and surmounted by a smaller cap. Pigmentation of the opaque body wall varies in living specimens from dark to light pink or pinkish white and in preserved specimens the color is changed to yellow. The metatrochal region and head are pale yellow and on the dorsal head there is a single pair of small red eyes. The lip is a simple thin lobe, slightly heart-shaped and flattened, but with no bifurcating ciliated groove. The body, when contracted, is "lemon-shaped" and from the blunt posterior end a thin, rod-shaped terminal attachment organ can be extended.

Planktotrophic pelagospheras may swim throughout the water by means of the prominent metatrochal collar (Fig. 25a–c), but most frequently under laboratory conditions their activity is carried out on or near the bottom. Young larvae of *Phascolosoma agassizi*, reared in the laboratory, have been reported to be chiefly bottom feeders although they are also able to feed on planktonic microorganisms (Rice, 1973). The larva may attach by the terminal organ and extend out from this point of attachment either parallel to the substratum or upward at any angle. With head upward, food may be directed into the mouth by the cilia of the ventral surface of the head, or, still attached, the larva may bend downward, applying the ventral surface of the head to the substratum with lip extended posteriorly, grazing over the bottom. The larva is able to release itself from its terminal attachment and glide along with terminal organ directed upward and head flattened against the bottom. Frequently the larvae lie in a quiescent state or they may crawl in the manner of an inchworm, presumably scraping material from the bottom. The continual eversion of the buccal organ during feeding probably aids in the removal of food from the substratum. This tough muscular organ is believed to function in breaking up material into small particles for feeding and to aid also in the rejection of unwanted material or in swallowing. The secretion of the lip glands is presumed to have some function in feeding, but the nature of the secretory product is unknown (Rice, 1973). A peculiar behavior pattern of uncertain significance is the insertion of the terminal organ or posterior tip of the larva into the mouth. Jägersten (1963) has suggested that by this action the larva may discharge some secretion from the terminal organ into the mouth.

The pelagosphera larva is a significant agent in the dispersal of species over wide geographical areas. Planktotrophic pelagospheras have been found in warm and temperate waters around the world. Studies in the North Atlantic Ocean of at least 5 species of pelagosphera larvae show that the larvae occur along the entire length of the major east-west currents and that, because of their long larval life (Section 4.4.3.5), may be transported between continents (Scheltema and Hall, 1965, 1975; Scheltema, 1975).

4.4.3.4 LARVAL ORGANOGENESIS

Derivation of tissues and organs has been studied in 7 species of sipunculans, representing all developmental patterns: *Golfingia minuta*, *Phascolion strombi* (Åkesson, 1958), *Phascolopsis gouldi*, *Golfingia vulgaris* (Gerould, 1907), *G. elongata* (Åkesson, 1961a), *Phascolosoma agassizi* (Rice, 1973), and *Sipunculus nudus* (Hatschek, 1883).

As in all protostomous coelomates, the endomesoderm is derived from the 4d cell and the coelom originates by a splitting of the mesodermal bands of the trochophore into splanchnic and somatic layers. The coelom forms before trochophoral metamorphosis in species with planktotrophic development and relatively later in species with more yolky eggs and lecithotrophic development (Section 4.4.3.2). In all species yolk granules are released from the regressing prototroch cells into the developing coelom where they furnish an important source of nutrition for the larva.

The stomodeum develops as an ectodermal invagination at the anterior site of the closed blastopore and pushes farther anteriorly into the region of the prototroch. At the time of metamorphosis of the trochophore, the stomodeum ruptures through the egg envelope to form the mouth and the ventral ciliated surface of the head. The esophagus is derived from the stomodeum and, in planktotrophic species, the buccal organ (in part) and the lip gland, both appendages of the mouth, are also derived from the stomodeum. Entomeres, frequently marked by distinctive pigmentation, give rise to the stomach and intestine. A stomach is present only in planktotrophic larvae and disappears at metamorphosis of the pelagosphera. The anus and rectum are derived from a proctodeal invagination which is located in a dorsal position in the middle or posterior trunk.

Larval retractor muscles, retained as the muscles of the adult, are reported to arise from ectomesoderm in all species except *Sipunculus nudus*. Hatschek (1883) proposed that in the latter species the retractors originate from somatic mesoderm, although he was unable to observe their early development. Reports of the origin of body wall musculature are conflicting. In *S. nudus* both circular and longitudinal muscles are supposed to form from endomesoderm (Hatschek, 1883). Circular muscles of *Golfingia vulgaris* and *Phascolopsis gouldi* are presumed to be derived from ectomesoderm, but no information is available on the source of the longitudinal muscles in these species (Gerould, 1907). Circular musculature of *G. minuta* and *Phascolion strombi* is believed to originate from ectomesoderm and longitudinal musculature from endomesoderm (Åkesson, 1958).

Descriptions of the derivation of the nephridia vary for different species. Observations on development in *Phascolion strombi, Golfingia minuta, G. vulgaris,* and *Phascolopsis gouldi* (Gerould, 1907; Åkesson, 1958) indicate that the nephridia are composite structures, conforming to the concept of mixonephridia by Goodrich (1945). In these species the nephridium proper has been reported to arise from an ectodermal invagination and the internal funnel from peritoneum. In *Sipunculus nudus* and *Phascolosoma agassizi*, on the other hand, a purely mesodermal origin of the nephridia has been suggested (Hatschek, 1883; Rice, 1973).

The larval cuticle in most species is formed by a transformation of the egg envelope at metamorphosis of the trochophore. However, in three species, *Phascolopsis gouldi, Golfingia vulgaris,* and *Sipunculus nudus,* the egg envelope is shed at the time of metamorphosis and the cuticle of the larva is a newly formed secretion of the epidermis (Section 4.4.3.2).

4.4.3.5 METAMORPHOSIS OF THE PELAGOSPHERA

The end of the pelagosphera stage is marked by the loss of metatrochal cilia in both lecithotrophic and planktotrophic pelagospheras. Lecithotrophic larvae, which live in the plankton for only a short time, change to the adult form gradually, passing through a vermiform stage and sometimes requiring several weeks to attain the form of the juvenile. During this time the coelomic yolk is entirely absorbed, the gut is completed, and that portion of the body anterior to the postmetatrochal sphincter is elongated to form the introvert with terminal ciliated tentacular lobes. In *Golfingia pugettensis* 7 weeks may elapse before these changes are completed; in *Themiste alutacea* the change may take 3–4 weeks.

Planktotrophic pelagospheras, after a prolonged period in the plankton, undergo a relatively rapid metamorphosis into the juvenile form. After a month in the plankton, the larva of *Sipunculus nudus* elongates and sinks to the bottom where metamorphosis takes place over a period of 1 or 2 days (Fig. 22c). The metatroch and associated organs of the mouth are lost, the mouth moves anteriorly, a tentacular lobe forms on either side of the mouth, and the head becomes proportionately smaller (Hatschek, 1883). Unidentified planktotrophic pelagospheras from the oceanic plankton have been reported to undergo a similar metamorphosis with regression of the lip and formation of tentacles on the rim of the mouth (Jägersten, 1963). Metamorphosis of "Baccaria citrinella" or Type A larvae is marked by an elongation of the introvert to a length three times that of the trunk with 35 rings of hooks at its anterior end. Over a period of several weeks, the juvenile develops anterior and posterior shields which characterize it as belonging to the genus *Aspidosiphon* (Hall and Scheltema, 1975).

Metamorphosis of oceanic pelagospheras, of unknown adult affinities, has been observed in the laboratory from 11 to 129 days after collection. Maximum duration of larval life has been calculated from approximations of age at the time of collection to be 254 days (Scheltema and Hall, 1975). Metamporphosis of planktotrophic pelagospheras reared from spawnings of known adults in the laboratory has not been observed, although larvae of *Phascolosoma agassizi* have been maintained in culture to an age of 7 months.

4.4.4 Summary of Developmental Patterns

The most common developmental patterns in the phylum Sipuncula are summarized in the schematic diagram (Fig. 26). As shown for Category I, the embryo may develop directly within the egg coats with no ciliated stage, hatching out as a small, crawling worm which gradually transforms into a juvenile, or it may develop into a lecithotrophic trochophore with ciliated prototroch and apical tuft which then develops further along one of three pathways, represented by Categories II, III, and IV. In Category II the trochophore transforms directly into the vermiform stage, ending the pelagic phase, or the trochophore may metamorphose, as in Categories III and IV, into a second larval stage, the pelagosphera with metatrochal band of cilia. The pelagosphera of Category III is lecithotrophic, remaining in the plankton only a short time

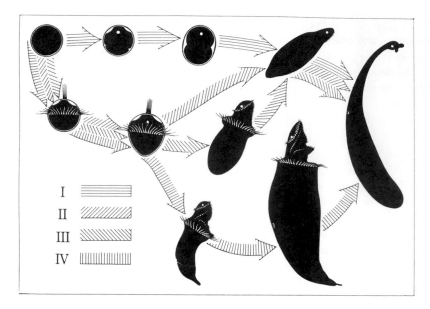

Fig. 26. Diagrammatic summary of chief developmental patterns in the Sipuncula. Category I. Direct development with no pelagic stages. Category II. Pelagic, lecithotrophic trochophore which transforms into vermiform stage. Category III. Pelagic, lecithotrophic trochophore metamorphoses into second larval stage, lecithotrophic pelagosphera, which then transforms into vermiform stage. Category IV. Pelagic, lecithotrophic trochophore metamorphoses into second larval stage, planktotrophic pelagosphera. After a prolonged period in the plankton the larva, having increased in size, undergoes a second metamorphosis into the juvenile form. (From Rice, 1975.)

before transforming into the vermiform stage, then gradually assuming the features of the juvenile. The planktotrophic pelagosphera of Category IV lives for a prolonged period in the plankton, increasing considerably in size, before undergoing a second metamorphosis directly into the juvenile form.

Acknowledgments

I am grateful to Harbor Branch Foundation Laboratory, Fort Pierce, Florida, where much of this chapter was written, for use of their facilities, and the cooperation of their staff in preparation of the figures. Mary Goldizen, Harbor Branch Foundation Laboratory, drew Figs. 17, 19, 20, 21, and 22. Douglas Putnam and William Davenport assisted in the photographic efforts.

Carolyn B. Gast, scientific illustrator in the Department of Invertebrate Zoology, National Museum of Natural History, Smithsonian Institution, is gratefully acknowledged for her illustrations which appear as Figs. 1, 3, 16, 23, and 26.

4.5 References

Åkesson, B. (1958). A study of the nervous system of the Sipunculoideae with some remarks on the development of two species *Phascolion strombi* Montagu and *Golfingia minuta* Keferstein. *Undersökningar över Öresund* 38, 1–249.

Åkesson, B. (1961a). The development of *Golfingia elongata* Keferstein (Sipunculidea) with some remarks on the development of neurosecretory cells in sipunculids. *Ark. Zool.* (Ser. 2), 13(2), 511–531.

Åkesson, B. (1961b). Some observations on pelagosphaera larvae. *Galathea Rept.* 5, 7–17.

Andrews, E. A. (1889). The reproductive organs of *Phascolosoma gouldii*. *Zool. Anz.* 12, 140–142.

Andrews, E. A. (1890). Notes on the anatomy of *Sipunculus gouldii* Pourtalès. *Johns Hopkins Univ. Biol. Lab. Stud.* 4, 390–430.

Awati, P. R. and Pradhan, L. B. (1936). The anatomy of *Dendrostoma signifer* Selenka and de Man. *J. Univ. Bombay* 4(5), 114–131.

Carlisle, D. B. (1959). On the neurosecretory system of the brain and associated structures in *Sipunculus nudus* with a note on the cuticle. *Gunma J. Med. Sci.* 8, 183–194.

Claparède, É. (1863). "Anatomie und Entwicklungsgeschicte wirbelloser Thiere an der Küste von Normandie angestellt." vii + 120 pp. Leipzig.

Clark, R. B. (1969). Systematics and Phylogeny: Annelida, Echiura, Sipuncula. In "Chemical Zoology" (M. Florkin and B. T. Scheer, eds.) Vol. 4, Chapter 1, pp. 1–68. Academic Press, New York and London.

Cole, J. B. (1952). The morphology of *Golfingia pugettensis*: A sipunculid worm. M. S. Thesis. 78 pp. University of Washington, Seattle.

Damas, H. (1962). La collection de *Pelagosphaera* du "Dana". *Dana Rept.* No. 59, 22 pp.

Dawydoff, C. N. (1930). Quelques observations sur *Pelagosphaera*, larve de sipunculide des côtes d'Annam. *Bull. Soc. Zool. Fr.* 55, 1–22.

Fisher, W. K. (1947). New genera and species of echiuroid and sipunculoid worms. *Proc. U.S. Nat. Mus.* 97 (3218), 351–372.

Fisher, W. K. (1952). The sipunculid worms of California and Baja, California. *Proc. U.S. Nat. Mus.* 102(3306), 371–450.

Franzén, Å. (1956). On spermiogenesis, morphology of the spermatozoan, and biology of fertilization among invertebrates. *Zool. Bidr. Uppsala* 31, 355–482.

Gabe, M. (1953). Données histologiques sur la neurosécrétion chez quelques Sipunculiens. *Bull. Lab. Dinard* 38, 3–15.

Gerould, J. H. (1903). Studies on the embryology of the Sipunculidae, I. The embryonal envelope and its homologue. Mark Anniversary Volume, pp. 439–452. Henry Holt and Co., New York.

Gerould, B. (1907). Studies on the embryology of the Sipunculidae, II. The development of *Phascolosoma*. *Zool. Jahrb. Abt. Anat. Ontog. Thiere* 23, 77–162.

Gonse, P. (1956a). L'ovogenèse chez *Phascolosoma vulgare*. I. Définition cytologique des stades de croissance des ovocytes. *Acta Zool.* 37, 193–224.

Gonse, P. (1956b). L'ovogenèse chez *Phascolosoma vulgare*. II. Recherches biométriques sur les ovocytes. *Acta Zool.* 37, 225–233.

Gonse, P. (1957a). L'ovogenèse chez *Phascolosoma vulgare*. III. Respiration exogène et endogène de l'ovocyte. Effet de l'eau de mer. *Biochim. Biophys. Acta* **24**, 267–278.

Gonse, P. (1957b). L'ovogenèse chez *Phascolosoma vulgare*. IV. Etude chromatique des sucres du plasma, action de différents substrats et du malonate sur la respiration de l'ovocyte. *Biochim. Biophys. Acta* **24**, 520–531.

Goodrich, E. S. (1945). The study of the nephridia and genital ducts since 1895. *Quart. J. Microsc. Sci.* **86** (342–344), 113–392.

Green, W. (1975). The annual reproductive cycle of *Phascolosoma lurco* (Sipuncula). *Proc. Intern. Symp. Biol. Sipuncula and Echiura,* in press. Naučno Delo Press, Belgrade.

Häcker, V. (1898). Die pelagischen Polychaeten- und Achaetenlarven der Plankton-Expedition. *Ergeb. Plankton-Exped. Humboldt-Stiftung* **2**, 1–50.

Hall, J. R. and Scheltema, R. S. (1975). Comparative morphology of open-ocean pelagosphaera. *Proc. Intern. Symp. Biol. Sipuncula and Echiura,* in press. Naučno Delo Press, Belgrade.

Hatschek, B. (1883). Ueber Entwicklung von *Sipunculus nudus. Arb. Zool. Inst. Univ. Wien Zool. Stat. Triest* **5**, 61–140.

Heath, H. (1910). *Pelagosphaera,* a larval Gephyrean. *Biol. Bull.* **18**, 61–140.

Hérubel, M. A. (1908). Récherches sur les sipunculides. *Mém. Soc. Zool. Fr.* **20**, 107–419.

Hyman, L. H. (1959). "The Invertebrates: Smaller Coelomate Groups," Vol. 5. McGraw-Hill, New York.

Jägersten, G. (1963). On the morphology and behaviour of *Pelagosphaera* larvae (Sipunculoidea). *Zool. Bidr. Uppsala* **36**, 27–35.

Keferstein, W. (1863). Beiträge zur Kenntnis der Gattung *Phascolosoma. Z. Wiss. Zool.* **12**, 35–51.

Mingazinni, P. (1905). Un Gefireo pelagico: *Pelagosphaera Aloysii* n. gen., n. sp. *Rend. Acad. Naz. Lincei* **14**, 713–720.

Murina, V. V. (1965). Some data on the structure of pelagospheres—sipunculid larvae. *Zool. Zh.* **44**(11), 1610–1619 (in Russian).

Pasteels, J. J. (1935). Recherches sur le déterminisme de l'entrée en maturation de l'oeuf chez divers Invertébrés marins. *Arch. Biol.* **46**, 229–262.

Rajulu, G. S. (1975). Asexual reproduction by budding in the Sipuncula. *Proc. Intern. Symp. Biol. Sipuncula and Echiura,* in press. Naučno Delo Press, Belgrade.

Rajulu, G. S. and Krishnan, N. (1969). Occurrence of asexual reproduction by budding in Sipunculida. *Nature (London)* **223**, 186–187.

Rice, M. E. (1966). Reproductive biology and development in Sipuncula. Doctoral Dissertation, University of Washington. Seattle. 322 pp.

Rice, M. E. (1967). A comparative study of the development of *Phascolosoma agassizii, Golfingia pugettensis,* and *Themiste pyroides* with a discussion of developmental patterns in the Sipuncula. *Ophelia* **4**, 143–171.

Rice, M. E. (1970). Asexual reproduction in a sipunculan worm. *Science* **167**, 1618–1620.

Rice, M. E. (1973). Morphology, behavior, and histogenesis of the pelagosphera larva of *Phascolosoma agassizii* (Sipuncula). *Smithson. Contrib. Zool. No.* **132**, 1–51.

Rice, M. E. (1974). Gametogenesis in three species of Sipuncula: *Phascolosoma agassizii, Golfingia pugettensis,* and *Themiste pyroides. La Cellule,* **70**(2), 295–313.

Rice, M. E. (1975). Observations on the development of six species of Caribbean Sipuncula with a review of development in the phylum. *Proc. Intern. Symp. Biol. Sipuncula and Echiura*, in press. Naucno Delo Press, Belgrade.

Sawada, N., Noda, Y., and Ochi, O. (1968). An electron microscope study on the oogenesis of *Golfingia ikedai*. *Mem. Ehime Univ. Sci. Ser. B*6(1), 25–39.

Scheltema, R. S. (1975). The frequency of long-distance larval dispersal and the rate of gene-flow between widely separated populations of sipunculans. *Proc. Intern. Symp. Biol. Sipuncula and Echiura*, in press. Naucno Delo Press, Belgrade.

Scheltema, R. S., and Hall, J. R. (1965). Trans-oceanic transport of sipunculid larvae belonging to the genus *Phascolosoma*. *Amer. Zool.* 5(2), 216 (Abstr.).

Scheltema, R. S. and Hall, J. R. (1975). The dispersal of pelagosphera larvae by ocean currents and its relationship to geographical distribution of sipunculans. *Proc. Intern. Symp. Biol. Sipuncula and Echiura*, in press. Naučno Delo Press, Belgrade.

Schleip, W. (1934). Die Regeneration des Rüssels *Phascolion strombi* Mont. (Sipunculidae). *Z. Wiss. Zool.* 145, 462–496.

Senna, A. (1906). Sulla struttura di alcune larve (Pelagosphaera) di Sipunculidi. *Pubbl. Inst. Stud. Superiori Pract. Perfezionamento Firenae; Sez. Sci. Fis. Nat.,* pp. 50–78.

Shipley, A. E. (1890). On *Phymosoma varians*. *Quart. J. Microsc. Sci.* 31, (1), 1–27.

Stephen, A. C. (1941). The Echiuridae, Sipunculidae and Priapulidae collected by the ships of the "Discovery" Committee during the years 1926 to 1937. *Discovery Rep.* 21, 237–260.

Stephen, A. C. and Edmonds, S. J. (1972). "The Phyla Sipuncula and Echiura." The British Museum (Natural History), London, 527 pp.

Thorson, G. (1946). Reproduction and larval development of Danish marine bottom invertebrates, with special reference to the planktonic larvae in the Sound (Øresund). *Medd. Kommn Danm. Fisk. Havunders. Plankton* 4, 1–523.

Towle, A. and Giese, A. C. (1967). The annual reproductive cycle of the sipunculid *Phascolosoma agassizii. Physiol. Zool.* 40, 229–237.

Wegener, F. (1938). Beitrag zur Kenntnis der Russelregeneration der Sipunculiden. *Z. Wiss. Zool.* 150, 527–565.

Williams, J. (1972). Development of a rock burrowing sipunculid inhabiting stony coral. *Amer. Zool.* 12(4), 723 (Abstr.).

Chapter 5

POGONOPHORA

Eve C. Southward

5.1 Introduction

The phylum Pogonophora was discovered comparatively recently and has only become well-known during the past 20 years. There are a few comprehensive accounts available: Hyman (1959) gives an excellent summary of knowledge of the group in the late 1950's and Ivanov (1963a), in the English edition of his monograph, includes all work on anatomy, embryology, distribution, and systematics up to that date, much of it having been published earlier by himself, in Russian. Translations and reviews of the embryological work have appeared in several languages (Alvarado, 1957; Manton, 1958; Ivanov, 1958a, 1960, 1963b, 1964a; Johansson, 1969; Kirsteuer, 1969).

Pogonophora are free-living, tubicolous, benthic marine invertebrates without mouth or internal digestive system. Because of the absence of a mouth in the adult and the absence of a blastopore in the embryo, it has been difficult to decide whether these animals belong to the Deuterostomia or Protostomia. In 1963 it was fairly generally agreed that Pogonophora should be classified among the Deuterostomia (Ivanov, 1963a; Southward, 1963). However, since 1963, various discoveries, including a previously unknown segmented posterior end (Webb, 1964a,b; Ivanov, 1964b, 1965), have induced some rethinking of the systematic position of the Pogonophora (see review by Southward, 1971a) and it now seems likely that they may be related to the Annelida and should be included among the Protostomia.

This has caused some confusion in the terminology of pogonophoran anatomy. Ivanov assumes (1963a and earlier) that pogonophores are Deuterostomia and therefore that their main nerve trunk is dorsal and the opposite side of the body, which bears the tentacles and papillae (Fig. 1), is ventral. This convention has been followed by nearly all authors. If pogonophores can be shown to be protostomian, as now seems probable, their dorsoventral orientation must be reversed. In addition, the division of the coelom into three segments in the embryo, as described by Ivanov (1957, 1963a) and the apparent division into three segments in the adult (Johansson, 1939; Ivanov, 1963a and earlier) gave rise to the use of the terms protosoma, mesosoma, and metasoma for the segments. The later discovery of a multisegmented posterior end in the adult was followed by a revised theory of the homologies between the segments of embryo and adult (Webb, 1964c, 1965), but the terms proto-, meso-, and metasoma were retained. A few years later Webb (1969b) reconsidered the segmentation of pogonophores and concluded that they must be polymeric, not trimeric, and suggested that the terms protosoma, mesosoma, and metasoma should be replaced by others based mainly on the functions of the various parts (see Table I). However, these new terms are rather unwieldy and there is a parallel system of terminology, which has been in use for many years (Ivanov, 1963a and earlier; Southward and Southward, 1966), in which the small anterior section of the body is called the "forepart" and the long section behind the diaphragm is called the "trunk" (Table I and Fig. 1). The small, segmented, posterior section should have a separate name and the term "opisthosoma," as suggested by Webb (1969a,b), seems suitable. I prefer to use these terms for the body regions and to avoid, as far as possible, using the terms dorsal and ventral. The two sides of the body can be referred to as "neural" (dorsal–Ivanov) and "tentacular" (ventral–Ivanov).

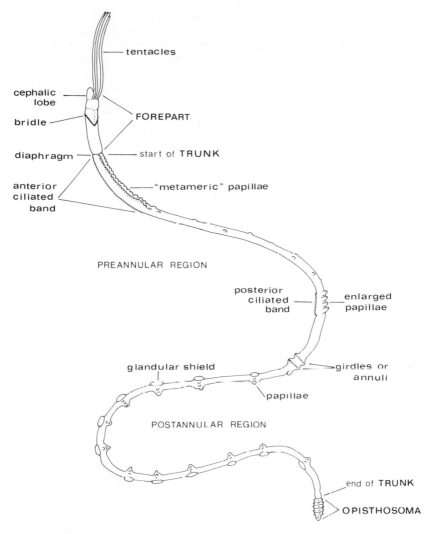

Fig. 1. Diagram of a typical pogonophore (much shortened), showing the anatomical terms used in this chapter.

5.2 Asexual Reproduction

Asexual reproduction is rare in the Pogonophora. One species, *Sclerolinum brattstromi*, is long and fragile, fragmenting readily; the fragments are capable of regenerating anterior and posterior ends, and so two or more shorter individuals can be produced, living in the same tube

TABLE I
TERMINOLOGY OF BODY SEGMENTS OF ADULT POGONOPHORA

Ivanov (1963a)	Webb (1964c)	Webb (1969b)	Present paper
Protosoma Mesosoma } Forepart	Protosoma	Tentaculomere }	Forepart
	Mesosoma {	Frenulomere {	
Metasoma Trunk		Gonomere	Trunk
— —	Metasoma	Opisthosoma	Opisthosoma

(Webb, 1964d; Southward, unpublished). It is not known whether the individuals produced in this way ever leave the original tube and start a new one of their own.

Other *Sclerolinum* species may possibly have the same capacity for regeneration, but there is no sign of such a thing in the many species of other genera that have been studied.

5.3 Sexual Reproduction

5.3.1 Sexual Dimorphism

In pogonophores the sexes are separate but the only external difference between them is the position of the genital apertures. The male, of most species, has two rather large rounded papillae at the anterior end of the trunk, immediately in front of the first two metameric papillae, with the genital apertures in their centers (Fig. 2). In some species of *Siboglinum* (e.g., *S. atlanticum* and *S. candidum*) there is only one papilla, between the anteriormost metameric papillae, and one opening, the two genital ducts uniting just before reaching the papilla (Southward and Southward, 1966). The female genital apertures lie, inconspicuously, on either side of the trunk in the posterior part of the preannular region, in front of the zone of enlarged papillae, if this is present.

In some species there is a difference in size and proportions between the sexes. In *Polybrachia canadensis*, for example, females are generally thicker than males of the same length, and eventually they grow longer than males (Southward, 1969). In a mature animal, full of gametes, the epidermis of the trunk is often transparent enough to allow the gametes to be seen and the sex determined without damage to the body wall.

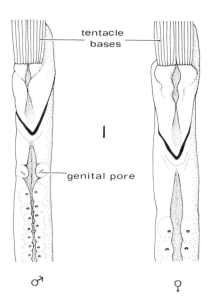

tentacle
bases

genital pore

♂ ♀

Fig. 2. *Lamellisabella coronata,* male and female: forepart and anterior part of trunk, viewed from tentacular side. Scale line, 0.1 mm.

5.3.2 Anatomy of the Reproductive System

In both sexes there is one pair of gonads in the trunk and their products are carried to the exterior by a pair of coelomoducts. The sexes differ in the position of the gonads as well as in the position of the genital apertures.

The male reproductive system has been studied by Ivanov (1958b, 1963a) in *Polybrachia, Lamellisabella,* and *Siboglinum* and is basically similar in all three genera. The testes are very long tubular organs which start somewhere near the posterior end of the postannular region of the trunk and continue all along this region, past the girdles, and a short way into the preannular region. They are quite wide canals which run close to or merge with the median mesentery between the blood vessels and occupy the neural side of each lateral coelomic space, or often the whole of the space. Close to the mesentery the testis wall is extremely thin but the outer side has a definite lining epithelium, partly ciliated. Outside this is a very thin circular muscle layer and a covering of peritoneal cells. The testes are filled with floating masses of developing germ cells. The anterior end of each testis opens into a sperm duct, which is apparently morphologically a coelomoduct. This duct makes a backward loop and then proceeds anteriorly (Fig. 3)

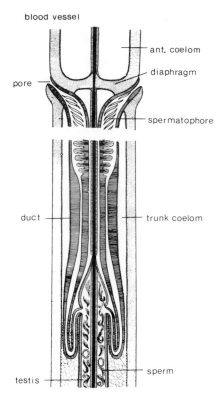

FIG. 3. Diagram of male reproductive system of a pogonophore. (From Ivanov, 1963a.)

along the preannular region of the trunk to open through a genital papilla just behind the diaphragm. The walls of the sperm ducts consist of thick internal epithelium, a thin muscle layer, and a covering of peritoneal epithelium. The posterior part of the duct is lined with tall ciliated cells; the following part has a very thick ridged epithelium, rich in gland cells, where the spermatophores are formed; finally, the long anterior section of the duct is lined with ciliated cells and serves as a storage place for the spermatophores.

The female reproductive system has been investigated in four genera, *Polybrachia, Lamellisabella, Spirobrachia,* and *Siboglinum,* and there is little difference between them (Ivanov, 1958b, 1963a). The paired ovaries are attached to the median mesentery at the anterior end of the trunk, just behind the diaphragm (Fig. 4). Like the testes, they are long tubular organs, but they start at the anterior end of the trunk and run backward through the lateral coeloms almost to the region

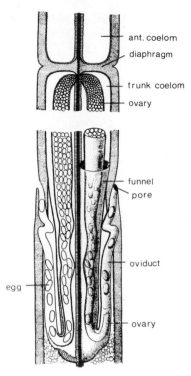

ant. coelom

diaphragm

trunk coelom

ovary

funnel

pore

oviduct

egg

ovary

FIG. 4. Diagram of female reproductive system of a pogonophore. (From Ivanov, 1963a.)

of enlarged papillae, or to within a short distance of the girdle region in species which lack enlarged papillae. The ovary walls consist of small flat cells, apparently derived from the peritoneal lining cells and are attached to the median mesentery only at the extreme anterior end. Blood vessels in the walls originate from one of the longitudinal blood vessels at the point of attachment (dorsal vessel in Ivanov's terminology). The anterior ends of the ovaries are packed with small oocytes. Farther back the oocytes gradually grow larger and become more loosely packed, the largest being found at the posterior end of the ovary. Here the ovary opens into an oviduct, which may surround the posterior end of the ovary with a thin-walled funnel. The oviducts are fairly short **U**-shaped tubes which open to the exterior through lateral pores a little in front of the zone of enlarged papillae. The walls of the oviducts are lined with nonciliated epithelium, outside which they have a well-developed layer of circular muscle and an outer layer of peritoneal cells. The eggs must be extruded by muscular contractions.

The reproductive system described above seems to be typical of most species of Pogonophora, but a few may have only one side of the female genital system fully developed. One unusual species has the opening of the oviduct close to the anterior end of the trunk, at the same level as the male genital aperture (Webb, 1969a).

5.3.3 Origin of Germ Cells and Gonads

The male germ cells presumably originate inside the testes from some kind of germinal epithelium which has not yet been observed (Ivanov, 1963a). The wall of the testis seems to be derived from the peritoneum. The female germ cells originate on the sides of the dorsal blood vessel (Ivanov terminology) at the anterior end of the trunk and develop as a swelling under the peritoneal lining. As they increase in number and size they carry the peritoneum with them to form the ovary wall and gradually the ovary extends back along the trunk (Ivanov, 1963a).

5.3.4 Cytodifferentiation of Gametes

5.3.4.1 SPERMATOGENESIS

The long testes are filled with masses of male germ cells at various stages of development. Ivanov (1958b, 1963a) has described several stages, starting from comparatively large rounded cells or spermatogonia which divide to form groups of smaller cells (Figs. 5 and 7A). These in turn divide to produce groups of many small cells (morulae). The small cells grow a little larger, develop tails, and become very much elongated as they mature, remaining grouped around a central mass of cytoplasm, the cytophore (Figs. 5 and 7C,D). The mature spermatozoa have a very long head region. In *Siboglinum atlanticum* (Fig. 6A), for example, the head region and tail region are both about 70 μm long (Southward and Southward, 1963 and unpublished). The fine structure of the spermatozoon of *Siboglinum ekmani* has been described by Franzén (1973). This spermatozoon has a helical acrosome about 3 μm long; a helical nucleus (30×0.3 μm), the posterior part of which is surrounded by three helically wound, filamentous mitochondria; a short connecting piece with two centrioles; and a flagellum about 30 μm long, the first half of which is stiffened by an outer sheath of granular material. Franzén has also followed the elongation of the nucleus and mitochondria through the spermatid stages. He notes that filamentous mitochondria surrounding the nucleus are rather unusual and that in other animals with elongated spermatozoa the mitochondria usually

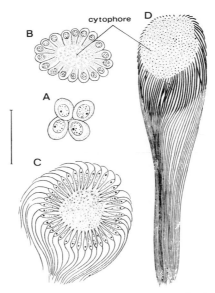

Fig. 5. Successive stages in spermatogenesis of *Lamellisabella zachsi*. (A) 4-cell stage; (B) spermatocyte morula; (C) spermatid mass; (D) mass of sperms. Scale line, 0.05 mm. (From Ivanov, 1963a.)

spread down the flagellum. Elongated spermatozoa are found in animals with internal fertilization; in *Siboglinum* fertilization must occur in the tube or internally, but it has not yet been observed. The spiral shape of the spermatozoon resembles the shape in the polychaete *Manayunkia*, some Gnathostomulida and some Mollusca (Franzén, 1956, 1958; Sterrer, 1969), but its significance is quite unknown.

Spermatophores are formed in a thick-walled glandular region of the sperm duct (Section 5.3.2). The spermatozoa are bundled together and encased in a thin envelope of polysaccharide material (Ivanov, 1963a; Southward and Southward, 1966). One end of the envelope is drawn out into an extremely long filament, which is coiled into a skein close to the spermatophore while it remains within the duct (Fig. 6C), but once in the water the filament uncoils and spreads out (Fig. 8C). The shape and size of the spermatophore are constant for each species and are useful in the systematics of Pogonophora. Spermatophores may be spindle-, cigar-, or leaf-shaped (Figs. 6B–H and 8A,B), and usually are symmetrical, but in *Galathealinum arcticum* they are remarkably unsymmetrical (Fig. 6H). They vary in length from about 40 μm to nearly 2 mm, the size being roughly correlated with the size of the species. The outside of the envelope and filament may be ornamented with tiny

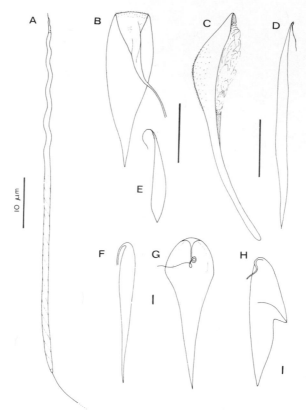

Fɪɢ. 6. (A) "Head" of spermatozoon of *Siboglinum atlanticum;* (B–H) spermatophores of: B, *Lamellisabella coronata;* C, *Polybrachia canadensis,* with filament;
D, *Siboglinum candidum;* E, *Siboglinum bayeri;* F, *Zenkevitchiana longissima* (After
Ivanov, 1963a); G, *Spirobrachia beklemishevi* (After Ivanov, 1963a); H, *Galathealinum arcticum.* Scale lines (B–H), 0.1 mm.

bristles or capitate hairs (Webb, 1963) (Figs. 6B,C,E and 8B). The
spermatophores are stored inside the sperm duct, neatly arranged side
by side, with the filament-bearing end forward. A large male contains
several hundred spermatophores, perhaps more than a thousand in some
species.

5.3.4.2 Oogenesis

Ivanov (1958b, 1963a) and Bubko (1971) have studied oogenesis
in pogonophores and have observed that oocytes are formed at the anterior end of the ovary and pushed backward as more and more are
formed. As they progress along the ovary they increase in size and

yolk content. Since there are no apparent follicle or nurse cells, nutriment must be obtained from the fluid in the ovary. The largest oocytes lie at the posterior end of the ovary and sometimes within the oviducts. In small pogonophores, such as *Siboglinum* species, the full-grown oocytes are elongated, but in larger species, which have more space in the ovaries and oviducts, they are more or less spherical. Their size has not received much attention, but in *Oligobrachia dogieli* and *O. ivanovi* they are about 400 μm in diameter (Ivanov, 1963a; Southward and Southward, 1963). *Polybrachia canadensis* has oocytes 160 μm in diameter and *Galathealinum arcticum*, which is a large pogonophore, has oocytes only 150 μm in diameter. The full-sized oocyte of the small species *Siboglinum caulleryi* is about 650 μm long and 130 μm in diameter, so there is no correlation between oocyte size and species size. There does, however, seem to be a tendency for members of the order Thecanephria to have smaller oocytes than members of the order Athecanephria, but this requires further investigation.

5.3.5 Gametogenic Cycles within Populations

Collection of pogonophores is not easy, because of their mainly deep-water habitats, and sampling of one population at intervals has not been attempted. Some large samples of *Polybrachia canadensis*, collected in the northeast Pacific in August and September, provided examples of a range of growth stages, and it was found that females with a preannular region more than 20 mm long contained recognizable oocytes. In slightly larger females some full-sized oocytes were present, and the number of full-sized oocytes increased as the animals grew longer. Thus in this species, full-sized oocytes are accumulated and seem to be retained for a long period (Southward, 1969). There could be a simultaneous spawning by the larger individuals. Males of most species accumulate spermatophores, so that eventually the preannular region becomes distended with them.

5.3.6 Factors Influencing Gametogenesis

Pogonophores generally live in environments with little seasonal variation in temperature, light, or salinity (Southward, 1963, 1971a), always below the intertidal zone, and usually in depths of more than 200 m. As stated above (Section 5.3.5), gametogenesis seems to result in the gradual accumulation of gametes over a long period, and there is no evidence as to whether or not there is any seasonal cycle.

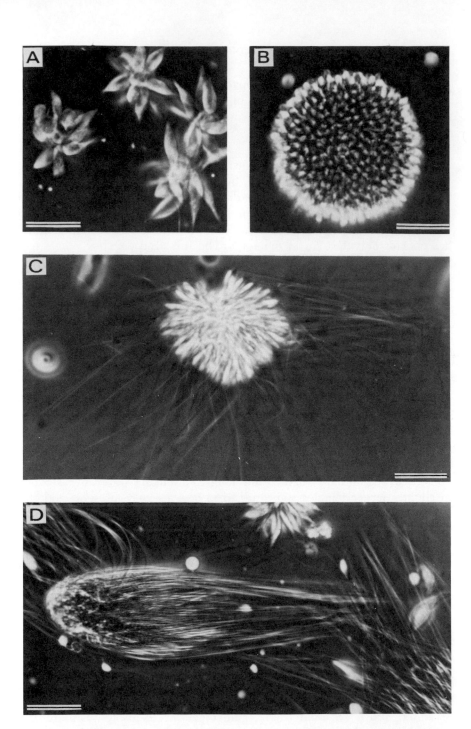

FIG. 7. Spermatogenesis in *Siboglinum ekmani:* (A) early morulae; (B) fully formed morula; (C) spermatid mass; (D) mass of sperms. Scale lines, 20 μm.

5.3.7 Reproductive Behavior

Pogonophores are sedentary tube dwellers, living partly buried in mud or other deposits. They are not mobile enough to migrate or aggregate for spawning. They often occur in quite dense populations, up to 200 per square meter, which is presumably sufficiently close together for reproduction.

5.3.8 Spawning and Mating

Whether or not any form of mating takes place in Pogonophora is still unknown. Large numbers of spermatophores, each containing hundreds of sperm, are stored by the males in their sperm ducts. The spermatophores are ejected one by one, filament first, from the genital apertures (Fig. 2); this can be observed under the microscope in slightly squeezed specimens (Fig. 8C). Webb (1963) has watched living *Siboglinum ekmani*, inside the tube, fold the tentacle backward over the metameric region of the trunk, pick up the ejected spermatophores, and draw them toward the front end of the tube. One animal was seen to do this several times, and as each fresh batch of spermatophores was brought forward, the earlier ones were pushed farther forward until eventually they were pushed out of the tube. Ivanov has suggested (1963a and earlier) that the spermatophores float about near the sea floor until their filaments become entangled with the tentacles of a female and can then be drawn into her tube. Carlisle (in Ivanov, 1963a, p. 98) suggests that waterborne transfer of spermatophores is likely to be unsuccessful because of excessive wastage. He considers that some form of transfer on the tentacles of the male may take place. However, this could only be achieved if the animals were within "tentacle length" of one another, and even the largest species have tentacles only a few centimeters long, so they would have to be very close together. It seems to me that this is unlikely to happen very often, but that waterborne spermatophores, with their very long filaments, do have a good chance of being captured by a female some distance from the male, or at least becoming entangled with her tube. The covering membrane of the spermatophore disintegrates after a few hours in seawater (Southward and Southward, 1963), and each contains enough spermatozoa to fertilize a large batch of eggs.

Fertilization takes place inside the tube of the female and the fertilized eggs come to lie in the top part of the tube, in front of the animal (Fig. 9). Fertilization has not been observed, but it could be either internal or external. The process of moving the eggs from the genital aperture to the front end of the animal has not been observed either.

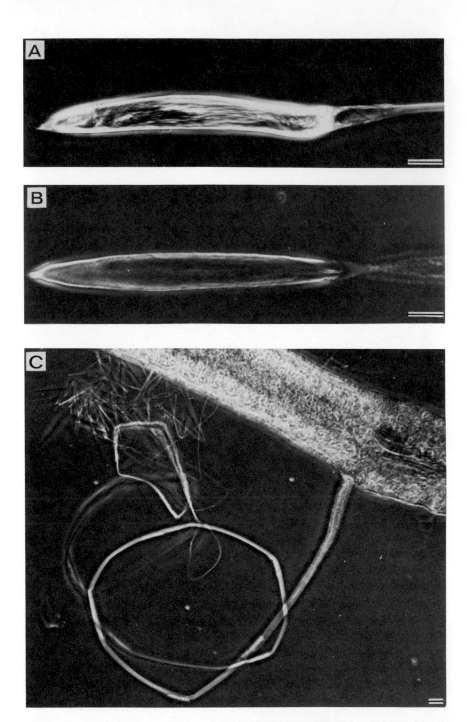

FIG. 8. (A) *Siboglinum fiordicum,* spermatophore with base of filament; (B) *Siboglinum ekmani,* spermatophore with base of filament; (C) *Siboglinum ekmani,* diaphragm region of male, with forepart on left and trunk on right, spermatophore being extruded through genital pore, filament first. Scale lines, 20 μm.

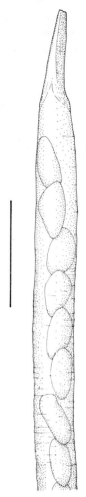

FIG. 9. *Siboglinum candidum:* early embryos in top end of tube. Scale line, 1 mm.

The number of eggs produced at one time varies from about five in some *Siboglinum* species to more than 100 in *Oligobrachia ivanovi* (Southward and Southward, 1963). Development is roughly synchronous in the whole batch of eggs. Zygotes or embryos have been found in the tubes of the following species: *Oligobrachia dogieli* (Ivanov, 1957); *O. ivanovi* (Southward and Southward, 1963); *O. webbi* (Brattegard, 1966); *Crassibrachia sandersi* (Southward, 1968); *Siboglinum fiordicum* (Jägersten, 1957; Webb, 1964c; Nørrevang, 1970a); *Siboglinum caulleryi* (Ivanov, 1957); *Siboglinum inerme* (Southward, in Ivanov, 1963a); *Sibo-*

glinum   holmei   (Southward,   unpublished);   Siboglinum   gosnoldae
(Southward, unpublished); *Siboglinum mergophorum* (Southward, unpublished); *Siboglinum candidum* (Southward, unpublished); *Siboglinum*-3 unnamed Antarctic species (Southward, unpublished).

All these species belong to the order Athecanephria. Members of the
order Thecanephria, including such genera as *Diplobrachia, Polybrachia,
Galathealinum,* and *Lamellisabella,* have never been found brooding
their young. This together with their apparently smaller eggs, suggests
that they may have a free-living developmental stage, but more evidence
is needed to support this speculation.

5.3.9 Breeding Period

There are not enough data to determine the breeding season for any
species, or the length of the embryonic phase. It is also not known
whether an individual reproduces more than once in its lifetime.

5.4 Development

5.4.1 Embryonic Development

The zygotes lie in the top part of the tube, in front of the female
but often alongside her tentacle. The species in which zygotes or embryos
have been found are listed in Section 5.3.8. Caullery (1944) was the
first to see embryos in a species of *Siboglinum* without being certain
what they were, but he suspected that they were embryos from their
resemblance to the oocytes. In 1957, Jägersten described the embryos
of another species of *Siboglinum* (probably *S. fiordicum*) and about
the same time Ivanov (1957) described several stages in the development
of *Siboglinum caulleryi* and *Oligobrachia dogieli*. Webb (1964c, 1965)
has described the external appearance of several embryonic stages of
S. fiordicum, and recently Nørrevang (1970,a,b) has examined the internal anatomy of four stages of the same species. Thus we have a
fairly complete idea of the development of *S. caulleryi* and *S. fiordicum,*
which are closely related species with embryos of much the same size
and appearance. There are some differences in the descriptions of their
internal anatomy, by Ivanov and Nørrevang, which are discussed below.
In the account which follows, the early stages of cleavage are taken
mostly from Ivanov's description, and the later stages from both Ivanov
and Nørrevang, with some additional information from Jägersten (1957)
and Webb (1964c, 1965).

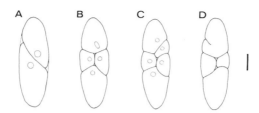

FIG. 10. *Siboglinum caulleryi*, early cleavage stages: (A) 2-cells; (B) 4-cells; (C) 5-cells, viewed from concave side; (D) 5-cells, viewed from convex side. Scale line, 0.1 mm. (After Ivanov, 1963a.)

The zygote is enclosed in a thin membrane which remains intact through the early stages of cleavage. In *Siboglinum caulleryi* and S. *fiordicum* the zygote is elongate-oblong with the nucleus about halfway along and close to one side. This side is concave in S. *caulleryi*. In *Oligobrachia* species the zygotes are more nearly spherical. The first cleavage in *Siboglinum* is oblique and produces two equal-sized blastomeres (Fig. 10A), which then divide simultaneously, each into a large and a small cell (Fig. 10B). The third cleavage is apparently not synchronous. Ivanov (1963a) has observed the incomplete formation of a fifth cell by the division of one of the large cells, but in a review article he has figured an 8-cell stage of an unnamed species of pogonophore with four small cells regularly arranged over four larger cells, and has stated that this shows radial cleavage (Ivanov, 1970). Ivanov has no information on the development between the 8-cell and the 80-cell stages. Bakke (1972) has seen the 8- and 16-cell stages of S. *fiordicum*, and in his opinion the cleavage pattern is closer to radial than spiral. I have found a stage of S. *fiordicum* with about 20 cells (Fig. 11). It has about seven macromeres on the side equivalent to the convex side of S. *caulleryi* and about 12 micromeres on the other side. There were still very large cells at each end and some of the cleavages were incomplete, but the micromeres were beginning to spread toward the

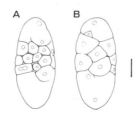

FIG. 11. *Siboglinum fiordicum*, 20-cell stage: (A) micromere side; (B) macromere side. Scale line, 0.1 mm.

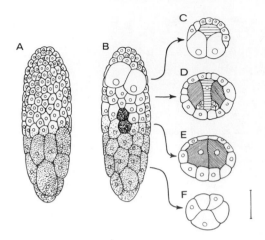

FIG. 12. *Siboglinum caulleryi* 130–140-cell stage: (A) from concave side; (B) from convex side; (C–F) transverse sections at the levels indicated. Horizontal hatch, anterior, median, internal cell; oblique hatch, pair of posterior internal cells. Scale line 0.1 mm. (From Ivanov, 1963a.)

anterior pole. The 80-cell stage of S. *caulleryi* has small cells almost covering the anterior end and larger cells forming the posterior end. Two large cells remain on the surface of the convex side near the anterior end. In a later stage, with 130 to 140 cells, the same general arrangement is found and the two large cells can still be seen (Fig. 12B). Internally there are three massive cells and two groups of very small cells (mesenchyme) lying near the concave side (Fig. 12D). This stage may be regarded as a gastrula (Ivanov, 1963a), gastrulation having taken place by epiboly. There is no blastocoel or blastopore, and nothing remains to show its primitive position. A 200-cell embryo is the earliest stage of S. *fiordicum* described by Nørrevang (Fig. 13). This has small cells covering the anterior half and larger cells forming the posterior half. Inside there are several large cells and a group of small cells (Fig. 13d) near the midline of one side (termed side B by Nørrevang and equivalent to the concave side of S. *caulleryi*).

The early ciliated embryo of *Siboglinum* has an epidermis composed of small cells, including two rings of ciliated cells (Fig. 14A). Large cells are arranged centrally around a very small oblong cavity, the archenteron (Fig. 15a), and there is a mass of small cells in front of the posterior ciliated ring between the ectoderm and endoderm, Nørrevang, 1970a). A patch of ciliated cells develops on one side of the body (Fig. 15), and later elongates to form a ciliated band. This ciliated side of the body is dorsal according to the deuterostome theory, but

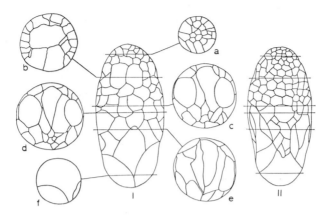

Fig. 13. *Siboglinum fiordicum,* 200-cell stage: I, side B; II, side A; a–f, transverse sections at levels indicated, side A upward. (From Nørrevang, 1970a).

ventral according to the protostome theory. Before this band appears, close to or joined with the anterior ciliated ring (Fig. 14B,C,), it is difficult to decide which end of the embryo is which (unless it can be studied inside the maternal tube, where it develops with the anterior end downward, toward the mother, in all species of *Siboglinum* studied). A comparison of the accounts of the internal anatomy at this stage, in *S. caulleryi* and *S. fiordicum,* indicates that either Ivanov or Nørrevang has reversed the anterior and posterior ends (Figs. 15 and 16A). Ivanov found the mesodermal tissue mainly near the anterior end, while Nørrevang found it near the posterior end. Ivanov shows the coelomic spaces developing anteriorly and laterally and he considers that they are derived from an anterior endodermal pouch which produces backwardly directed pockets which become separated from the anterior unpaired coelom.

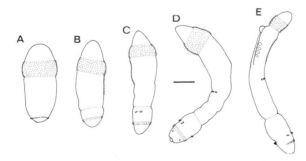

Fig. 14. *Siboglinum fiordicum,* successive ciliated embryonic stages. Scale line, 0.1 mm. (After Webb, 1964c.)

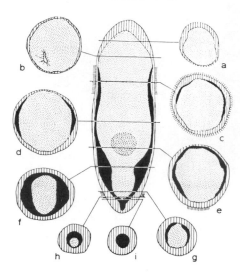

Fig. 15. *Siboglinum fiordicum*, early ciliated embryo. Diagrammatic reconstruction of horizontal section from transverse sections, a–i, at levels indicated. Vertical hatch, ectoderm; dots, endoderm; solid black, mesoderm. (From Nørrevang, 1970a.)

Nørrevang describes the mesoderm at the same stage in *S. fiordicum* as being still solid, and concentrated at the sides and posterior end of the embryo (Fig. 15).

A later stage in development is described by both authors as having a constriction near the posterior end, and a ciliated band behind the anterior ciliated ring (Figs. 16B and 17). Thus, Ivanov and Nørrevang agree on the orientation at this stage. Looking back at their illustration of the previous stage (Figs. 15 and 16A) it appears that Nørrevang oriented the embryo correctly and Ivanov was mistaken. For this reason I prefer to follow Nørrevang's description of the development of the mesoderm and coelom. He finds that by the time the posterior constriction is formed, the mesoderm is well differentiated (Fig. 17). The development of the coelom is schizocoelic, not enterocoelic as Ivanov thought. There are one or two thin-walled coelomic cavities at the anterior end and two pairs of lateral cavities a little farther back. At the level of the ciliated band there are several flat coelomic cavities on either side, but it is difficult to distinguish the exact number. Behind the level of the ciliated band the mesodermal layer becomes thicker but lateral cavities may be present. Two longitudinal blood vessels can be detected. Just behind the septum the mesoderm is very thin, but at the posterior end of the embryo there is a mass of mesodermal tissue with the cells arranged in arcs separated by distinct lumina (Fig. 17f). Nørrevang

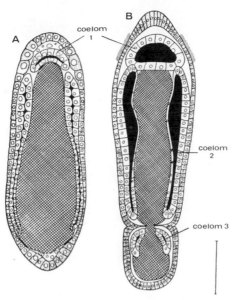

FIG. 16. *Siboglinum caulleryi*, embryos. Diagrammatic horizontal sections. (A) Prior to development of cilia. (B) During formation of posterior septum. Dots, ectoderm; white, mesoderm; cross hatch, endoderm; solid black, coelom. Scale line, 0.1 mm. (From Ivanov, 1963a.)

regards this posterior zone of mesoderm formation as independent of the anterior mesoderm and comparable with the postlarval segments of annelids.

The posterior coelomic compartments of the embryo seem to be the rudiments of the opisthosome segments of the adult pogonophore (see Webb, 1964c, 1969b; Ivanov, 1964b, 1965). Ivanov's theory of a tricoelomate embryo seems to be based on a misconception of the orientation of the early ciliated stage, and the pogonophore embryo apparently consists of a somewhat indefinite number of "larval segments" in front of the septum and a small number of "postlarval" segments which develop later behind the septum.

Externally, the embryo at the stage of formation of the posterior septum usually has several small rod-shaped bristles or setae, commonly arranged in four pairs, a little behind the septum (Figs. 14C and 19B), growing out of epidermal pouches (Fig. 18). The further development of the embryo consists of gradual elongation of the region between the anterior ciliated ring and the posterior septum. A tentacle rudiment develops behind the anterior ciliated ring and another set of four pairs of small bristles develops in the middle region of the embryo (Fig.

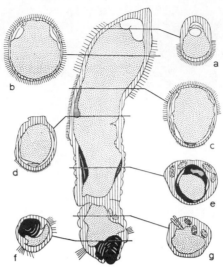

FIG. 17. *Siboglinum fiordicum,* embryo at stage of formation of posterior septum. Diagrammatic horizontal section reconstructed from transverse sections, a–g, at levels indicated. Vertical hatch, ectoderm; horizontal hatch, blood space; dots, endoderm; solid black, mesoderm; white, coelom. (From Nørrevang, 1970a).

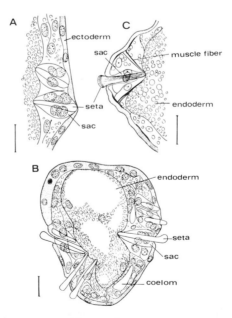

FIG. 18. Transverse sections of late embryos, showing developing posterior setae: (A and B) *Oligobrachia dogieli;* (C) *Siboglinum caulleryi.* (From Ivanov, 1963a). Scale lines, 0.02 mm.

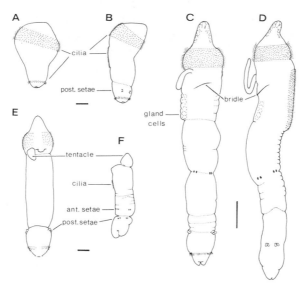

Fig. 19. Pogonophore embryos. (A and B) Successive stages of *Oligobrachia ivanovi;* (C) late stage of *Siboglinum candidum;* (D) late stage of *Siboglinum gosnoldae;* (E) late stage of *Oligobrachia dogieli* (after Ivanov, 1963a); (F) late stage of *Crassibrachia sandersi.* Scale lines, 0.1mm.

14D) (Jägersten, 1957; Webb, 1964c; Southward, 1968). Epidermal glands become visible as swollen clear cells, and the ciliated band becomes longer. In *Siboglinum* species the tentacle begins to elongate (Fig. 19C,D) but in *Oligobrachia dogieli* a second tentacle rudiment develops beside the first (Fig. 19E) (Ivanov, 1963a).

5.4.2 Organogenesis

About the time of the development of the ciliated rings the endoderm loses its central cavity and becomes a liquid mass which is gradually used up as the embryo develops, and there is no trace of an alimentary canal. The mesoderm gives rise to two longitudinal blood vessels while the coelomic cavities are forming and by the time the tentacle rudiment appears, red blood can be seen in the living embryo of *Siboglinum fiordicum* (Southward, unpublished). Other mesoderm cells which line the coelomic cavities become muscular and the worm-shaped embryo is capable of both longitudinal and circular contraction. Both ectoderm and mesoderm contribute to the formation of the tentacles. The epidermis becomes differentiated at an early stage, producing first the ciliated rings and band, then the first set of bristles, and later the second set

of bristles and unicellular epidermal glands. Ivanov (1963a) describes
the formation of the posterior bristles, which emerge from multicellular
epidermal sacs (Fig. 18) and he suggests that several cells are involved
in the formation of each. The cuticle seems to be very thin in the embryo,
but two thickenings which develop laterally behind the tentacle rudiment
are the first signs of the bridle keels (Fig. 19C,D). The posterior septum
is largely mesodermal and divides the opisthosoma from the trunk
(Webb, 1969b; Nørrevang, 1970a) while the diaphragm separating the
forepart from the trunk (see Fig. 1) is a later development, not seen
in the embryo.

The oldest embryos found in adult tubes are worm-shaped, with anterior and posterior ciliated rings still present (Figs. 14E and 19C,D).
They have a longitudinal ciliated band, a short tentacle, faint lateral
bridle keels, and two rings of bristles or setae. The anterior bristles
have enlarged rounded heads with small teeth, like the toothed platelets
of the adult girdles, while the posterior ones are rod-shaped without
enlarged heads. Webb (1965) has found, in some embryos of *Siboglinum
fiordicum*, a third set of two setae behind the posterior ciliated ring
(Fig. 12D). A small depression at the posterior end of the embryo is
termed the sucker by Webb (1964c), who suggests that it is used in testing the substrate.

The youngest individuals found living on their own are much larger
than the biggest known embryos. In *Polybrachia canadensis* the smallest
juveniles found are about 20 mm long, and have developed the diaphragm between forepart and trunk and the metameric papillae, though
they have still only one tentacle (adult has 30 to 40) and incompletely
formed girdles (Southward, 1969).

5.4.3 Factors Influencing Successful Development

The embryos have no mouth or gut and are not able to feed. At
first they must be dependent on their stores of yolk, but it is possible
that they can take up dissolved organic compounds from the water
in the tube which diffuses through its walls. Adult pogonophores are
capable of taking up dissolved amino acids, glucose, and fatty acids
from dilute solutions (Little and Gupta, 1968, 1969; Southward and
Southward, 1968, 1970, 1972), and one experimental observation on embryos of *Siboglinum fiordicum* suggests that they may have a higher
rate of uptake than that of the adult (Southward and Southward,
unpublished).

The rate of development and growth is not known. When some embryos of *Oligobrachia ivanovi* were kept in the laboratory (at 5°C in

the dark) they survived several weeks without much further development (Southward and Southward, 1963). Those that were kept in their original tubes survived much longer than those that were removed and kept in bowls of similar seawater. The embryos outside the tubes became fouled with mucus and bacteria and died. Thus the tube seems to protect the embryo from bacteria as well as from predators.

5.4.4 Settlement

The embryos are kept inside the tube by either a terminal seal (Fig. 9) or the folding over of the flimsy tube tip. It is not known how they make their way out. When embryos are taken out of the tube the beating of their cilia carries them forward, rotating on a longitudinal axis (Webb, 1964c; Southward and Southward, 1963). They have a strong tendency to sink to the bottom and cannot be said to be good swimmers. The worm-shaped embryos can crawl along the bottom and appear to test the substratum with their posterior sucker (Webb, 1964c). They can also stick themselves down with the secretion of their unicellular glands (Webb, 1965; Southward, unpublished). Inside the maternal tube, their ciliary currents probably help to circulate water among them and help to prevent their sticking together (Southward and Southward, 1963). The worm-shaped embryo is small and light enough to be swept along by water currents and this is presumably the chief method of dispersal (Southward, 1971b); see also new work by Bakke (1974).

5.4.5 Embryology and the Systematic Position of Pogonophora

The systematic position of Pogonophora is still a matter for argument, since the discovery of the segmented posterior end reopened the discussion (see Southward, 1971a). A study of the adult morphology is not very helpful, since it is so much modified by the absence of mouth and alimentary canal that one cannot decide which is the dorsal side. The embryology should give valuable clues. For instance, if the pattern of cleavage was obviously spiral or radial it would provide a pointer toward protostomian or deuterostomian affinity. Bakke (1972) has studied cleavage in *Siboglinum fiordicum* and thinks it is closer to radial than spiral. However, the situation is distorted by the elongated shape of the egg and the large quantity of yolk. The cleavage of a spherical egg is more likely to show the primitive pattern, and it is to be hoped that suitable material of *Oligobrachia* or some Thecanephrian pogonophore with small eggs, such as *Polybrachia,* may be found soon.

A second clue lies in the way the coelom is formed. Ivanov (1963a and earlier) considered coelom formation to be enterocoelic and comparable to coelom formation in some hemichordates. Livanow and Porfirewa (1967) have queried this after examining Ivanov's illustrations, and they have suggested schizocoelic development. More recently, Nørrevang has provided evidence of schizocoelic development in *Siboglinum fiordicum*, which points to a protostomian affinity rather than a deuterostomian one.

The development of the posterior region of the embryo into the adult opisthosoma, with its coelomate segments and setalike bristles, suggests some relationship to the Annelida (George and Southward, 1973), though it is true that setae of annelid type are found in the not very closely related Brachiopoda.

There is a growing body of evidence to show that pogonophores are more nearly related to Annelida than to any deuterostome group, and Nørrevang (1970a) has suggested a close relationship to Archiannelida. I think the relationship is more distant, because of the singularity of many pogonophore features and because of the antiquity of the group; Pogonophora seem to be sufficiently distinct to remain in a separate phylum with status similar to Echiura and Sipuncula.

5.5 References

Alvarado, R. (1957). Desarrollo embrionario de los Pogonoforos. *Bol. Soc. Espan. Hist. Natur., Inf. Sci.* **55**, 495–496.

Bakke, T. (1972). *Siboglinum fiordicum* Webb (Pogonophora): Larvenes utvikling, bunnfelling og behaviour i akvarium. Unpublished thesis, University of Bergen, Norway.

Bakke, T. (1974). Settling of the larvae of *Siboglinum fiordicum* (Pogonophora) in the laboratory. *Sarsia* **56**, 57–70.

Brattegard, T. (1966). A new species of multitentaculate Pogonophora from northern Norway. *Sarsia* **22**, 55–63.

Bubko, O. (1971). Some evidence on the oogenesis of a Pogonophora species. *Cytologia* **13**, 1461–1467 (in Russian).

Caullery, M. (1944). *Siboglinum* Caullery 1914 type nouveau d'invertébrés, d'affinités à préciser. *Siboga Exped. Monogr.* **25**bis, 1–26.

Franzén, Å. (1956). On spermiogenesis, morphology of the spermatozoon, and biology of fertilization among invertebrates. *Zool. Bidr. Uppsala* **31**, 355–482.

Franzén, Å. (1958). On sperm morphology and acrosome filament formation in some Annelida, Echiuroidea and Tunicata. *Zool. Bidr. Uppsala* **33**, 1–28.

Franzén, Å. (1973). The spermatozoon of *Siboglinum* (Pogonophora). *Acta Zool.* **54**, 179–192.

George, J. D. and Southward, E. C. (1973). A comparative study of the setae of Pogonophora and polychaetous Annelida. *J. Mar. Biol. Ass. U.K.* **53**, 403–424.

Hyman, L. H. (1959) "The Invertebrates," Vol. 5. McGraw Hill, New York.

Ivanov, A. V. (1957). Materials on the embryonic development of Pogonophora. *Zool. Zh.* 36, 1127–1144 [in Russian].

Ivanov, A. V. (1958a). Ein Beitrag zur Embryonalentwicklung der Pogonophoren. *Sowjetwissenschaft Naturwiss. Beitr.* 10, 1068–1086.

Ivanov, A. V. (1958b). Structure of the genital system of Pogonophora *Zool. Zh.* 37, 1363–1374 [in Russian]. (Transl. in *Syst. Zool.* 10, 24–34, 1961.)

Ivanov, A. V. (1960). Embranchement des Pogonphores. "Traité de Zoologie" (P. P. Grassé, ed.), Vol. 5, part 2, pp. 1521–1622. Masson, Paris.

Ivanov, A. V. (1963a). "Pogonophora." Academic Press, London.

Ivanov, A. V. (1963b). Les pogonophores-leur structure, développement et position systématique. *Trav. Mus. Hist. Nat. "Grigoire Antipa"* 4, 119–130.

Ivanov, A. V. (1964a). Pogonophora, Bau, Embryonalentwicklung und Stellung im System. *Biol. Rundschau* 1, 145–154.

Ivanov, A. V. (1964b). On the structure of the hind region of the body in Pogonophora. *Zool. Zh.* 43, 581–589 [in Russian].

Ivanov, A. V. (1965). Structure de la région postérieure sétigère du corps des Pogonophores. *Cah. Biol. Mar.* 6, 311–323.

Ivanov, A. V. (1970). Verwandtschaft und Evolution der Pogonophoren. *Z. Zool. Syst. Evolutionsforsch.* 8, 109–119.

Jägersten, G. (1957). On the larva of *Siboglinum. Zool. Bidr. Uppsala* 32, 67–80.

Johansson, K. E. (1938). *Lamellisabella zachsi* Uschakow, a representative of a new class of animals, Pogonophora. *Zool. Bidr. Uppsala* 18, 253–268. [in German].

Johansson, K. E. (1969). Pogonophora. "Handbuch der Zoologie (Kükenthal)," Bd 3, Hälfte 2, Lief. 18 (6), 1–50.

Kirsteuer, E. (1969). Morphologie, Histologie und Entwicklung der Pogonophora, Hemichordata und Chaetognatha. *Fortschr. Zool.* 20, 129–172.

Little, C. and Gupta, B. L. (1968). Pogonophora: uptake of dissolved nutrients. *Nature (London)* 218, 873–874.

Little, C. and Gupta, B. L. (1969). Studies on Pogonophora. III. Uptake of dissolved nutrients. *J. Exp. Biol.* 51, 759–773.

Livanow, N. A. and Porfirewa, N. A. (1967). Die Organisation der Pogonophoren und der Beziehungen den Polychäten. *Biol. Zentrabl.* 86, 177–204.

Manton, S. M. (1958). Embryology of Pogonophora and classification of animals. *Nature (London)* 181, 748–751.

Nørrevang, A. (1970a). On the embryology of *Siboglinum* and its implications for the systematic position of the Pogonophora. *Sarsia* 42, 7–16.

Nørrevang, A. (1970b). The position of Pogonophora in the phylogenetic system. *Z. Zool. Syst. Evolutionsforsch.* 8, 161–172.

Southward, A. J. and Southward, E. C. (1963). Notes on the biology of some Pogonophora. *J. Mar. Biol. Ass. U.K.* 43, 57–64.

Southward, A. J. and Southward, E. C. (1968). Uptake and incorporation of labelled glycine by pogonophores. *Nature (London)* 218, 875–876.

Southward, A. J. and Southward, E. C. (1970). Observations on the role of dissolved organic compounds in the nutrition of benthic invertebrates. Experiments on three species of Pogonophora. *Sarsia* 45, 69–95.

Southward, A. J. and Southward, E. C. (1972). Observations on the role of dissolved organic compounds in the nutrition of benthic invertebrates. III. Uptake in relation to the organic content of the habitat. *Sarsia* 50, 29–46.

Southward, E. C. (1963). Pogonophora. *Oceanogr. Mar. Biol. An. Rev.* 1, 405–428.

Southward, E. C. (1968). On a new genus of pogonophore from the Western Atlantic, with descriptions of two new species. *Bull. Mar. Sci.* **18**, 182–190.

Southward, E. C. (1969). Growth of a pogonophore: a study of *Polybrachia canadensis* with a discussion of the development of taxonomic characters. *J. Zool. London* **157**, 449–467.

Southward, E. C. (1971a). Recent researches on the Pogonophora. *Oceanogr. Mar. Biol. Ann. Rev.* **9**, 193–220.

Southward, E. C. (1971b). Pogonophora of the northeast Atlantic: Nova Scotia to Florida. *Smithsonian Contrib. Zool.* **88**, 1–29.

Southward, E. C. and Southward, A. J. (1966). A preliminary account of the general and enzyme histochemistry of *Siboglinum atlanticum* and other Pogonophora. *J. Mar. Biol. Ass. U.K.* **46**, 579–616.

Sterrer, W. (1969). Beiträge zur Kenntniss der Gnathostomulida. I. Anatomie und Morphologie des Genus *Pterognathia* Sterrer. *Ark. Zool.* **22**, 1–125.

Webb, M. (1963). A reproductive function of the tentacle in the male of *Siboglinum ekmani* Jägersten (Pogonophora). *Sarsia* **13**, 45–49.

Webb, M. (1964a). The posterior extremity of *Siboglinum fiordicum* (Pogonophora). *Sarsia* **15**, 33–36.

Webb, M. (1964b). A redescription of *Siboglinum ekmani* (Pogonophora). *Sarsia* **15**, 37–48.

Webb, M. (1964c). The larvae of *Siboglinum fiordicum* and a reconsideration of the adult body regions (Pogonophora). *Sarsia* **15**, 57–68.

Webb, M. (1964d). Additional notes on *Sclerolinum brattstromi* (Pogonophora) and the establishment of a new family Sclerolinidae. *Sarsia* **16**, 47–58.

Webb, M. (1965). Additional notes on the adult and larva of *Siboglinum fiordicum* and on the possible mode of tube formation. *Sarsia* **20**, 21–34.

Webb, M. (1969a). *Lamellibrachia barhami* gen. nov., sp. nov. (Pogonophora), from the northeast Pacific. *Bull. Mar. Sci.* **19**, 18–47.

Webb, M. (1969b). Regionation and terminology of the pogonophoran body. *Sarsia* **38**, 9–24.

Chapter 6

CHAETOGNATHA

M. R. Reeve and T. C. Cosper

6.1 Introduction

Within recent times, the phylum Chaetognatha has been discussed in monographs by Kuhl (1938), Hyman (1959), de Beauchamp (1960), and Alvariño (1965). The most recent substantial review of their biology was by Ghirardelli (1968), which dwelt heavily on their reproduction and included many of his own personal observations accumulated over many years of active research in this field. Burfield (1927) and John (1933) authored monographs on *Sagitta* and *Spadella*, respectively. It would serve little purpose here to do more than summarize these contributions and add pertinent information from our own experience. Because it is only recently that chaetognaths other than the benthic genus *Spadella* have been adapted to laboratory conditions, there is still very

157

little information available in those areas which necessitate observation and experimentation with live animals.

The Chaetognatha are a phylum of "small, bilaterally symmetrical enterocoelous marine animals, of mostly planktonic habits, without circulatory or excretory systems and with a slender, transparent torpedo-shaped body, provided with one or two pairs of lateral horizontal fins, and terminating anteriorly in a rounded head armed on each side with a group of grasping spines, posteriorly in a horizontal tail fin" (Hyman, 1959). Although there are only approximately 70–80 species in about eight genera (according to the authoritative estimate of David, 1963), chaetognaths are generally considered to constitute one of the most important groups of marine zooplankton, second only to copepods.

6.2 Asexual Reproduction

Asexual reproduction is not known to occur in chaetognaths.

6.3 Sexual Reproduction

6.3.1 Sexual Dimorphism, Sex Determination, Hermaphroditism

All chaetognaths are hermaphrodites.

6.3.2 Origin of Germ Cells and Gonads

Information available on this subject is covered in Sections 6.4.1 and 6.4.2.

6.3.3 The Reproductive System

All chaetognaths are hermaphrodites whose body is divided into three distinct regions (Fig. 1). The head is separated from the trunk by a transverse coelomic septum; the trunk is separated by a postanal (caudal) coelomic septum from the tail or caudal region.

6.3.3.1 THE MALE SYSTEM

The male reproductive system is located totally in and on the caudal region of the body (Fig. 1C). Three paired structures, the testes, vasa deferentia, and seminal vesicles, are separated by a longitudinal septum

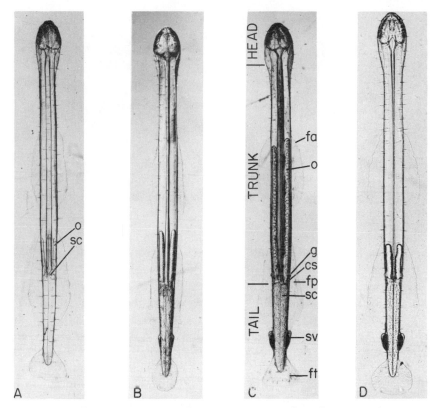

F IG . 1. The development of sexual maturity in *Sagitta hispida*. (A) Early immature
(5 mm long); (B) late immature (7 mm long); (C) fully mature (9 mm long);
(D) fully mature after having laid a batch of eggs (10 mm long). Abbreviations:
cs, caudal septum; fa, anterior fin; fp, posterior fin; ft, tail fin; g, gonopore; o, ovary;
sc, sperm cell in tail coelom; sv, seminal vesicle.

which begins at the caudal septum and runs dorsoventrally to the distal
end of the tail. This one dividing transverse septum is found in all
chaetognaths, but in *Sagitta* there are two subsepta, one to either side
of, and parallel with, the main septum. However, the secondary septa
begin "a little behind the caudal septum . . . and end at the level of
the seminal vesicles" (Ghirardelli, 1968). Therefore, in *Sagitta* there
is still communication between all parts of each side of the larger dividing septum, across which there is no apparent communication. The genital products in the mature stages completely fill the caudal segment,
with the notable exception of *S. inflata* (Fig. 2), where the male products
are confined to small regions near the tail fin (Ghirardelli, 1968).

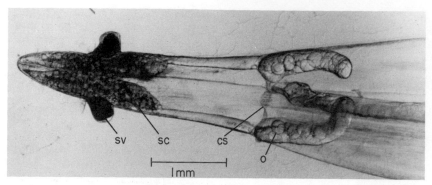

FIG. 2. Reproductive organs of *Sagitta inflata* (abbreviations as in Fig. 1).

The testes were described by Burfield (1927) as solid bodies lying anteriorly in the tail cavity and "attached to the outer wall in the region of the lateral line and ventral longitudinal muscle band." This band of cellular material is broader anteriorly and is covered with an endothelium which is continuous with the body wall at the site of attachment.

Spermatogenesis has been studied in detail by, among others, John (1933) for *Spadella cephaloptera*, Burfield (1927) for *Sagitta bipunctata*, and Ghirardelli and Arnaud (1966) for *Sagitta setosa*. In a maturing animal, masses of diploid cells ($2N = 18$) are budded from the testes, and meiosis according to the classical scheme occurs while the cells are being circulated in the caudal coelomic fluid. The rounded masses of cells come to fill each caudal compartment and are kept in constant, regular movement within the coelomic fluid, presumably by ciliation of the lateral line and medial septum areas (Burfield, 1927) or perhaps by the action of ripe sperms (Bordas, 1912). The research of Ghirardelli and Arnaud (1966) tended to support the former by showing that all the sperms develop at nearly the same rate and at the same time, although this was contradicted by Burfield (1927) and by Stevens (1910). Whatever the mechanism or physiological benefit of this movement, it is quite evident and regular in all stages of the maturing male products. The procession is toward the transverse septum along the body wall, medially along that septum, and caudally along the longitudinal septum. A diagram of this progression was afforded by Burfield (1927) for *S. bipunctata* and by Jägersten (1940) for *S. setosa*. John (1933) noted generic differences and the condition for *Spadella cephaloptera*. Ghirardelli (1968) stated that their "movement is more or less rapid depending on the species and the degree of fullness of the coelom." Movement is slower in *Spadella*. Continued development to the mature state yields spermatozoa which are filiform in shape. Afzelius (1963) has shown unusual filament conformations in the sperm tails of *Sagitta elegans* which do not conform to the usual $9 + 2$ arrangement. Some sperm tails were normal and others showed either a $9 + 4$ or a $10 + 4$ arrangement. The latter had a tenth filament which was simple and without the "arms" possessed by other filaments.

The ripe filiform sperm moves through the vas deferens, a duct beginning at the posterior end of the testis, traversing the body wall, and emptying into the anterior portion of the seminal vesicle. The mouth of each vas deferens is ciliated, which facilitates sperm movement into the seminal vesicles (Burfield, 1927).

The seminal vesicles are located laterally on the caudal portion of the body just anterior to the caudal fin. In *Sagitta* there is a pronounced lateral protrusion and, since their morphology changes with their degree of fullness, tightly packed seminal vesicles have been regarded by some workers as good taxonomic characters because of their species-specific morphology (Tokioka, 1939, 1955; Sund, 1959, Alvariño, 1965). In mature organisms the seminal vesicles are tightly packed with filiform spermatozoa. The method of sperm emission is by rupture of the seminal vesicle wall. Ghirardelli (1968) stated that in *Spadella* sperms are contained in "the form of a spermatophore whose dimensions are always the same as those of the vesiculae seminalis." Formed spermatophores have been reported by Murakami (1959) for *S. crassa* and by Dallot for *S. ferox*, *S. neglecta*, and *S. hedoti* (see Ghirardelli, 1968). We have observed compacted aggregations of sperms of the shape and size of the seminal vesicle in *S. hispida* which appear to serve as spermatophores. This subject is treated further in Section 6.3.4.

6.3.3.2 THE FEMALE SYSTEM

The female reproductive system (Fig. 1C) is located totally in the trunk cavity and is separated from the male reproductive system by the caudal septum. The ovaries are elongated solid bodies, round in cross-section, lying posteriorly in the trunk section. They are attached laterally to the body wall via a short mesentery which is "continuous on the outer side with the lining of the trunk cavity in the region of the lateral line" (Burfield, 1927).

Several authors have published accounts of aspects of the female reproductive system and oogenesis of certain species, including Ghirardelli (1953) for *Pterosagitta draco;* Vasiljev (1925) and John (1933) for *Spadella cephaloptera;* Bordas (1912, 1920), Stevens (1903, 1905) and Burfield (1927) for *Sagitta bipunctata;* and Stevens (1910) for *Sagitta elegans.*

In juveniles the ovary occupies a small area of the posterolateral trunk. Cross sections of the immature ovary reveal a crescent-shaped germinal epithelium along the lateral to lateromedial perimeter of the ovary. Smaller oocytes are still in contact with the germinal epithelium. With increasing growth an accessory cell is extruded from the egg membrane (Stevens, 1905), separating the egg from the ovary wall. More mature eggs are always more medial in position, and in organisms with many ripe eggs the intestine may be flattened because of this mechanical intrusion. The degree of maximum forward extension of the ovaries in the trunk cavity varies considerably interspecifically. The ovary of S. *hispida* may extend to the region of the ventral ganglion, while the mature ovary of S. *inflata* (Fig. 2) is only a small fraction of the total trunk length. The ducts of the female reproductive system will be discussed in the next two sections.

6.3.4 Fertilization

The fact that the entire phylum is hermaphroditic is one of its most characteristic attributes. Yet there is little more that can be said in terms of the significance of this fact. Ghiselin (1969), in his review on the evolution of hermaphroditism amongst animals, acknowledged three possible explanations of which only one—the "classical low-density model"—was applied to chaetognaths. The ability to reproduce without the aid of another animal would greatly increase the reproductive potential of a population at low density, or of low mobility.

The hermaphrodite, under these circumstances, would be self-fertilizing. This, in turn, assumes simultaneous availability of mature sperm and eggs in the same animal. Both of these circumstances are subject to contradictory opinions in the literature. A further assumption is the existence of a most extreme form of inbreeding. Even the basic assumption, that of spatial dispersion to the point of serious difficulty of meeting another member of the same species, is a matter of intuition. Although they are carnivores, and consequently less abundant than their natural prey, copepods, chaetognaths are second only to them in terms of abundance in the macroplankton. There is little quantitative information on their swimming speed and acuity of sensory perception. It might be argued that at extremes of low density the case for this model of hermaphroditism becomes obvious. In inshore and oceanic near-surface waters, however, we do not have the data which would permit an opinion on their degree of dispersion relative to a need for hermaphroditism, considering that the planktonic oceanic amphipods (an analagous carnivore group of probably lower density) are dioecious.

In any case the extent of self-fertilization is far from clear. The scattered observations throughout the literature concerning fertilization conflict to the extent that two recent major reviewers arrived at opposite interpretations. Hyman (1959) stated that "self-fertilization is thus apparently the rule in *Sagitta*," while Alvariño (1965) concluded for chaetognaths as a whole that ". . . it is generally accepted that cross-fertilization by copulation is the rule."

The benthic genus *Spadella*, which has been successfully maintained in aquariums for many years, has been clearly demonstrated to engage in a behavioral sequence in which cross-fertilization occurs. Although John (1933) believed that the process was not reciprocal, Ghirardelli (1968) has summarized his long researches into the reproductive morphology of cross-fertilization in S. *cephaloptera*, which is of a reciprocal nature. After a succession of quick movements and after touching certain parts of the upper body of the other organism, two organisms orient themselves in opposing directions, the bodies almost touching. A spermatophore from each individual is placed on the neck of the partner on "the dorsal median line directly behind the ciliary loop." The posterior portion of the spermatophore dissolves, and a stream of spermatozoa move caudally. Before reaching the caudal septum, the stream bifurcates, resulting in some sperms from one spermatophore being deposited in each of the seminal receptacles. Experimental placement of spermatophores has proved successful only if the spermatophore is placed in the median dorsal line behind the corona. This experimental placement of the spermatophore will cause successful spermatozoan movements, regardless of whether the spermatophore comes from the same or from another organism. Ghirardelli believed that self-fertilization could not be effected in nature because of the specificity of spermatophore placing.

Unlike the planktonic genera, *Spadella* is functionally adapted to contact with and attach to surfaces, and consequently, there can be some assurance that when it is confined to small volumes and inspected microscopically there is greater chance of it behaving naturally. Very few planktonic species have been observed alive and in good condition, and from our own observation their impending death (other things being equal) is a direct function of their degree of confinement. There exist, therefore, few and conflicting opinions on the mode of fertilization in the other genera.

Grassi (1883) believed fertilization to be a reciprocal mating process. Van Oye (1931) worked with an unnamed species of *Sagitta* and described a cross-fertilization process wherein a full seminal vesicle released a spermatophore coated with a mucoid substance. The spermatophore moved to the caudal fin, which formed a curved repository for it until the meeting of two organisms in this condition. The animals then arranged themselves so the heads were in opposing directions; the spermatophore of the one would come into contact with the expanded genital orifice of the other. By holding the received spermatophore with its lateral fins, the organism allowed the sperms to penetrate the gonopore into the seminal receptacle (i.e., the extended terminal portion of the duct running anteriorly alongside the ovary). The exchange of spermatophores was always mutual. Ghirardelli (1968) pointed out that fins contain no musculature, and we have not observed any fin-curling ability in these animals. Nevertheless, an egg brood pouch formed by a similar mechanism has been described (see Section 6.3.6). Although Van Oye considered self-fertilization to be unlikely on morphological grounds, the pathways of sperm movements that he described resemble self-fertilization mechanisms reported by Jägersten (1940) and Dallot (1968).

Murakami (1959) observed *Sagitta crassa* with their heads oriented in opposite directions, presumably copulating. He believed that cross-fertilization occurs, at least in this species.

Bordas (1920) supported Stevens' (1910) belief that self-fertilization was possible, since sperms from the seminal vesicles of *Sagitta bipunctata* could enter the seminal receptacles by moving along the tail.

Jägersten (1940), working with *Sagitta setosa*, saw spermatozoa migrating from a ruptured seminal vesicle toward the caudal septum. Periodic fillings and ruptures of the seminal vesicles occurred. In different specimens the mode of migration differed, as did the destination of the sperm. Contents from one side of the organism could enter the seminal receptacle on either side of the animal.

Dallot (1968) also never observed cross-fertilization, and his investigations supplied evidence of self-fertilization. He maintained isolated individuals of *S. setosa* from which about 50% of the spawnings gave fertile eggs. Although not excluding mating without cross-fertilization as a possibility, he felt that the survival of the species would not depend on it. His observations are detailed further in Section 6.3.6.

Ghirardelli (1968) was impressed by the fact that the seminal vesicles often bear structures (such as the serrated-edged cup surmounting the vesicle of *Sagitta bipunctata*) which separate together with the spematophore and which could be attached experimentally to the body of another individual of the same species. Indeed, he felt that they might be considered as copulatory structures which would fit only into female genital orifices of the same species. These tubular structures are particularly prominent on spermatophores from *Eukrohnia bathyantarctica*. David (1958, 1965), who reported their occurrence, believed that these saclike spermatophores originated from the seminal vesicles of other individuals. Commenting on these specimens, Alvariño (1968) suggested that they "might be brooding sacs beginning to develop" (see also Section 6.3.6).

Ghirardelli (1968) noted that he was unable to induce spawning in either S. *inflata* or S. *bipunctata*.

Both Jägersten (1940) and Ghirardelli (1968) showed that sperm from a spermatophore which was placed close to the seminal receptacle migrated into that organ. Dallot (1968) observed the migration of sperms from the seminal vesicles to a seminal receptacle of the same animal. We have not observed behavior of the type described by Murakami as indicative of pairing in our laboratory populations of *Sagitta hispida*. Extensive observations (Reeve and Walter, 1972b), however, have showed that a copulatory act does occur in this species. It consists of a behavioral sequence in which two animals attach to each other by means of their head spines, their bodies flexing and twisting about each other as they move upward in a spiraling motion, followed by intervals of rest in which they sink downward. In more than 50 observed attachments, which lasted from a few seconds to a few minutes, the encounters usually ended with both animals gaining and losing a spermatophore. The site of attachment was always on the lateral trunk wall (rather than dorsally, as described by Ghirardelli for *Spadella*) and sperms immediately began streaming toward the gonopore on that side. On reaching it, some traversed the body to the gonopore on the other side and penetration of the seminal receptacle (i.e., expanded terminal portion of the duct running anteriorly alongside the ovary) began. The whole process was completed in little more than 10 minutes. Originally, Reeve and Walter (1972b) mistook the "attacks" as attempted cannibalism, and since they have observed spermatophores attached to animals in the process of ingesting a victim, they suggested that cannibalism may sometimes follow copulation, especially if one animal is smaller or weaker than the other. Acts of self-insemination were also seen on rare occasions in *Sagitta hispida*, although only 2% of animals individually confined in small dishes produced eggs. Self-insemination was easily induced by breaking open the seminal vesicle; animals which survived the operation usually laid fertile eggs. Animals which were isolated after being inseminated would lay only one batch of fertile eggs, any subsequent batches being infertile. It appears, therefore, that *Sagitta hispida* cannot store viable sperms and requires reinsemination for each batch of eggs. Since self-fertilization could occur, although it was not usual, Reeve and Walter suggested that it could provide a short-term survival mechanism for populations temporarily at very low densities. These observations are the only ones which demonstrate that both self- and cross-fertilization occur in the same species, but neither Dallot nor Ghirardelli excluded the possibility. Since the general morphological characteristics of all planktonic chaetognaths are similar, it may not be speculative to expect that most may be capable of both forms of fertilization. If self-fertilization is a regular phenomenon, which becomes more likely when opportunities for cross-fertilization are minimal, it might be expected to occur more frequently in species habitually in lower densities, such as typically oceanic and bathypelagic ones.

It appears well-established that fertilization occurs while the egg is still in position in the ovary. Sperms migrate from the seminal receptacle along the "seminal pouch" or "sperm duct" (but see next section) close to the outer edge of the ovary. They pass through a fine duct formed by accessory fertilization cells to the egg. Ghirardelli (1968) recorded his own detailed researches on this subject, reviewed the opinions of others, and provided photomicrographs and diagrams to illustrate the process.

6.3.5 Egg Laying

The term "egg laying" rather than "spawning" is used here to denote the release of the fertilized zygote rather than gametes. Although unfertilized eggs can be extruded (see last section above), there appears to be no provision for their subsequent fertilization.

The investigations of Stevens (1910) into the process of egg laying in *Sagitta bipunctata* were confirmed by Ghirardelli (1968), who studied the same species. After fusion of the egg and sperm, the egg (zygote) actively pushes its "way between the oviduct wall and the sperm duct receptacle, and by its own contractions or by shifting of material within the egg membrane," moves into the temporary oviduct (Stevens, 1910). Once housed in the temporary oviduct, the eggs are compressed against each other toward the genital papilla. Since the temporary oviduct is blind, the eggs issue to the outside through a temporary opening near the gonopore. This extrusion is caused either by inner contractions of the ovarian wall (Conant, 1896; Dallot, 1968) or possibly by the pressure of the eggs themselves (see figures of Ghirardelli, 1968).

Egg laying in *Spadella* is a modification of the process described for *Sagitta*. Ghirardelli (1968) stated that after the egg is in the temporary oviduct, the egg is pear-shaped "and its more pointed end slips between the epithelial cells of the wall of the vagina which are pushed aside in all directions, so that the egg seems to penetrate into the lumen of the vagina." The process took 30 seconds or less, the eggs were laid in clusters of 4 to 12, and were attached to submerged objects nearby.

Reeve and Lester (1974) reviewed the range of often contradictory opinions in the literature regarding the number of female reproductive ducts, and concluded that in *Sagitta hispida,* at least, there is a single duct running forward from the gonopore and seminal receptacle along the outer border of each ovary. The sperm pass anteriorly along this tube from the exterior and the eggs pass posteriorly along this tube to the exterior, in both cases by way of the gonopore. This view, supported by a series of photomicrographs (see Reeve and Lester, 1974) contradicts that of Conant (1896) who reported the presence of an additional temporary oviduct in the same species.

The eggs leave the ovary by passing through the germinal epithelium (as described by Ghirardelli, 1968; and Reeve and Lester, 1974) and become compressed in the duct leading to the exterior so that their original shape is now almost cuboidal. Usually within minutes of this very characteristic visual appearance of the cuboidal eggs, they are passed out through the gonopore. The whole ovary remains shrunken with their departure and only gradually increases to its former size as subsequent eggs enlarge. The extrusion of the eggs is probably accomplished (as Conant, 1896, believed) at least partly by active contraction of the ovary, and Conant, Dallot (1968), and we have seen waves of contraction in the ovary wall. The body wall musculature also appears to play an important part in egg extrusion in *Sagitta hispida.* We have seen, both in photomicrographs of live animals (Fig. 3A) and in histological sections of the bodywall, evidence of muscle contractions in the region of the gonopores at the moment of egg laying. Reeve and Lester (1974) provided photomicrographs of the entire process.

6.3.6 Factors Influencing Egg Laying

There are few direct observations on the effects of temperature or salinity on egg laying or breeding, although temperature especially must play an important role in seasonal periodicities (see below). Murakami (1959, 1966) was able to maintain *Sagitta crassa* over part of its life cycle and noted that the period from egg laying to hatching was affected by both temperature and salinity, a minimum time of 15 hours being recorded above 27°C. Reeve (1970a) noted that survival of mature individuals after egg laying was inversely related to temperature, although total egg production over most of the temperature range was directly proportional.

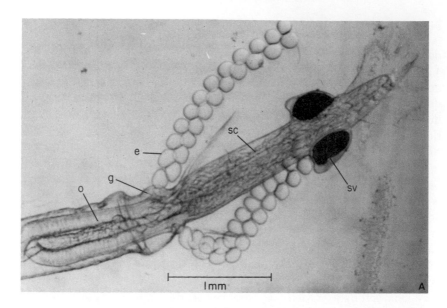

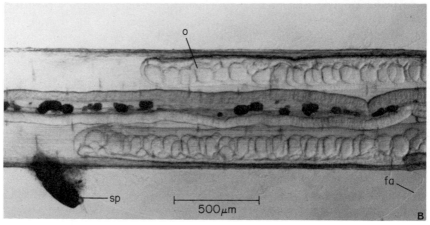

FIG. 3. (A) *Sagitta hispida* in the act of egg laying (caudal region). (B) Spermatophore attached to body wall of trunk region of *Sagitta hispida* at the level of the anterior fin. e, Eggs; sp, spermatophore; other abbreviations as in Fig. 1.

Egg laying is not necessarily a single event which is the culmination of the life of the animal. In *Sagitta setosa* (Dallot, 1968) and *S. hispida* (Reeve, 1970a) at least, a succession of batches of eggs may be laid. The latter species, for instance, lays an average of 50 eggs per batch at first. This number is highly variable from batch to batch and between animals but tends to decline with age. The somatic weight and length of animals does not increase and often declines over this period until death.

A serious problem of such studies is the need to inspect microscopically the live animal at least daily and to confine it individually in small volumes to facilitate searches, eliminate cannibalism, and replicate experiments. In 50-ml volumes, batches were laid until death, which occurred in a few days. In 250-ml volumes at this temperature, animals survived much longer and by the 30th day when over 30% still remained, egg laying had virtually ceased. Further discrepancies, which will doubtless arise as more experimental data become available, are difficult to resolve in terms of the natural environment. Effects of predation and food supply, for instance, may well curtail natural egg production below that in laboratory experiments.

One might consider the successive batches of eggs as one brood, with the possibility that after a period of dormancy a second phase of egg laying might commence. This was not shown for S. *hispida*. All the animals had died after 45 days, due in part, it is assumed, to deficiencies of laboratory culture. Ghirardelli (1968) summarized evidence for longer periods between egg-laying activity such as that for S. *inflata*, which appears to pass through several maturity phases during a season (see following section).

The production of successive batches of eggs was related to seminal vesicle function by Dallot (1968), who observed an alternating cycle involving both seminal vesicles and ovaries in *Sagitta setosa*. The mature vesicle released its sperms which traveled to the seminal receptacle. Egg laying took place shortly thereafter. The ovary was reduced in length as the vesicle filled again. This pattern of events would be repeated. The likelihood of unfertilized eggs being laid could even be predicted by observing departures from the sequence. Reeve's daily observations of egg laying (1970a) did not extend to larval viability, but he recorded daily changes in ovary length, body length, and the contents of ovaries, seminal vesicles, and tail, and egg-laying frequency.

Food supply unquestionably affects egg laying, as well as all other aspects of the organism's metabolism. Starvation of a mature *Sagitta hispida* results in cessation of reproductive activity and loss of weight. Reeve *et al.* (1970) showed that starved animals lost 4% of their dry weight per day and their tissues became more watery. Starvation could be borne for a period of several days, equivalent to at least one-quarter of the minimum life-cycle duration at this temperature. Although from Ghirardelli's (1968) data (as well as those of others) S. *inflata* continues to increase in size markedly after the onset of maturity, S. *hispida* does not. The food intake of the latter over this period is, therefore, utilized largely for its own maintenance metabolism and for the production of eggs and sperm. The proportion of its ingested nitrogen utilized for gamete production was 41% (Reeve, 1970a), which was at least as high a ratio as that incorporated into its own body tissue during its early growth.

Chaetognaths usually lay eggs between sunset and sunrise. Ghirardelli (1968) reported that the eggs of *Spadella* can be laid at any time, although there is a preference for the night hours, but Vasiljev (1925) observed only night egg laying. Dallot (1968) reported that *Sagitta setosa* laid eggs at sunrise. Also cited as having sunrise laying patterns are *S. hispida* and *S. hexaptera* (Conant, 1896), while *S. bipunctata* was found to lay at sunset. Conant found that egg laying of *S. hispida* could be delayed by several hours if the water was cooled and that hours of egg laying in this species would fluctuate somewhat with the period of year. Stevens (1910) suggested that the hour of egg laying was a species-specific characteristic in *Sagitta*, *S. elegans* possibly laying eggs at anytime during the day, *S. bipunctata* only at sunset. Ghirardelli (1968) reported that *S. inflata* laid eggs between "2300 hours and the first hours of morning."

Alvariño (1968), in a review of aspects of egg laying, stated "that eggs of the species of chaetognaths belonging to the genus of highest evolutionary rank are probably better equipped for survival." Such a statement is strictly speculative at this time, though it appears reasonable enough. There are very few records of egg laying. Conant (1896) recorded that *Sagitta hispida* eggs, covered by a layer of "gelatinous substance," were extruded as two linear rows. In aquarium tanks this species deposits its eggs both on the side and bottom surfaces (Reeve, personal observations). The eggs usually adhere together in a string or clump but are occasionally separate. They are negatively buoyant and so would sink if laid in the water column. Some eggs are undoubtedly deposited directly by animals on the bottom. Egg clumps on the walls must also have been deposited directly, and this species and *S. helenae* can attach to vertical aquarium surfaces. The two rows of eggs are usually double (making four distinguishable rows in all) and are often in a more or less straight line, but may be twisted and curled in a variety of ways. We do not know whether some of their eggs are laid in the water column in nature and adhere together, although since both species inhabit inshore shallow waters (and can themselves adhere to objects), we believe that most of their eggs are directly attached to surfaces such as sea grasses (see Reeve and Lester, 1974) and are thus prevented from being carried offshore.

Others (see Alvariño, 1968) have noted the laying of free eggs in various species of the genus *Sagitta*, and one of us (Cosper) has observed that eggs laid by *S. inflata* in aquariums accumulate singly on the bottom. Sanzo (1937) noted that the eggs of *Pterosagitta draco* from the water column clumped in gelatinous masses within which the developing young remained for a time. There are several suggestions of marsupia or brood pouches in *Eukrohnia* which appear to be of two types. One is formed by a curling of the lateral fins in which eggs and even larvae (e.g., MacGinitie, 1955) were found; the other is a sac attached in the region of the female gonopore (e.g., Alvariño, 1968). Alvariño challenged David's (1958) interpretation of a structure appended to the oviduct of *E. bathyantarctica* which the former believed was not a seminal vesicle functioning as a spermatophore, but a brooding sac. Ghirardelli (1968), however, to whom David sent his specimens, clearly confirmed the original observations. *Spadella* lays its eggs (Ghirardelli, 1968) in clusters of 4 to 12, each of which is attached by an elastic peduncle to the substrate.

A variety of parasites have been found within the trunk cavity of chaetognaths, and an assumption might be made that they would exert adverse effects on reproduction, especially with regard to the ovaries. These include a cercaria (Hutton, 1954) and one or more nematodes (e.g., Russell, 1932a). In occasional infections of *Sagitta hispida*, one of us (Reeve) has seen both types. They can occupy a considerable portion of the trunk coelom, pushing the ovaries aside as they move. Weinstein (1973) extensively reviewed chaetognath parasites.

There are many observations in the literature (reviewed by Weinstein, 1973) of some species of chaetognaths, including *Sagitta elegans*, performing "ontogenetic" descents, where the older and more mature individuals are found deeper than the younger ones. At some times of the year the deepest samples contained only mature animals. Weinstein suggested that such a phenomenon could aid a neritic species like *Sagitta elegans* in maintaining its preferred range by retaining breeding stocks in hollows and bottom topographic irregularities. He always found animals when using hyperbenthic sampling gear, designed to fish very close to the bottom. Data for *Sagitta hispida* populations in the shallow waters of Biscayne Bay indicate that surface samples rarely contain mature animals, which are only encountered from hyperbenthic samples (Reeve, unpublished). In this species, the relationship with the bottom has been carried further, since eggs are laid in clusters and presumably attached to surfaces as they are in the laboratory aquarium. Such a behavioral pattern appears even more likely to aid in the maintenance of a breeding population within a restricted neritic locality.

6.3.7 Seasonal Breeding Periodicities

There are three orders of egg laying periodicities on which some information exists for chaetognaths. Two—the laying of successive batches over a period of a few days, and breeding activity within the same individual separated by longer periods of inactivity—have been noted in the last section. There is a larger body of literature on the seasonal breeding and the appearance of successive generations of pelagic chaetognaths drawn from examination of preserved material and the construction of size/frequency plots. It has been reviewed by Hyman (1959) and Ghirardelli (1968), and extensively documented by Alvariño (1965). Weinstein (1973) has provided an exhaustive review of the breeding cycles of *Sagitta elegans,* the species for which most information is available, and which reflect the now well-established principles that latitude (i.e., temperature) affects the number of generations per year, and that there is an inverse relationship between temperature and the size of the animal at maturity.

Dunbar (1962) showed that *Sagitta elegans* in Canadian eastern arctic waters had a biennial life cycle. The population appeared to be composed of two distinct breeding groups, one isolated reproductively from the other, since they bred on alternate years. He concluded that low temperature rather than food availability imposed the particular growth patterns encountered. Weinstein (1973) also found a biennial cycle in the Gulf of St. Lawrence. McLaren (1966) found that in Ogac Lake at a temperature of 8°C the species grew more rapidly and reached maturity at a smaller size. He felt that "where generation length is set by marked seasonality of food supply, as in the arctic, high fecundity and associated large size and slow development may be selected for." Dunbar (1940), comparing development of *S. elegans* in the Arctic with that in the Plymouth area, found that ovaries began to develop sooner (relative to testis development) than in temperate waters, an "automatic result of the slowing down of all growth processes in the colder water."

In the north temperate waters of the North Sea off Flamborough, Wimpenny (1936) showed that *Sagitta elegans* passed through three generations per year, while a little farther south off Plymouth it went through five or more (Russell, 1932a,b, 1933). Pierce (1941) reported a single extended (January–May) breeding period in Port Erin Harbour (Isle of Man).

Clarke *et al.* (1943) found that on Georges Bank the main breeding period for *Sagitta elegans* varied from April to May. There was evidence for a second breeding period in late summer or autumn. Sherman and Schaner (1968) found that only one annual generation of *S. elegans* was produced in the coastal waters of the Gulf of Maine, with breeding lasting from spring until fall. Redfield and Beale (1940) found a late spring-summer breeding season for *S. elegans* in the Gulf of Maine, while Huntsman and Reid (1921) gave April–September as the breeding period for *S. elegans* in the Bay of Fundy.

Working with *S. crassa*, Murakami (1959) reported an annual cycle of three generations (spring-summer and summer-autumn short-period generations; and autumn-spring generation with a life span of 5 months). He also suspected another generation period (summer-winter long-period generation), which he later substantiated (1966).

David (1955) found that the "small northern" race of *Sagitta gazellae* from antarctic waters spawned "throughout the spring, summer, and autumn, and perhaps to a small extent in winter." He was not able to investigate the breeding periodicities of the larger southern race because of inadequate data from that area covering a complete season.

It was formerly believed that chaetognaths died after egg laying (Kuhl, 1938), and in the slower growing cold-water forms this may be the case. In warm-water species, however, the situation is undoubtedly more complicated. Generation times may be very short (see next section), breeding may be more or less continuous throughout the year, and animals may go through more than one breeding phase. Attempts to elucidate the course of breeding activities are often impossible from the examination of preserved samples.

Thomson (1947) believed that the gonads of *Sagitta inflata* from southeastern Australian waters matured periodically over the course of a year (see also Michael, 1919, for the same species from the Philippines). Furnestin (1952) suggested that *S. inflata* from the Mediterranean had several sexual cycles per individual throughout the year. Ghirardelli (1962) confirmed by cytological examination of the ovary that there are at least two sexually mature forms of ovary in *S. inflata*. In one form the ovaries do not extend "beyond the anterior end of the posterior fins," and the eggs in these short-ovary forms are "massed without any definite order;" in the long-ovary forms the eggs are regularly arranged in a line and the eggs extend beyond the posterior fins. The short-ovary form is usually the first stage of ovarian activity, and it is followed by the long-ovary stage.

Stone (1966) investigated *Sagitta inflata* of the Agulhas Current, finding that individuals from neritic waters contained more eggs than those from oceanic water. The neritic area had both more food per unit volume and a lower temperature, both of which he recognized could be influencing egg number.

Owre (1960) worked with preserved *Sagitta inflata* from the Florida

Current area and found that it was "difficult to recognize separate broods and follow them through to maturity because, in contrast to comparable data for arctic and temperate species, the periods of heavy breeding are superimposed on what appears to be continual breeding throughout the year." She also found *Krohnitta pacifica*, *Pterosagitta draco*, and *Sagitta serrodentata* to be year-round breeders. Not all Florida Current representatives shared this characteristic. She also reported that *S. bipunctata* breeds once a year in the late winter or early spring, that *S. minima* breeds mostly in the summer (there is evidence for fall breeding), and that *S. decipiens* and *K. subtilis* have two breeding periods, one in the summer and one in the winter. In addition, Pierce (1951) found that *S. hispida*, *S. helenae*, and *S. tenuis* from the same latitude breed "to some extent the year round."

Sagitta hispida probably breeds in Biscayne Bay (Miami) throughout the year, but its numbers are extremely reduced during the height of the summer, when water temperatures reach 32°C and above. This may be related to a failure to raise larvae at temperatures of 33°C and above in the laboratory. It seems that in Biscayne Bay, where temperatures exceed those of the tropical ocean surface, there is a reversal of the trend of more intensive breeding activity with increasing temperature and an accompanying reversal of optimum breeding season.

Dallot (personal communication) is compiling extensive biometrical data for *Sagitta hispida* and other species from various latitudes which clearly demonstrate the inverse relationship between increasing temperature and size of the animal at maturity.

6.4 Development

6.4.1 Embryonic Development

The embryology of chaetognaths received early attention in the work of several authors and is well reviewed by Hyman (1959) and Ghirardelli (1968). Much of the earlier work had as its goal the elucidation of the phylogenetic relationships of this uniquely isolated phylum and the origin and tracing of the germ cell determinant. All chaetognath eggs undergo equal holoblastic cleavage (Fig. 4). The blastula develops into a gastrula of which the blastopore is the future posterior end of the organism. Germ cells are segregated at an early stage, the enlarged one-cell (*Spadella*) or two-cell (*Sagitta*) primordium occupying an anterior position in the archenteron. These primordia become detached

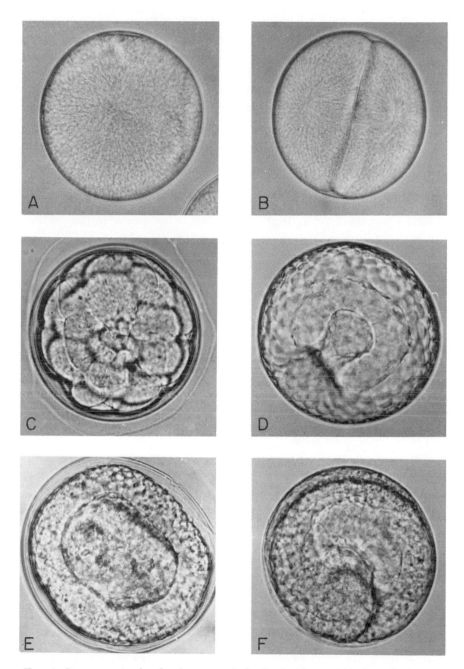

Fɪɢ. 4. Six stages in the development of the fertilized egg (A) through cleavage (B and C) and gastrulation (D) until immediately before hatching (E and F), where the embryo is curled around the circumference of the egg (200 μm diam.).

before the formation of the coelom and lie free in the archenteron cavity. The primordia are pushed backward by two folds which appear in the anterior archenteron and progress posteriorly. The blastopore closes and the folds reach the posterior wall of the archenteron, defining two lateral coelomic spaces and a centrally located intestine. A stomodaeal invagination meets and fuses with the gut, thereby defining the mouth and pharynx. Each coelomic sac is cut anteriorly into two head sacs which form the future head musculature.

While the embryo continues to elongate, the one primordial germ cell in each lateral coelom divides once in *Sagitta,* the single cell in *Spadella* divides to become two binucleate cells. Continued growth and elongation of the embryo results in a disappearance of all lateral coelomic spaces. Thus, the germ cells are embedded in mesoderm in the posterior trunk.

6.4.2 Larval and Juvenile Development

Several authors represent the mature embryo (e.g., Sanzo, 1937) as being curled around the circumference of the egg within the membrane (see Fig. 4), with its head tucked toward the center. One of us (Cosper) has observed hatching in *Sagitta inflata* which starts with contractions beginning at each end of the embryo and extending to its midregion, having the effect of uncoiling the organism and forcing its tail and head against the membrane. The contractions become more pronounced 10–15 minutes before hatching, and immediately before hatching the contractions become an uninterrupted series of violent jerking movements. After hatching, the larva swims with the same series of darts which characterize the adult organisms. The lateral extremities of the larva are curled ventrally upon hatching, but this curled posture disappears within an hour after hatching. Hatching times are within 26–68 hours of egg laying.

Spadella larvae do not swim about, but immediately attach themselves to the substrate where the eggs were laid. Attachment is by adhesive cells on the ventral trunk and by adhesive projections on the sides of the head.

Three to four days after hatching, the primordial germ cells (2 on each side in *Sagitta,* now 8 in *Spadella*) migrate across the "reformed trunk cavities to the lateral body wall" (Hyman, 1959). While this migration proceeds, a tail-trunk septum is being laid down, leaving half the germ cells in the trunk cavity and half in the caudal region. The ovaries arise from the germ cells in the trunk cavity, while the testes arise from the corresponding cells in the tail cavity.

There are very few illustrations of the larvae of chaetognaths (see Doncaster, 1902; Sanzo, 1937) but they conform to our own observations for *Sagitta inflata*, *S. helenae*, and *S. hispida*. Immediately after hatching, the alimentary system is relatively undeveloped and the fins are represented by little more than a stubble of rays projecting from the tail. Development is very rapid (at least in these warm-water species at 20°–30°C), and within a few hours the rays of the tail fin are well formed and those of the posterior laterals quickly follow. Over the next few hours the interstitial tissues of the fins grow out progressively from the body wall. The newly hatched animal already possesses seizing spines and within 2 to 3 days begins to feed. As it lengthens, its bodily proportions take on those of the older animal. Initially its tail is equal in length to the trunk and head combined. The "tangoreceptors," at first few in number and projecting (relatively) much farther from the body wall, increase in number and decrease in relative size. After several days (see Reeve, 1970b), the anterior lateral fins begin to develop (see Section 6.3).

It is only very recently (Reeve, 1970b) that larvae of a pelagic chaetognath have been raised to maturity in the laboratory, and even in the relatively hardy benthic genus *Spadella* this has been reported only for *S. cephaloptera* (Ghirardelli, 1959), which, nevertheless maintains itself in large recirculating aquarium systems, such as that at Plymouth Marine Laboratory. Murakami (1959, 1966) reported studies on *Sagitta crassa* from the Seto Inland Sea in which he maintained larvae hatched from eggs for 20 days and adults up to 3 months. He felt that development stopped for lack of suitable food. Dallot (1968) also obtained larvae of *S. setosa* which lived for a few days in the laboratory.

There are several major problems to be overcome in the successful rearing of pelagic chaetognaths. The first is the collection of undamaged live specimens. It is no accident, as Ghirardelli (1968) pointed out, that success has so far only been achieved with species habituating inshore waters which are undoubtedly able to withstand greater mechanical, temperature, salinity, and other shocks. With current collection methods, offshore species such as *Sagitta inflata* are usually dead by the time they reach the deck of the ship. Operations involving transport to and maintenance in the laboratory must also be designed to ensure continued care in this direction. In raising the young through their larval stage, particularly, the provision of suitable food is one of the major difficulties. Reeve and Walter (1972a) reported that when ciliates, rotifers, and mixed microplankton (largely copepod nauplii) were offered, only the latter sustained growth through the larval period. At times when natural microplankton is not readily available, laboratory cultures of planktonic copepods must be supplied, although growth of these is quite an accomplishment at present. Animals at all stages of development in aquariums are very susceptable to violent aeration, water turbulence, entrapment on filters, etc., as well as cannibalism, especially in over-crowded conditions.

Data on growth rates from the newly hatched to mature animal (Reeve and Walter, 1972a) demonstrated the wide range of conditions tolerated by *Sagitta hispida* and the rapidity with which a life cycle could be completed in this warm-water species. Growth rates were directly proportional to temperatures between 17° and 31°C, and at the highest temperature 50% of the population required only 18 days to reach maturity (as defined below) from hatching. As suggested by Dallot (personal communication) from biometrical studies, there was an inverse relationship between size at maturity and temperature, the mean dry weights and lengths of newly mature animals at 17° and 31°C being 263 and 119 μg and 11.5 and 8.4 mm, respectively. Populations were able to reach maturity in salinities ranging from 25 to 40‰, as well as in inshore and oceanic seawater and that constituted artificially from a commercial preparation of sea salts.

6.4.3 Maturity Classification Schemes

As young animals grow to maturity, various characters of the mature reproductive system gradually develop. In order to quantify and compare breeding cycles between species, several workers have proposed schemes, or relied on those of others, which defined advancing stages of sexual maturity. Such work was often limited by its inevitable dependence on preserved material and an assumption (see Kuhl, 1938) that death followed egg laying. It is becoming increasingly evident (Owre, 1960; Dallot, 1968; Ghirardelli, 1968; Reeve, 1970a) that maturation and spawning can be cyclic processes rather than single events in one lifetime.

The most recent and comprehensive scheme is that of Ghirardelli (see his 1968 review) which takes into account the appearance of both male and female characteristics viewed at 100 magnifications through a dissecting microscope and from the cytological appearance of sections. Other schemes which have used both male and female development are those of Russell (1932a), Kramp (1939), Pierce (1951), and Colman (1959), while Thomson (1947) and Owre (1960) used ovarian development only. Reeve (1970a) assigned stages of development to maturing testicular products, seminal vesicles, and ovaries in *Sagitta hispida* on the basis of the observation of live animals at 50 magnifications. He noted that the ovary changed markedly in length on a cyclic basis as successive batches of eggs enlarged to maturity and were subsequently laid. On the basis of ovarian characteristics alone, it was not possible to determine whether an animal possessing small ovaries containing only small eggs had yet to lay eggs for the first time (Fig. 1B), or had already laid one or more times (Fig. 1D). Successive ovarian cycles have been seen also in S. *inflata* by Furnestin (1957) and Ghirardelli (1962), suggested as likely by Owre (1960), and observed in the living S. *setosa* by Dallot (1967, 1968). Ghirardelli noted that, cytologically, such secondarily shortened ovaries show the same features as those of specimens which have for the first time reached a younger maturity stage. These facts could be inferred in the case of S. *inflata* because the animal continued to increase in size giving rise to larger "immature" forms. In the case of S. *hispida*, however, whose growth rate levels off at maturity, only sequential observation of the live animal could reveal its cyclic changes.

A more constant indication of maturity in *Sagitta hispida* was the appearance of seminal vesicles. Reeve (1970a) found that the development and filling of the seminal vesicles and spawning of the first batch of eggs could all take place within 48 hours. Subsequently, seminal vesicles were rarely seen completely emptied (though this might occur briefly immediately following loss of a spermatophore) and never disappeared. Our less detailed (unpublished) observations of S. *helenae* and S. *inflata* do not conflict with this criterion of maturity. In the absence of any observations on living cold-water forms where slower developmental rates (to the extreme of the biennial cycle of S. *elegans*, Dunbar, 1962) might materially change time scales and cyclic periodicities, we do not wish to generalize on criteria of maturity at this point. Weinstein (1973) reported, for instance, that he sometimes found large *Sagitta elegans* with no seminal vesicles or sperm in the tail, and long empty ovaries, which he considered to be "spent" animals, prior to death.

It is possible for the sake of convenience to divide the entire life history of *Sagitta hispida* into four stages (Reeve, 1970b) while keeping in mind that development in the phylum is direct and continuous, with no abrupt transformations. The first of these may be termed the larva (see Hyman, 1959), which shortly after hatching, possesses only one of two pairs of lateral fins and whose tail and trunk are of approximately equal lengths (Fig. 5A, B). The beginning of the next (juvenile) phase may be considered as that time at which the full compliment of fins is attained as bodily proportions approach much more closely those of the older animal (Fig. 5C). The third (immature) stage commences with the first sign of the onset of sexual maturity at 50 magnifications (Fig. 1A) which corresponds roughly to maturity stage II of Ghirardelli

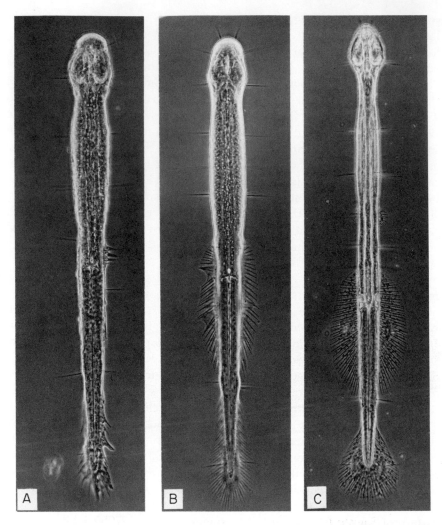

Fig. 5. Early postembryonic development of *Sagitta hispida*. (A) Newly hatched larva (0.8 mm long); (B) 5-day-old larva (1.5 mm long); (C) 10-day-old animal in transition to juvenile stage as anterior fins begin to appear.

for the testes ("first germ elements free in the caudal coelom") and his maturity stage I for the ovary ("little developed . . . slightly elongated . . ."). The fourth life history stage (Fig. 1C), and most clearly pinpointed, is the mature stage when the seminal vesicles appear as noted in the previous paragraph.

The end of the larval phase cannot be indicated by the appearance of an additional pair of lateral fins except in the genus *Sagitta*, as the other genera do not possess more than one pair. Since virtually nothing is recorded concerning the early development of the other genera, once again no generalities can be drawn. A figure of the larva of *Pterosagitta draco* (Sanzo, 1937), however, suggests that at first its tail and lateral fins are continuous and only later become separate, a characteristic which might be utilized as the end of the first phase.

Reeve (1970b) noted that in terms of life history stages defined above, the larval stage lasted 10–15 days (0.9–4.0 mm), the juvenile 4–8 days (4.0–6.5 mm), and the immature 6–10 days at 23.5°C. Maturity was attained within 33 days (50% maturity—25 days), and the last animal died on the 57th day.

6.5 Conclusion

The culture of marine zooplankton is a field which has developed only within the last decade and even now is currently restricted to a few copepod species. The culture of other zooplankton groups, including chaetognaths, is still in its infancy. We expect that great strides will be made in this direction over the next decade, shedding light on some of the many unanswered questions concerning reproduction in chaetognaths, which can only be resolved by the systematic study of a range of species living under controlled conditions.

Acknowledgments

Supported by National Science Foundation Grant GB 12377. Contribution of the University of Miami, Rosenstiel School of Marine and Atmospheric Science.

6.6 References

Afzelius, B. A. (1963). Cilia and flagella that do not conform to the 9 + 2 pattern. I. Aberrant members within normal populations. *J. Ultrastruct. Res.* 9, 381–392.

Alvariño, A. (1965). Chaetognaths. *Oceanogr. Mar. Biol. Ann. Rev.* 3, 115–194.

Alvariño, A. (1968). Egg pouches and other reproductive structures in pelagic Chaetognatha. *Pac. Sci.* 22, 488–492.

Bordas, M. (1912). La Spermatogénèse dans le *Sagitta bipunctata*. *La Cellule* 28, 167–211.

Bordas, M. (1920). Estudio de la ovogénesis en la *Sagitta bipunctata*, Quoy et Gaim. *Trab. Mus. Nac. Cienc. Nat. Madrid. Ser. Zool.* 42, 5–119.

Burfield, S. T. (1927). Sagitta. *Proc. Trans. Liverpool. Biol. Soc.* 41, 1–104.

Clarke, G. L., Pierce, E. L., and Bumpus, D. F. (1943). The distribution and reproduction of *Sagitta elegans* on Georges Bank in relation to the hydrographical conditions. *Biol. Bull.* **85**, 201–226.

Colman, J. S. (1959). The "Rosaura" Expedition 1937–38. Chaetognatha. *Bull. Brit. Mus. Natur. Hist. Zool.* **5**, 219–253.

Conant, F. S. (1896). Notes on the Chaetognaths. *Johns Hopkins Univ. Circ.* **15**,(126), 82–85.

Dallot, S. (1967). La reproduction du Chaetognathe planktonique *Sagitta setosa* Müller, en été, dans la rade de Villefranche. *C. R. Acad. Sci.* (*Paris*) **264**, 972–974.

Dallot, S. (1968). Observations préliminaires sur la reproduction en élevage du Chaetognathe planctonique *Sagitta setosa* Müller. *Rapp. Commun Int. Mer Medit.* **19**, 521–523.

David, P. M. (1955). The distribution of *Sagitta gazellae* Ritter-Zahony. *Discovery Rept.* **27**, 235–278.

David, P. M. (1958). A new species of *Eukrohnia* from the Southern Ocean with a note on fertilization. *Proc. Zool. Soc. London* **131**, 597–606.

David, P. M. (1963). Some aspects of speciation in the Chaetognatha. p. *In* "Speciation in the Sea" (J. P. Harding and N. Tebble, eds.), pp. 129–143. The Systematics Association, London.

David, P. M. (1965). The Chaetognatha of the South Ocean. *In* "Biogeography and Ecology in Antartica" (J. Mieghem, P. Van Oye, and J. Schell, eds.), pp. 296–323. Junk, The Hague.

de Beauchamp, P. (1960). Classe des Chétognathes (Chaetognatha). *In* "Traité de Zoologie" (P. P. Grassé, ed.), Vol. 5, part 2, pp. 1500–1520. Mason, Paris.

Doncaster, L. (1902). On the development of *Sagitta;* with notes on the anatomy of the adult. *Quart. J. Micros. Sci.* **46**, 351–398.

Dunbar, M. J. (1940). On the size distribution and breeding cycles of four marine planktonic animals from the Arctic. *J. Anim. Ecol.* **9**, 215–226.

Dunbar, M. J. (1962). The life cycle of *Sagitta elegans* in arctic and subarctic seas, and the modifying effects of hydrographic differences in the environment. *J. Mar. Res.* **20**, 76–91.

Furnestin, M. L. (1952). Chaetognathes récoltés en Méditerranée par le "PRESIDENT THEODORE TISSIER" aux mois de Juin et Juillet 1950. *Trav. Stat. Aquiculture Peches Cast.* **4**, 277–317.

Furnestin, M. L. (1957). Chaetognathes et zooplancton du secteur atlantique marocain. *Rev. Trav. Office Sci. Tech. Peches Maritimes* **21**, 113–356.

Ghirardelli, E. (1953). Appunti sulla morfologia dell' apparecchio riproduttore femminile e sulla biologia della riproduzione in *Pterosagitta draco* Krohn. *Monit. Zool. Ital.* **61**, 71–78.

Ghirardelli, E. (1959). Habitat e biologia della riproduzione nei Chetognati. *Arch. Oceanogr. Limnol.* **11**, 1–18.

Ghirardelli, E. (1962). Ambiente e biologia della riproduzione nei Chaetognati. Metodi di valutazione degli stadi di maturità e loro importanza nelle ricerche ecologiche. *Pubbl. Staz. Zool. Napoli Suppl.* **32**, 380–389.

Ghirardelli, E. (1968). Some aspects of the biology of the chaetognaths. *Advan. Mar. Biol.* **6**, 271–375.

Ghirardelli, E. and Arnaud, J. (1966). Contribution à l'étude de la spermatogénèse chez les Chaetognathes. *Arch. Zool. Ital.* **51**, 309–325.

Ghiselin, M. T. (1969). The evolution of hermaphroditism among animals. *Quart. Rev. Biol.* **44**, 189–208.

Grassi, G. B. (1883). I Chaetognati. *Fauna Flora Golf. Neapel.* **5**, 1–126.

Huntsman, A. G. and Reid, M. E. (1921). The success of reproduction in *Sagitta elegans* in the Bay of Fundy and the Gulf of St. Lawrence. *Trans. Roy. Can. Inst.* **13**, 99–112.

Hutton, R. F. (1954). *Metacercaria owreae* n. sp., an unusual trematode larva from Florida Current chaetognaths. *Bull. Mar. Sci. Gulf Carib.* **4**, 104–109.

Hyman, L. H. (1959). Phylum Chaetognatha. "The Invertebrates" Vol. 5. McGraw-Hill, New York.

Jägersten, G. (1940). Fur Kenntnis der Physiologie der Zeugung bei *Sagitta. Zool. Bidr. Uppsala* **18**, 397–413.

John, G. (1933). Habits, structure, and development of *Spadella cephaloptera. Quart. J. microsc. Sci.* **75**, 625–696.

Kramp, P. (1939). Chaetognatha. *Medd. Grønland* **80**, 1–40.

Kuhl, W. (1938). Chaetognatha. *In* "Klassen und Ordnungen des Tierreichs" (II. G. Bronn, ed.), Bd. IV, Abt IV, Buch 2, Teil 1, pp. 1–226.

MacGinitie, G. E. (1955). Distribution and ecology of the marine invertebrates of Point Barrow, Alaska. *Smithson. Misc. Collect.* **128**, 1–201.

McLaren, I. A. (1966). Adaptive significance of large size and long life of the chaetognath *Sagitta elegans* in the Arctic. *Ecology* **47**, 852–855.

Michael, E. L. (1919). Report on the Chaetognatha collected by the United States Fisheries Steamer "ALBATROSS" during the Philippine Expedition, 1907–1910. *Bull. U.S. Nat. Mus.* **1**(4), 235–277.

Murakami, A. (1959). Marine biological study on the planktonic chaetognaths in the Seto Inland Sea. *Bull. Maikai Reg. Fish. Res. Lab.* **12**, 1–186. (Engl. summary)

Murakami, A. (1966). Rearing experiments of a chaetognath, *Sagitta crassa. Inform. Bull. Planktol. Japan* **13**, 62–65. (Engl. abstr.)

Owre, H. (1960). Plankton of the Florida Current. Part VI. The Chaetognatha. *Bull. Mar. Sci. Gulf Carib.* **10**, 255–322.

Pierce, E. L. (1941). The occurrence and breeding of *Sagitta elegans* Verrill and *Sagitta setosa* J. Müller in parts of the Irish Sea. *J. Mar. Biol. Ass. U.K.* **25**, 113–124.

Pierce, E. L. (1951). The Chaetognatha of the West Coast of Florida. *Biol. Bull.* **100**, 206–228.

Redfield, A. C. and Beale, A. (1940). Factors determining the distribution of populations of chaetognaths in the Gulf of Maine. *Biol. Bull.* **79**, 459–487.

Reeve, M. R. (1970a). The biology of Chaetognatha. I. Quantitative aspects of growth and egg production in *Sagitta hispida. In* "Marine Food Chains" (J. H. Steele, ed.), pp. 168–189. Oliver and Boyd, Edinburgh.

Reeve, M. R. (1970b). Complete cycle of development of a pelagic chaetognath in culture. *Nature (London)* **227**, 381.

Reeve, M. R. and Lester, B. (1974). The process of egg-laying in the chaetognath *Sagitta hispida. Biol. Bull.* **147**, 247–256.

Reeve, M. R. and Walter, M. A. (1972a). Conditions of culture, food size selection and the effects of temperature and salinity on growth rate and generation time in *Sagitta hispida* Conant. *J. Exp. Mar. Biol. Ecol.* **9**, 191–200.

Reeve, M. R. and Walter, M. A. (1972b). Observations and experiments on methods of fertilization in the chaetognath *Sagitta hispida. Biol. Bull.* **143**, 207–214.

Reeve, M. R., Raymont, J. E. G., and Raymont, J. K. B. (1970). Seasonal biochemical composition and energy sources of *Sagitta hispida*. *Mar. Biol.* **6**, 357–364.

Russell, F. S. (1932*a*). On the biology of *Sagitta*. The breeding and growth of *Sagitta elegans* Verrill in the Plymouth Area. *J. Mar. Biol. Ass. U.K.* **18**, 131–146.

Russell, F. S. (1932*b*). On the biology of *Sagitta*. II. The breeding and growth of *Sagitta setosa* J. Müller in the Plymouth Area, 1930–31, with a comparison with that of *S. elegans* Verrill. *J. Mar. Biol. Ass. U.K.* **18**, 147–160.

Russell, F. S. (1933). On the biology of *Sagitta*. IV. Observations on the natural history of *Sagitta elegans* Verrill and *Sagitta setosa* J. Müller in the Plymouth Area. *J. Mar. Biol. Ass. U.K.* **18**, 559–574.

Sanzo, L. (1937). Colonia pelagica di uova di Chetognati (*Spadella draco* Krohn). *Mem. Regio. Comitato Talassogr. Ital.* **239**, 1–6.

Sherman, K. and Schaner, E. G. (1968). Observations on the distribution and breeding of *Sagitta elegans* (Chaetognatha) in coastal waters of the Gulf of Maine. *Limnol. Oceanogr.* **13**, 618–625.

Stevens, N. M. (1903). Ovogenesis and spermatogenesis of *Sagitta bipunctata*. *Zool. Jahrb. Abt. Anat.* **18**, 227–240.

Stevens, N. M. (1905). Further studies on the ovogenesis of *Sagitta*. *Zool. Jahrb. Abt. Anat.* **21**, 243–252.

Stevens, N. M. (1910). Further studies on reproduction in *Sagitta*. *J. Morphol.* **21**, 279–319.

Stone, J. H. (1966). The distribution and fecundity of *Sagitta enflata* Grassi in the Agulhas Current. *J. Anim. Ecol.* **35**, 533–541.

Sund, P. N. (1959). A key to the Chaetognatha of the tropical eastern Pacific Ocean. *Pac. Sci.* **13**, 269–285.

Thomson, J. M. (1947). The Chaetognatha of south-eastern Australia. *Bull. Coun. Sci. Ind. Res. Melbourne* **222**, 1–43.

Tokioka, T. (1939). Chaetognaths collected chiefly from the bays of Sagami and Suruga, with some notes on the shape and structure of the seminal vesicle. *Rec. Oceanogr. Wks Japan* **10**, 123–150.

Tokioka, T. (1955). Notes on some chaetognaths from the Gulf of Mexico. *Bull. Mar. Sci. Gulf Carib.* **5**, 52–65.

Van Oye, P. (1931). La fécondation chez les Chaetognathes. *Bull. Mus. Roy. Hist. Nat. Belg.* **7**, 1–7.

Vasiljev, A. (1925). La fécondation chez *Spadella cephaloptera* Lgrhs. et l'origine de corps déterminant la voie germinative. *Biol. Gen.* **1**, 249–278.

Weinstein, M. (1973). Studies on the relationship between *Sagitta elegans* Verrill and is endoparasites in the southwestern Gulf of St. Lawrence. Ph.D. thesis, McGill University, Montreal.

Wimpenny, R. S. (1936). Distribution, breeding and feeding of some important plankton organisms of the southwest North Sea in 1934. Part I. *Calanus finmarchicus* (Gunn), *Sagitta setosa* (J. Müller), and *Sagitta elegans* (Verrill). *Fish. Invest. London Ser.* 2, **215**, (3), 1–53.

Chapter 7

HEMICHORDATA

Michael G. Hadfield

7.1 Introduction

Comprehensive reviews of the literature pertaining to research on the Hemichordata are not lacking. The treatments by Dawydoff (1948), in the "Traité de Zoologie," and Hyman (1959) contain excellent descriptions of the Hemichordata. More recent good discussions of the group appear in Beklemishev (1969) and Barrington (1965). The last two works are, however, bound to disappoint the student of invertebrate reproduction because of their scant attention to this subject. The most comprehensive treatment ever given the hemichordates is that of C. J. van der Horst in Bronn's Klassen und Ordnungen des Tier-Reichs (1932–1939). I have relied heavily on this work for its discussions of the classical literature. I have also used van der Horst's assignments of species to genera.

The illustrations contained in this chapter are new. They are derived from my own investigations of the reproduction and early development of an Hawaiian population of *Ptychodera flava*. The Hawaiian form has been placed in as the subspecies *P. flava laysanica* (Spengel). The population sampled in this study is located near shore on the south end of Waikiki Beach, Oahu, Hawaii, at a depth of 1 to 3 m. Monthly samples were collected, preserved, and 10–20 animals embedded and sectioned to determine the condition of their gonads. Animals were also maintained for 6 to 8 weeks in aquariums which were half filled with sand from their native habitat. Spawning was observed several times in these aquariums.

Because the anatomy, life habits, and developmental modes differ so greatly between the two classes of Hemichordata, the Enteropneusta and Pterobranchia are considered separately. Planctosphaeroidea, considered a third class by many authors, will be treated here as an enteropneust larva and is discussed in Section 7.2.3.4. This treatment is, I feel, justified by the excellent discussion of Damas and Stiasny (1961).

7.2 Enteropneusta

7.2.1 Asexual Reproduction

7.2.1.1 Occurrence and Types

Asexual propagation has been well documented in the Enteropneusta, although thoroughly studied in only three species: *Balanoglossus capensis* by Gilchrist (1923), *B. australiensis* by Packard (1968), and *Glossobalanus crozieri* by Petersen and Ditadi (1971). The processes of vegetative reproduction are apparently identical in these species. An adult worm breaks into two parts, separating immediately anterior to the area of the body containing the hepatic caeca. An elongate piece breaks off from the posterior region of the anterior half and then fragments into many (<12 in *B. australiensis*) smaller tubular fragments each 2–15 mm long. Each fragment (called "regenerands" by Packard) consists of outer body wall, coelomic compartments, mesenteries and, often, gonads. The small fragments undergo morphogenesis, without growth, to reconstitute all of the missing parts of an adult worm. Apparently both large remaining segments of the adult, the anterior branchiogenital half and the posterior hepatointestinal half, also regenerate missing parts. This mode of asexual propagation can thus, at one time, provide 14 individuals from one (*B. australiensis*).

None of the above-named authors gives a thorough description of the histological or cellular processes involved in asexual propagation and thus the following comments are based on external examination of the regenerating fragments.

1. Both ends heal; the gut apertures remain open.

2. At the anterior end of the "regenerand" a small proboscis differentiates, apparently dorsal to the new mouth.

3. A new "stomochord" evaginates from the anterodorsal roof of the buccal cavity.

4. In the developing collar region, a dorsal longitudinal invagination becomes the new neurochord.

5. Gill slits perforate the body wall beginning just behind the posterior collar groove and their development progresses posteriorly.

6. Genital wings grow outward in the branchial region.

7. Hepatic saccules evaginate from the gut wall in the region behind the pharynx.

While reconstitution of missing parts is progressing in the "regenerands," the gonadal elements disappear from all but the future genital region. No description of this process is provided, but both Gilchrist and Packard stress the value of the gonadal yolk reserves for the nourishment of the reconstituting fragments.

Packard suggests that complete specific form is reached by redifferentiation of existing tissues. The information cited, however, is not sufficiently detailed or extensive to prove this point and the possibility remains that cellular proliferations may account for some new tissues.

The phenomenon of somatic regeneration, probably very widespread, if not universal, in the Enteropneusta, cannot be clearly separated from that of asexual propagation. When a worm is broken into two halves, if each half regenerates the missing parts, reproduction has occurred. Fortunately, there are numerous studies of regeneration in enteropneusts and some of these give particularly enlightening detail to the general discussions cited above.

Dawydoff (1902, 1907a,b, 1909), Rao (1955b), and Tweedel (1961) have extensively studied regeneration in *Glossobalanus minutus, Ptychodera flava,* and *Saccoglossus kowalevskii,* respectively. Dawydoff and Rao found that posterior halves rapidly regenerate anterior structures while anterior halves do not rebuild posterior structures. However, Dawydoff (1948) later cited papers by Cori (1902) and Kuwano (1902) as providing evidence that posteriorward regeneration can occur but is a much slower process. Tweedel (1961) found, in *S. kowalevskii,* that direction and type of regeneration was dependent on the level of section along the longitudinal axis of the worm. Tweedel's studies established the potential for, although with low percentages of laboratory success, regeneration of anterior parts of animals sectioned through the proboscis, collar, or the anteriormost branchial region. Only animals sectioned immediately posterior to the branchial region demonstrated

posteriorward regeneration. In the latter case, however, the posterior segment was incapable of anteriorward regeneration. These observations
thus throw doubt on the suggestion that the capacity for regeneration
automatically implies a capacity for asexual multiplication of a species.

Development of proboscis, collar, and branchiogenital region from
the anterior ends of cut worms usually begins with the fusion of the
gut wall and the body wall. On the anteriodorsal surface a "regeneration
bud" then forms. The bud contains mainly elements of the coelom which
will produce the coeloms and musculature of proboscis, collar, and heart.
The bud is penetrated by an anterodorsal diverticulum of the new buccal
cavity; it becomes the new "stomochord." The collar nerve cord arises
first as a condensation and thickening of existing neural elements, followed by invagination.

In general, two processes are noted in regeneration and these are
undoubtedly identical with processes occurring in asexual reproduction
by transverse fission. They are: (1) formation of a regeneration bud
which contains dedifferentiated mesodermal elements. The bud basically
provides the materials only for the building of a new proboscis and
its contents. (2) *In situ* dedifferentiation of existing structures provides
cellular elements for the construction of new ones. The buccal gut epithelium, the pharynx and gill slits, and the hepatic diverticula arise in this
way. Dawydoff calls this process "morphalaxy," and it is this process
which provides the missing parts posterior to the proboscis.

Asexual reproduction, in the strictest sense of the term, has been demonstrated only in three members of one family, the Ptychoderidae. Regeneration studies on three more species include only two non-ptychoderids,
the harrimaniids *Saccoglossus serpentinus* (Assheton, 1908) and S.
kowalevskii (Tweedel, 1961). It thus remains to be learned whether
or not all species which are capable of regeneration are also characterized
by regular asexual propagation and whether or not these potentialities
are limited to the members of just one enteropneust family.

7.2.1.2 Factors Influencing Asexual Reproduction

Gilchrist's (1923) study of *Balanoglossus* (*Ptychodera*) *capensis* from
South Africa showed that the animals bearing this name were, in fact,
conspecific with animals bearing the name *B. proliferans*. The former
species was known to reproduce sexually, the latter asexually. Interestingly, *B. capensis* was found to occur only in the winter months and
B. proliferans only in the summer. Gilchrist's studies finally revealed
that the asexual offspring of *B. proliferans* invariably mature into *B.
capensis* and that, at the end of summer, all remaining large *B. proliferans* transform into *B. capensis*. Gilchrist suggests that this species may

TABLE I
SPECIES KNOWN TO REPRODUCE ASEXUALLY OR TO REGENERATE
MISSING PARTS

Species	Regular asexual propagation	Regeneration shown	Source
Ptychoderidae			
Balanoglossus australiensis	+		Packard, 1968
Balanoglossus capensis	+		Gilchrist, 1923
Balanoglossus clavigerus		+	Cori, 1902
Balanoglossus misakiensis		+	Kuwano, 1902
Glossobalanus minutus	+	+	Dawydoff, 1902, 1907a,b, 1909; Packard, 1968
Glossobalanus crozieri	+		Petersen and Ditadi, 1971
Ptychodera flava		+	Rao, 1955b
Harrimaniidae			
Saccoglossus kowalevskii		+	Tweedel, 1961
Saccoglossus ruber (*serpentinus*)		+	Assheton, 1908

show a true "alternation of generations," but this point can certainly not be accepted for certain without knowledge of the form of settled, sexually produced larvae. What can be stated for certain is that reproductive habits in this species are seasonally mediated and that, in some way, the onset of summer stimulates the initiation of asexual propagation.

No other enteropneusts are known to separate so completely their sexual and asexual reproductive processes. My studies of *Ptychodera flava*, a species noted by Rao (1955b) to have great restitutive powers, have given no indication of any tendency toward regular reproduction via transverse fission at any time of the year. Table I shows the distribution of species known to reproduce asexually or which have been studied during regeneration.

7.2.2 Sexual Reproduction

7.2.2.1 SEXUAL DIMORPHISM AND HERMAPHRODITISM

No instances of hermaphroditism have been reported for the Enteropneusta. In most cases, the sex of an individual worm can only be determined by examination of the sexual products, and then only when the animal is ripe. Some reports, however, describe external indications of sex which are, at best, slight color differences in the genital region. Burdon-Jones (1952) notes that the genital region of males of *Saccoglossus horsti* is gray, while that of females is salmon-pink. Hill (1894) declared that the testes of *Balanoglossus australiensis* were deep yellow to orange, the ovaries light yellow to white. Caullery and Mesnil (1904) note that the gonads of *Protobalanus koehleri* show through the body wall and that the testes are milk-white in color and the ovaries rose-white. The genital region of *Saccoglossus ruber* may appear more apricot-orange in males and citrine-drab in females (Burdon-Jones and Patil, 1960). The situation in *Ptychodera flava* is apparently variable according to locality. Willey (1899) said that the males of this species were distinguished by the presence of dark brown pigment inside the genital wings. Rao (1954a) claimed for this species, "The sexes are dimorphic, the males being readily identifiable by the presence of dark brown pigment patches on the genital plurae." My own observations on *P. flava* from Hawaii do not confirm Willey's and Rao's findings. The males and females of Hawaiian *P. flava* are colored externally exactly alike. A practiced eye can usually distinguish between the sexes of extremely ripe animals because the sperm show through the body wall of the genital wings as a dull gray, the eggs as yellow. These are not the only differences we note between members of the *P. flava* population in Hawaii and descriptions of those in the Indian Ocean.

The literature contains no information relating to sex determination in Enteropneusta. It would be most interesting to learn if asexually produced progeny of a single adult always retain the sex of their progenitor (see Section 7.2.1.1).

7.2.2.2 Anatomy of the Reproductive System; its Constituents and Origin

The gonads of enteropneusts are located in the anterior region of the trunk. The exact extent of the gonads varies in different species; it coincides with the branchial region in *Stereobalanus* and in *Harrimania maculosa;* in *Saccoglossus pygmaeus* it lies entirely behind the branchial region; in *Balanoglossus australiensis* the gonads run even into the hepatic region.

The gonads may be more or less visible externally, but only in ripe animals. In Ptychoderidae the gonads occupy broad flaps of tissue which originate ventrolaterally to curl dorsally enclosing the branchial chamber (Figs. 1 and 2). These flaps are referred to as genital wings or pleurae and they exist in immature as well as mature individuals. In *Saccoglossus* the gonads lie in two sets of winglike longitudinal structures of which

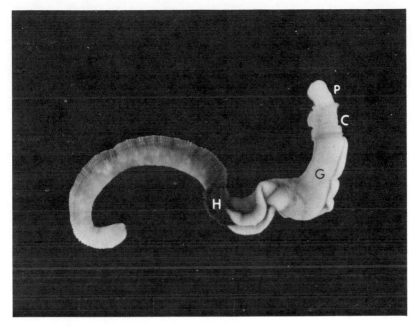

Fig. 1. *Ptychodera flava*; the living animal. C, Collar; G, genital wings containing gonads; H, hepatic sacculations; P, proboscis (life size).

one pair is dorsal to the branchial slits, the other ventral. Both pairs of folds project laterally.

Internally, enteropneust gonads lie within the trunk coelom but are always covered by coelomic epithelium. The origin and adult structure of the gonads clearly show that they are retroperitoneal. They arise (Morgan, 1894) as mesenchymal cellular aggregations in spaces derived from the blastocoel, between the basement membranes of the body wall epidermis and the peritoneum. As cellular proliferation proceeds, the gonads bulge into the coelom, always retaining their covering of basement membrane and coelomic epithelium.

The position of gonads in the coelom and the location of the gonopores relative to the branchial pores are variable. Gonads of enteropneusts are described as being dorsal or lateral, but these terms are confusing because in some forms they refer to the coelomic location of the gonads and in others to the relative position of the gonopores. In the Ptychoderidae the coelom of the branchiogenital region is subdivided by an additional pair of mesenteries which are lateral (dorsal and ventral mesenteries, suspending the gut from the body wall, are found in all

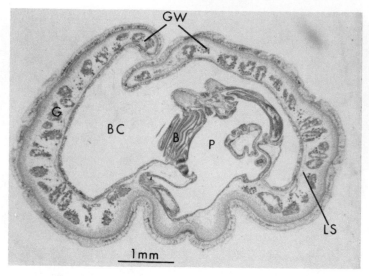

Fig. 2. Cross section through the branchiogenital region of *Ptychodera flava*. B, Branchial slits; BC, branchial chamber; G, gonads; GW, genital wings; LS, lateral septum; P, pharynx.

forms). These lateral septa extend from the gut to the lateral body wall in the dorsally directed crests of the genital wings (see Fig. 2). In most ptychoderids, gonads lie within the ventrolateral chambers of the coelom, as divided by the lateral mesenteries, and open dorsally into the branchial chamber created by the folds of the genital wings. These gonads are referred to as "lateral" both because of their position in the ventrolateral coelomic compartments and because their pores open lateral to the gill slits. In *Balanoglossus*, also in the Ptychoderidae, "dorsal gonads," lying dorsal to the lateral septa, are found in addition to lateral gonads. Their genital pores are, however, still in the lateral position.

In members of other enteropneust families, dorsal gonads may exist, but are labeled so because they open dorsal or medial to the branchial pores. The use of the term "dorsal gonad" in such forms does not imply the existence of lateral mesenteries.

The structure of enteropneust gonads is relatively simple. They are composed of numerous ovoid sacs each consisting of a layer of germinal epithelium surrounded by a basement membrane which is frequently richly vascularized and the whole is enclosed in a layer of coelomic epithelium (Figs. 3 and 5).

In the simplest gonads each opens to the outside by an individual pore. The ducts are very small and in *Ptychodera flava* appear to be closed in nonreproductive animals (personal observation). In some species secondary gonads are produced via separation of lobes from the primary gonads. The ducts of these are called, accordingly, secondary gonoducts. In other species where the gonads are highly lobate, individual lobes may develop gonoducts which are then referred to as accessory gonoducts (Dawydoff, 1948). The ducts open into the branchial chamber (Ptychoderidae) or into distinct channels, the branchiogenital grooves. In general, the branchial slits lie in the inner extremities of these grooves while the gonopores open along their dorsal or ventral walls (see Fig. 5A).

The exact origin of the germ cells is debated. Spengel stated in numerous works that he thought that they arose primarily from mesenchyme. Willey (1899), while not debating the origin of primary gonads, reported that secondary or accessory gonads have an ectodermal origin. Bateson (1886) felt that gonads were ectodermal in origin and Morgan (1894) felt that the gonads were clearly mesodermal in origin. Dawydoff (1948) shows the gonads forming retroperitoneally, in *Glossobalanus minutus*, from cells of unknown origin. Willey (1899) states, in his discussion of *Spengelia porosa*, that, "It seems not impossible that the gonads of the primary series may have a different origin from the accessory gonads when the latter can be shown to be distinct neoformations, as in the present species." A similar confusion concerning origin of the germ cells is still present in other large deuterostome groups (see Hyman, "Echinodermata," 1955, p. 308, for instance).

Whatever the source of the germinal epithelium in the Enteropneusta, it is responsible for the proliferation of more than just primary auxocytes. In many species, large yolk cells arise from the germinal epithelium and in others follicle cells may be proliferated in females. It is interesting to note that where yolk cells are present (Ptychoderidae, Spengelidae, and *Saccoglossus*) they fill the gonadal cavities in both sexes. In *Ptychodera flava* the large yolk cells are present most of the year, decreasing in volume and number only in the rather restricted reproductive season. The nutritional role of the yolk cells in gamete maturation appears unquestionable. In *Protobalanus* and most Harrimaniidae, where yolk cells are not found, the eggs themselves are four to ten times larger than in other species. A most interesting problem remains to be researched concerning the nature of the association between yolk cells and gametes in the Enteropneusta.

7.2.2.3 CYTODIFFERENTIATION OF THE GAMETES

There have been, to date, no published studies of the cytology of gametogenesis in any hemichordate. A few facts are available in older anatomical accounts concerning appearance of the gonads and gametes.

The morphology of sperm and egg surfaces has been illustrated at the ultrastructural level for one enteropneust (see Colwin and Colwin, 1967).

Spermatogenesis. Hill (1894) appears to be the first to have examined sperm of an enteropneust. He described the structure of the gonad of *Balanoglossus australiensis* and noted that the germ layer was not distinct, ". . . but certain small rounded cells lying irregularly internal to the limiting membrane in young gonads doubtlessly belong to it." Hill figured the sperm crudely and noted that the heads were rounded and the tails long. Willey's (1899) sketch of a testis of *Ptychodera flava* may have been meant to indicate that maturing sperm are found in the centers of the follicles.

Caullery and Mesnil (1904) have examined the testes of *Protobalanus koehleri* and noted that the spermatocytes lie in cords directed into a central follicular lumen. Spengel (1893) observed a similar disposition of germinal elements in *Harrimania kupfferi* and van der Horst (1932–1939) saw a similar arrangement in *Glandiceps talaboti*. Figure 33 of Caullery and Mesnil indicates that similar stages of spermatogenesis are clumped together at different locations in the testes.

Detailed illustrations of enteropneust sperm have been given by van der Horst (1932–1939) for *Baloglossus carnosus* and Burdon-Jones (1952) for *Saccoglossus horsti*. Their figures show the sperm to be about 55 μm long with a spherical head, a small rounded or pointed acrosome, and four basal mitochondrial bodies. By contrast, Spengel's (1893) illustration of a sperm of *Harrimania kupfferi* shows a narrow, elongate and pointed sperm head and does not indicate the presence of chondriosomes or a midpiece.

In a long series of papers dealing with fertilization in *Saccoglossus kowalevskii*, A. L. and L. H. Colwin (see review, 1967) have presented numerous electron micrographs of the tips of sperm of this species. Their studies have, however, been confined to the reactions of the acrosome, the details of which will not be reviewed here.

The sperm of *Ptychodera flava* from Hawaii resemble those of other enteropneust species in having a rounded head, four mitochondrial bodies, and a slender tail.

Spermatogenesis in individuals of an Hawaiian population of *Ptychodera flava* is restricted to a brief period which culminates in spawning in late November and December. In January only residual mature sperm are seen in the testes and in the period February through April it is impossible to detect males even via microscopic examination of the gonads. All follicles are gradually filling with yolk material. In May, June, and July a male may begin to proliferate small patches of very small spermatogonia which lie tightly against the basement membrane of the gonad. By August spermatogenesis has become an active process; the sites of spermatocyte proliferation broaden and increase in number and maturing stages extend in bands toward the center of each follicle (Fig. 3). It is difficult to tell whether or not there are fully matured sperm at this stage. Gonads sectioned in October and November show a steady increase in the volume of each testicular follicle occupied by sperm, and a concomitant decrease in the volume of yolk. By late November and December the follicles are completely engorged with ripe sperm, spermatogenesis has come to an end, and yolk is nearly or entirely absent. Spawning ensues.

All males gathered at any one time will not show an identical amount of sperm deposition in the testes. However, within an individual all follicles show the same degree of ripeness. Willey (1899) noted, as do we, that occasional individuals of *Ptychodera flava* may have gonads in only one of the genital pleurae. Such individuals make up less than 3% of the *P. flava* population which I have been sampling. Out of 205 individuals examined, only six were found to have gonads confined to one genital wing; two were male, three female, and one indeterminate.

Oogenesis. Dawydoff (1948) notes that the mature gonad of an enteropneust contains three sorts of cellular elements: sexual cells, follicle cells, and yolk cells. Follicle cells are probably limited to females and yolk cells are not present in some species; they are lacking in Harrimaniidae (except *Saccoglossus*). Dawydoff asserts that follicle cells are not present in most Spengelidae and Ptychoderidae. Judging from personal observations on *Ptychodera flava*, however, follicle cells may form such a thin layer about the oocytes as to be very difficult to detect with the light microscope (see Figs. 4 and 7A).

During growth of the oocytes in forms where yolk cells are abundant, the yolk gradually disappears until, in ripe specimens, no yolk may be visible outside the eggs. Oocytes are proliferated on the germinal epithelium and remain attached to it during most of their growth. As they grow, they protrude progressively into the lumens of the ovarian follicles and presumably occupy the space made available by the decline in volume of the yolk cells (van der Horst, 1932–1939, on *Saccoglossus otagoensis;* Dawydoff, 1948, on *Saccoglossus* sp., personal observations on *Ptychodera flava*).

The function and fate of the large yolk cells which fill the gonadal follicle of many enteropneust species throughout the year are enigmatic. In *Ptychodera flava* the yolk cells build very rapidly after the spawning period and only decline in volume during the phase of rapid oocyte proliferation and growth (Figs. 5 and 6). It would thus seem that the

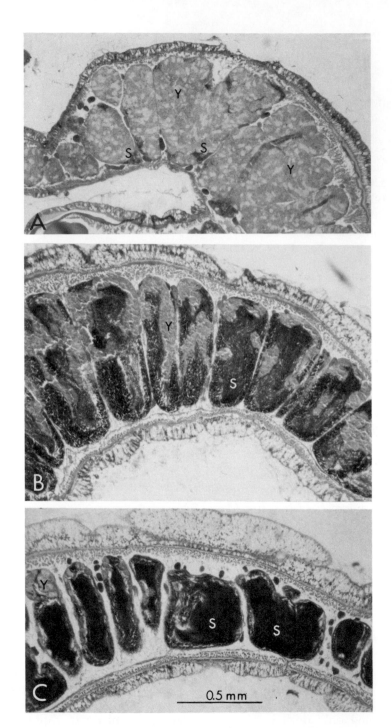

FIG. 3. Sections of genital wings of male *Ptychodera flava*, collected in: A, August; B, October; C, December. S, Sperm; Y, yolk.

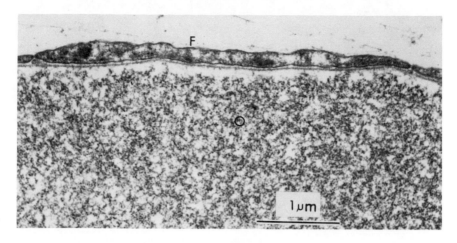

FIG. 4. Electron micrograph of an ovarian oocyte of *Ptychodera flava*. F, Follicle cell; O, oocyte cytoplasm.

contents of the yolk cells are utilized in gamete growth. Willey (1899) saw two functions in the yolky material: nutrition of the growing gametes and later, ". . . providing an albuminous medium to preserve the germ cells under the best possible physiological conditions during the final crucial stages of maturation." That the yolk cells provide a mechanism for the storage of nutritional materials, gained through a year of feeding, for use during a limited reproductive period seems incontestable.

De Jorge and Petersen (1968) have studied the glycogen content of mature and immature *Balanoglossus gigas* in Brazil. They note that the major accumulation of glycogen in immature animals is in the hepatic region, while in mature animals it is in the genital region. Their figures show that the increase of glycogen in the genital region from young animals to maturity is about 3.3 times (mg glycogen/100 gm fresh wt.). It would be most interesting to know if the glycogen is a part of yolk storage in the gonads and if the level varies throughout the year in mature animals. De Jorge and Peterson note, however, that, ". . . no sharp breeding cycle has been established for the *Balanglossus* species living in the Brazilian coast, mature animals being obtained in different seasons."

Gilchrist's (1923) investigation of *Balanoglossus capensis* showed that yolk cells were present in the gonads of all but very young stages. The yolk is not used up by the time of ripening. In the winter *B. capensis* "proliferans-stage" (see Section 7.2.1.1) yolk cells constitute the entire filling of the gonadal follicles.

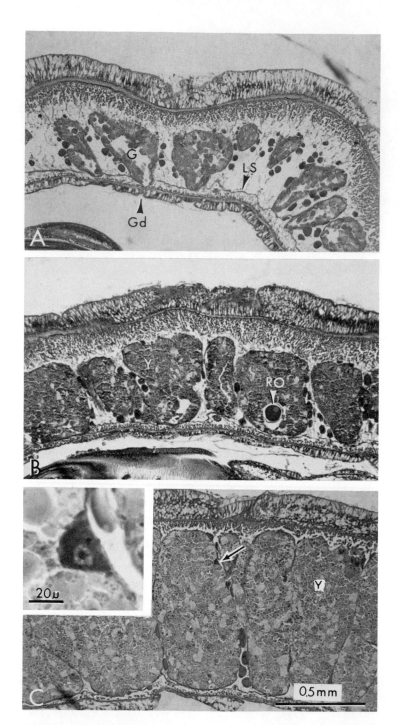

Fig. 5. Sections of genital wings of female *Ptychodera flava* showing progressive accumulation of yolk cells in the ovarian follicles. A, February; B, April; C, June. Inset shows very early oocyte indicated by arrow in C. G, Gonad; Gd, gonoduct; LS, lateral septum; RO, residual oocyte; Y, yolk.

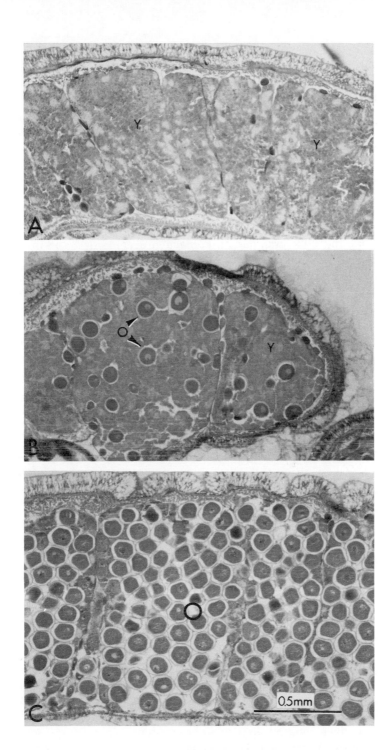

Fig. 6. Sections of genital wings of female *Ptychodera flava* showing progressive proliferation and growth of oocytes in: A, August; B, October; C, December. O, Oocytes; Y, yolk.

In enteropneust species where yolk cells exist, the mature oocytes are relatively small (see Table IV) and the cytoplasm of the oocytes is finely granular. By contrast, in those species without yolk cells, the cytoplasm of eggs contains large yolk spheres and the eggs are considerably larger. As shall be seen later, these differences also correspond to strikingly different modes of development.

The structure and function of the yolk cells are not clear. Willey (1899) maintained that they lacked nuclei. Dawydoff (1948) figured "cellules vitellines" with nuclei and we have observed nuclei in the yolk cells of *Ptychodera flava*, the same species studied by Willey. There is no doubt that the yolk granules of *P. flava* are contained within cellular membranes. An interesting problem remains to be studied concerning how yolk material, synthesized in the yolk cells, becomes incorporated into gametes. In the ovary this material must cross the barriers of the yolk cell membranes, the follicle cells, and the membranes of the oocytes.

The ripening of oocytes in Hawaiian *Ptychodera flava* follows the pattern described above for males from this same population (Figs. 5 and 6). Minute oocytes are not visible in the yolk-filled ovaries until early summer and their growth is slow until August or September. During the period from late September through mid-November oocyte growth is rapid, culminating in ovarian follicles being nearly devoid of yolk and solidly filled with full-grown oocytes by late November. All follicles within a single female demonstrate the same stage of ripening at any one time. Older studies of changes in nuclear appearance of oocytes during growth add little to modern understanding of the process in Enteropneusta (Caullery and Mesnil, 1904).

7.2.2.4 GAMETOGENIC CYCLES

That populations of at least some enteropneusts are synchronized in their reproduction is indicated by reports of epidemic spawning in *Saccoglossus horsti* (Burdon-Jones, 1951). Other reports of restricted reproductive periods, based on laboratory observation of spawning of animals in aquariums, concern the following: *Saccoglossus kowalevskii* (Colwin and Colwin, 1953; Costello *et al.*, 1957); *Ptychodera bahamensis* (Payne, 1937); *Ptych. flava* (Rao, 1954a; personal observation).

I have examined the gonads of 10 to 20 *Ptychodera flava laysanica* on a monthly basis for 24 consecutive months (January, 1970–January, 1972). Only the larger 50% of animals collected are reproductively active and, in a given month, all such animals show a similar gonadal state. As previously noted, gonadal volume is not an adequate measure of reproductive state since the gonads of even immature animals are filled with yolk cells which build during the year and decline only in a brief, about 3-month long, period of rapid gamete proliferation and growth (Figs. 3, 5, and 6).

Burdon-Jones (1951) notes that in *Saccoglossus horsti* ripe animals can be found at all months of the year, but spawning is restricted to the period May through July. He believes that gametogenesis can probably occur at any time, but that its completion is arrested until the actual spawning season and then the final stage of gamete maturation can occur quite rapidly. This is a most interesting suggestion and one which should be investigated more thoroughly through an annual study of gonad histology.

In *Ptychodera flava* we note that mature-appearing sperm can be seen in sections of the testes as early as September even though the testes do not become filled with ripe sperm until the onset of the spawning season in late November. There must obviously be some mechanism for retaining the earliest-formed sperm in a good physiological state for a period extending up to 2 months. By contrast, while new oocytes may appear over a 2-month period in this species, their arrival at full growth is a simultaneous process and one which occurs only just before spawning time. This observation leads to the interesting conclusion that the growth rate of individual oocytes cannot be a constant.

7.2.2.5 Factors Influencing Gametogenesis

External influences on gametogenesis in enteropneusts may only be implied from knowledge of strictly seasonal spawning in some species (see Table II). Whether such periodicity is induced by seasonal changes in temperature, light, tides, or other ecological parameters can not be stated. Nutrition may play a dominant role since the gametes of most enteropneusts are formed at the expense of large yolk-storage cells in the gonad, cells which grow during the nonreproductive periods of the year (personal observations on *Ptychodera flava*). It is interesting to note that a strict seasonality appears to exist in tropical as well as temperate enteropneust species. If factors influencing gametogenesis are the same from species to species then one would not expect them to relate to temperature and light which are seasonally variable in shallow temperate waters but negligible in at least parts of the tropics.

We know little about the nature of gonadal yolk stores in acorn worms. That they are rich in glycogen may be inferred from the studies of De Jorge and Petersen (1968). Of several elements studied by De Jorge and Petersen (1970) in *Glossobalanus crozieri*, only potassium appears to be in much greater concentration in the genital region than in other body sectors. This is not true of water, sodium, calcium, magnesium, phosphorus, chloride, nitrogen, or iodine.

7.2.2.6 Reproductive Behavior

Since most enteropneusts occur in dense beds in sandy to muddy intertidal flats, their numbers are usually sufficiently great to ensure fertilization in the milieu if epidemic spawning occurs. For instance, Burdon-Jones (1951) found populations of *Saccoglossus horsti* to have a density of 500 animals per square meter in a sandy flat on the British Solent. It is most unlikely that such fragile and slowly moving animals could manage sufficient migration to accommodate gamete aggregation. However, we must consider the interesting case of two species of *Glossobalanus* from Japan (Ikeda, 1908) and Java (Spengel, 1909). In both cases, the animals were found swimming, in great masses, at the surface of the sea. While both authors saw no suggestion of spawning, Dawydoff (1948) rightly points out, "Cette vie errante favorise la dissémination des produits génitaux." It is difficult to arrive at any other explanation for such extraordinary behavior on the part of these normally sluggish, burrowing worms.

Burdon-Jones (1951) did indeed observe epidemic spawning in *Saccoglossus horsti*. He noted that the females spawned first, followed approximately 20 minutes later by the males.

7.2.2.7 Spawning

Mechanisms. Saccoglossus horsti breeds in early summer in England. Burdon-Jones (1951) observed the spawning of this species in the field on three different occasions, always about 30 minutes after low tide exposure. His report states that spawning began when the water currents, which normally flow into the mouth of the burrow, reverse and flow outward, carrying out a stream of viscous mucus. Eggs then appeared in a mucous cord about 3 mm in diameter, issuing from the burrow opening. An egg mass containing 2500–3000 eggs was thus eliminated from the tube during the 5–10 minutes of spawning. Each egg mass covered an area of 7 to 8 cm². Spawning of the population continued for about 2½ hours.

Approximately 20 minutes after oviposition began, males commenced spawning. The process was much the same as seen in females except that the mucus surrounding the sperm was thinner and apparently more soluble, allowing clumps of sperm to disperse in such surface water as was present.

Burdon-Jones concludes that in spawning, first, the animal moves upward toward the mouth of the burrow, second, there is an increased secretion of mucus, probably mostly from the collar, and third, gametes are released through the genital pores. The animals may move back and forth in the burrows, with the forward movements pushing the gamete-laden mucus out of the burrows.

Rao (1954a) saw spawning of *Ptychodera flava* in India. He noted that the proboscis and collar protrude from the mouth of the burrow and that a milky substance is exuded. Strong peristaltic waves apparently carry the gametes forward in the burrows and movements of proboscis and collar distribute them. Rao does not mention mucus production during spawning, but in *Ptychodera flava* from Hawaii we have observed that animals commence spawning by shoving the proboscis out of the sand and then proliferating great quantities of mucus. The eggs are contained within this mucus and are not readily dispersed until some 20–30 minutes have passed. Sperm are also shed into mucus, but do not appear to adhere to it. They quickly dispersed in the aquarium.

There is another spawning mechanism described in the literature. Kirk (1938) said that *Saccoglossus otagoensis* free their eggs into a tough, stringy secreted mucous "cocoon" by rubbing off the entire gonadal ridges. The eggs are freed within the cocoon by the degeneration of the tissues surrounding them. Sperm are released (apparently not observed) in spermatophoral packets; these packets reach the cocoons and fertilize the eggs. The female eventually withdraws from the cocoon, leaving the eggs within it where they develop to free-swimming ciliated larvae.

A similar process of gamete release by dehiscence was reported by Bateson (1884) for *Saccoglossus kowalevskii*, but the studies of later workers (Colwin and Colwin, 1953, 1962b) do not confirm this. Stiasny (1913) claimed that *Balanoglossus clavigerus* releases its sexual products by rupture of the body wall along the free edges of the genital wings.

It appears that most burrowing enteropneust species release their eggs in a thin or thick mucus which flows from the mouth of the tube. Two notable exceptions are *Saccoglossus pusillus* (Davis, 1908) and *Balanoglossus clavigerus* (Stiasny, 1913) which deposit their eggs inside their burrows, but not in cocoons as does *Saccoglossus otagoensis*. Fertilization and early development occur within the burrow. Van der Horst (1932–1939) notes that fertilization probably also occurs in the open water in such species. It may be that only a few eggs remain in the burrow and are fertilized there.

Synchronization and Coordination. There are various observations recorded in the literature which indicate that spawning tends to be epidemic; when one animal spawns in an aquarium or bowl, others in the container begin to spawn. Epidemic spawning thus seems to be a general rule in the enteropneusts. As mentioned previously, we have at hand only Burdon-Jones' (1951) observation of spawning, in the field, of *Saccoglossus horsti*. In addition to the fact that females apparently spawn first and that the presence of eggs in the water induces males to spawn, Burdon-Jones noted the following: ripe specimens will spawn over a wide temperature range, from 13.5° to 21°C, but a prolonged, 2- to 3-day temperature rise always produces spawning in ripe animals; most animals spawned when the temperature was 16° to 17°C (60%). In the field, spawning was observed only when three factors converged: ripe animals, a fairly sudden rise in air temperature to at least 16°C, and spring tide low water.

Obviously, the factors leading to sexual ripening must be of a broader seasonal nature than those which produce spawning. Light, temperature, and food availability may all play a part. Once ripe, however, spawning is probably synchronized by a warming of the sand surface by warm air at low tide. Induction of males to spawn by female sexual products in the water assures fertilization.

If temperature serves as the trigger to spawning in the Hawaiian population of *Ptychodera flava*, it must be in a more gradual manner and in the reverse of the situation discussed above. The population I have studied lies at a minimum depth of 1.5 m at low water, where it would be little affected by daily variation in air temperature. Further, the spawning period, late November to late December, corresponds to a sharp seasonal decline of approximately 4°C in local nearshore sea temperature (Bathen, 1968). Ripe animals, brought into the lab, will usually spawn within a few hours; whether spawning is, in this case, precipitated by temperature changes or just the shock of collecting is not known.

Breeding Periods. Breeding seasons of enteropneusts appear to be characteristic of individual species and offer little on which to base generalizations (see Table II). We know of two species, *Saccoglossus horsti* and *Ptychodera flava,* which, respectively, appear to correlate their reproductive seasons with spring warming and winter cooling of waters. It is thus tempting to ascribe a summer breeding season to temperate forms, a winter one to tropical species. Such an attempt is thwarted by the midwinter breeding of *Balanoglossus australiensis* in New Zealand and the summer spawning of *Ptychodera bahamensis* in the Dry Tortugas, Florida.

Table II summarizes available information on breeding seasons of Enteropneusta. Obviously, the precise limits of the reproductive periods have yet to be worked out for most species.

TABLE II
BREEDING PERIODS IN THE ENTEROPNEUSTA

Species	Locale	Months	Field data	Source
Balanoglossus clavigerus	Adriatic	May–June	Yes	Heider, 1909 Stiasny, 1913, 1914a,b
Balanoglossus australiensis	Auckland	August	No	Packard, 1968
Balanoglossus capensis	So. Africa	"Winter"	No	Gilchrist, 1923
Glossobalanus minutus	Naples	Feb.–May	Yes	Dawydoff, 1928
Ptychodera flava	Madras	Dec.–Feb.	No	Devanesan and Varadarajan, 1940
Ptychodera flava	Madras	?March	No	Rao, 1954a
Ptychodera flava	Hawaii	Nov.–Dec.	No	Hadfield, un-published
Ptychodera bahamensis	Tortugas, Fla.	July–?	No	Payne, 1937
Saccoglossus pusillus	So. Calif.	Nov.–Feb.	Yes	Davis, 1908
Saccoglossus horsti	England	May–June	Yes	Burdon-Jones, 1951
Saccoglossus kowalevskii	Woods Hole	"Summer"	No	Bateson, 1884 Colwin and Colwin, 1962a, Costello *et al.*, 1957
Saccoglossus otagoensis	New Zealand	July–Aug.	Yes	Kirk, 1938
Saccoglossus cambrensis	Wales	June–July or cont.	Yes	Burdon-Jones, 1951
Protoglossus koehleri	?Wales	?May–July	No	Burdon–Jones, 1951

7.2.3 Development

7.2.3.1 FERTILIZATION

With the possible exception of *Xenopleura vivipara*, reported by Gilchrist (1925) to be viviparous, fertilization is external in the Enteropneusta. Eggs are freed into the maternal burrows and either fertilized there by floating-swimming sperm clumps or passed into the seawater above the habitat where they unite with freely spawned sperm.

The female sex cells are spawned in the primary oocyte stage and do not complete maturation until after fertilization. The eggs of many enteropneusts are irregularly shaped at spawning, a result of their packing within the ovary, but round up quickly on contact with seawater

(Burdon-Jones, 1952). In *Saccoglossus horsti* (Burdon-Jones, 1952) the germinal vesicle disappears after egg laying, but before fertilization. The oocytes of *Saccoglossus kowalevskii* apparently begin maturation, still within the ovary, about 4 hours before spawning. However, eggs spawned in the germinal vesicle stage may still advance to the first meiotic metaphase after spawning (Colwin and Colwin, 1953). The production of two or more polar bodies after fertilization in *Ptychodera flava* (Rao, 1954b; personal observation) indicates that this species, too, sheds oocytes in the primary stage (Fig. 7B). Colwin and Colwin (1953), working on *Saccoglossus kowalevskii*, and Payne (1937), studying *Ptychodera bahamensis*, note that eggs are fertilized when the first polar spindle is in metaphase.

After spawning, enteropneust oocytes are seen to be invested with a jelly coat which is often referred to as an "outer membrane." In *Saccoglossus horsti* (Burdon-Jones, 1952) this layer is 5–10 μm thick; it is seen also around the oocyte of *Ptychodera bahamensis*, but was not measured (Payne, 1937). In *Ptychodera flava* the jelly membrane is 20 μm thick immediately after oviposition, but swells to 70 μm in thickness during development (Rao, 1954b) (Fig. 7D).

The events of fertilization have been described in minute, fine-structural detail in *Saccoglossus kowalevskii* (Colwin and Colwin, 1963a, b). The Colwins found that the acrosome, lying at the apex of the sperm, is activated when it comes into contact with the jelly coat surrounding the oocyte. Activation results in dehiscence of the plasma membrane of the sperm tip and the construction and extrusion of an elongate acrosomal tubule which penetrates the jelly (said to be two distinct layers) to contact the plasma membrane of the oocyte. The membranes of acrosomal tubule and oocyte fuse, the tube widens, and the nucleus, mitochondria, centrioles, and axoneme of the sperm move into the egg cytoplasm. The fertilization cone, seen with the light microscope, occurs at the point of fusion of the two gamete membranes.

The formation of a fertilization membrane has been reported for numerous enteropneust species. Payne (1937) noted that the fertilization membrane of *Ptychodera bahamensis*, ". . . is formed in large part from the darkly staining particles within the cytoplasm which migrate to the periphery of the egg." Indeed, Colwin and Colwin (1963b) describe the extrusion of cortical granules in the formation of the fertilization membrane: ". . . the cortical granules release their contents into the perivitelline space, the space enlarges, two layers of cortical material join the inner egg envelope or 'vitelline membrane,' transforming it into the fertilization membrane . . ." This the Colwins define as the fertilization reaction and it resembles very closely the picture which emerged

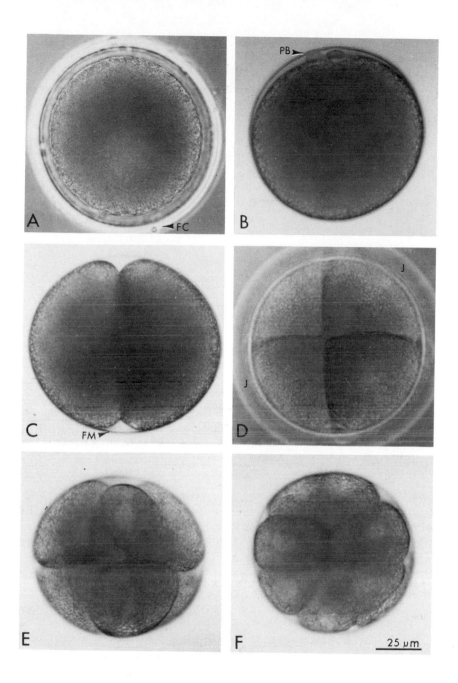

Fig. 7. The oocyte, zygote, and early cleavage in *Ptychodera flava*. (A) Oocyte teased from ovary, retaining follicle cells; (B) zygote with two polar bodies; (C) 2-cell stage; (D) 4-cell stage; (E) 8-cell stage; (F) 16-cell stage. Note complete equality of blastomeres in all stages. FC, follicle cell nuclei; FM, fertilization membrane; J, jelly coat; PB, polar bodies.

from numerous studies of fertilization in the sea urchin (Endo 1961a,b). In a few electron micrographs which I have obtained of ovarian oocytes of *Ptychodera flava* no structures resembling cortical granules are seen (see Fig. 4).

Maturation commences as soon as a sperm has successfully penetrated an oocyte. The meiotic spindle, located at the animal pole of the zygote, completes the first meiotic division, producing a polar body. The second meiotic division ensues, producing a second polar body, and division of the first often results in the presence of three polar bodies lying between the zygote surface and the fertilization membrane (see Fig. 7B).

7.2.3.2 CLEAVAGE AND EARLY EMBRYOLOGY

Cleavage has been observed in numerous species of Enteropneusta. It is holoblastic, radial, and nearly equal (see Fig. 7). *Saccoglossus horsti* (Burdon-Jones, 1952) was reported to have a slightly unequal third cleavage so that in 8-celled embryos the cells of the animal tier are slightly smaller than the vegetal four. Burdon-Jones says that the fourth cleavage in this species produces an even smaller vegetal quartette. In the 16-cell stage of *Saccoglossus horsti* an animal quartette of mesomeres caps two medial rows of four macromeres each and a vegetal quartette of micromeres. A similar inequality succeeding third cleavage was reported for *Saccoglossus kowalevskii* (Colwin and Colwin, 1953) and *Ptychodera flava* (Rao, 1954b), but I have been unable to confirm Rao's observations utilizing Hawaiian *P. flava* (see Fig. 7).

According to the Colwins, fourth cleavage in *Saccoglossus kowalevskii* results in three tiers of cells: an upper tier of 8 cells and two 4-celled tiers with the most vegetal quartette being composed of the smallest cells. The embryonic axes are laid out in the 16-cell stage. The two parallel rows of four cells at the animal pole are the presumptive dorsal and ventral sides, the ends of the rows are right and left. The animal pole of the embryo is the future anterior end, the vegetal pole the posterior end. These facts were determined by the Colwins with the use of vital stains. They report much variation in the size of homologous blastomeres in the 16-cell stage.

The times of events in early cleavage and embryology vary enormously from species to species (see Table III). First cleavage may occur as early as 1 hour after fertilization as in *Saccoglossus horsti* (Burdon-Jones, 1952) and as late as 4½ hours after in *Ptychodera flava* (Rao, 1954b). These differences are not simply reflected in the seawater temperature in which the embryos were developing since the latter species was cultured at 29°C and the former at 17°C.

TABLE III

DEVELOPMENTAL TIME SEQUENCES IN ENTEROPNEUSTA

	Saccoglossus kowalevskii[a] (21°–25°C) Fertilization to:	Saccoglossus horsti[b] (17°C) Fertilization to:	Ptychodera flava[c] (29°C) Spawning to:	Ptychodera bahamensis[d] (T not known) Fertilization to:
1st polar body	10 Minutes	30 Minutes		
2nd polar body	40–50 Minutes	50 Minutes		
1st cleavage	1¾–2½ Hours	1 Hour 5–10 minutes	4 Hours 30 minutes	1 Hour 35 minutes
2nd cleavage	2½–3¼ Hours	1 Hour 20–30 minutes	5 Hours 15 minutes	2 Hours 15 minutes
3rd cleavage	3¼–3¾ Hours	1 Hour 35–45 minutes	6 Hours	2 Hours 50 minutes
4th cleavage	3¾–4½ Hours		6½ Hours	3½ Hours
5th cleavage	4½–5 Hours		7½ Hours	4 Hours 10 minutes
6th cleavage	5 Hours			4 Hours 45 minutes
7th cleavage	5–8 Hours		9½ Hours	
Blastula	6–15 Hours		14¼ Hours	8 Hours 45 minutes
Gastrula	14–24 Hours	10+ Hours	22½ Hours	13 Hours
Blastopore closed			25 Hours	21¼ Hours
Rotation	19+ Hours	12 Hours		27 Hours
Telotroch		24 Hours		
Hatching	7th Day	30–36 Hours	42½ Hours	28 Hours
Tornaria			2½–3 Days	3 Days

[a] Colwin and Colwin, 1953.
[b] Burdon-Jones, 1952.
[c] Rao, 1954b.
[d] Payne, 1937.

There are, in the Enteropneusta, two apparently different modes of development, usually referred to as direct and indirect. These two patterns correlate neatly with taxonomic grouping, egg size, and, in some cases, with constitution of the cellular components of the ovary (see Table IV). Those species with eggs of 150 μm or less in diameter undergo indirect development via a larva called a "tornaria." Those species with eggs greater than 250 μm in diameter have direct development, with or without a planktonic stage. Dawydoff (1948) notes that yolk cells are present in the ovaries in all Ptychoderidae, Spengelidae, and in *Saccoglossus* species. The gonads of the remaining species of Harrimaniidae lack yolk cells, produce very large eggs, 400 to 1300 μm in diameter, and produce far fewer eggs at one time. It may be concluded that directness of development is thus strongly related to the quantity of yolk stored in the ova; those enteropneust species with small eggs produce the obligatorily planktotrophic larva, the tornaria (plural, tornariae).

The expression of directness of development is not truly apparent until later stages, generally after hatching. Since the early events of embryogenesis are similar in all forms these will be discussed first.

The coeloblastula of all enteropneusts undergoes gastrulation by a simple and direct invagination from the future posterior pole of the embryo (Fig. 8A). The invaginated archenteron, in most cases, comes to lie so tightly against the ectoderm as to nearly or completely obliterate the blastocoel for some time. The blastopore closes but usually a groove or indentation remains to mark the site of this closure and it is at this site that the anus later reopens. There can be little doubt that the blastopore gives rise to the anus in the Enteropneusta.

During gastrulation, external ciliation appears and by the time of blastopore closure, the embryo has begun to rotate within the enclosing membranes. The cilia soon grow longer at the anterior end of the embryo forming an apical tuft. The embryo elongates in the anteroposterior axis and a wide band of elongate cilia, the telotroch, becomes noticeable encircling the posterior end of the embryo. At this stage there may already be some external evidence of formation of the groove which separates the collar from the proboscis of the adult. In *Saccoglossus horsti* the larvae hatch at this stage (Burdon-Jones, 1952), as do those of *S. pusillus* (Davis, 1908) and *S. otagoensis* (Kirk, 1938). The embryos of *S. kowalevskii*, however, do not hatch until a much later stage of development is achieved (Colwin and Colwin, 1953). All indirectly developing species (*Balanoglossus* and *Ptychodera* species, mainly) hatch about this point.

When the external ciliation is complete and the blastopore closed, the protocoel (proboscis coelom) begins to arise by an anterior swelling

TABLE IV
DEVELOPMENTAL PATTERNS OF ENTEROPNEUSTA

Family/species	Egg diameter (μm)	Development direct/ indirect	Ovarian yolk cells	Source
Ptychoderidae				
Balanoglossus australiensis	100/160	?Dir.		Willey, 1899; Packard, 1968
Balanoglossus carnosus	150	?		Willey, 1899
Balanoglossus clavigerus		Ind.	+	Heider, 1909; Stiasny, 1913
Glossobalanus indicus		Ind.		Rao, 1955a
Glossobalanus minutus		Ind.		Dawydoff, 1928, 1948
Glossobalanus ruficollis	90	?		Willey, 1899
Glossobalanus sarniensis		Ind.		Knight–Jones, 1954
Ptychodera bahamensis		Ind.		Payne, 1937
Ptychodera flava	60/100/110	Ind.	+	Willey, 1899; Rao, 1954b; Hadfield (unpublished)
Spengelidae				
Glandiceps bengalensis		Ind.		Rao, 1955a
Glandiceps stiasnyi		Ind.		Rao, 1953
Glandiceps talaboti		?	+	van der Horst, 1932–1939
Harrimaniidae				
Harrimania kupfferi	1300 × 1000	?	−	Spengel, 1893
Protoglossus koehleri	400	?	−	Caullery and Mesnil, 1904
Saccoglossus cambrensis	350[a]	?		Brambell and Cole, 1939
Saccoglossus horsti	250	Dir.		Burdon-Jones, 1952
Saccoglossus kowalevskii	375	Dir.	+	Bateson, 1884, 1885, 1886; Colwin and Colwin, 1953
Saccoglossus otagoensis	250	Dir.	+	Kirk, 1938
Saccoglossus pusillus	264	Dir.		Davis, 1908
Stereobalanus canadensis	250	?	−	Spengel, 1893

[a] Ovarian.

of the archenteron (Fig. 8B). This lobe then gradually pinches off and becomes a separate cavity from the archenteron. The formation of the protocoel is thus indisputably, in all enteropneusts, a clearly enterocoelous formation. After separation from the archenteron, an anterior prolon-

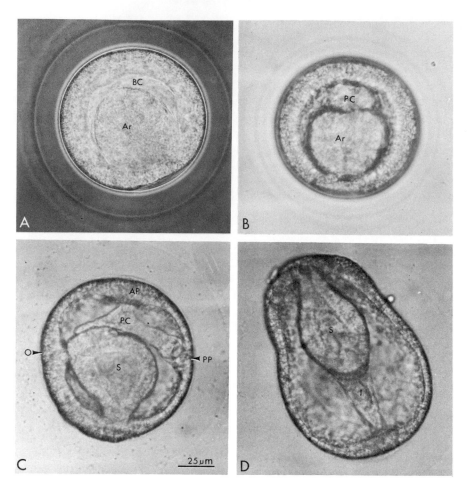

Fig. 8. Early development of *Ptychodera flava*. (A) Gastrula; (B) late gastrula with proboscis coelom; (C) stage with proboscis canal and pore formed; (D) hatching. AP, apical plate; Ar, archenteron; BC, blastocoel; I, intestine; O, oral plate; PC, proboscis coelom; PP, proboscis pore; S, stomach.

gation of the protocoel establishes a thin connection to the apical plate ectoderm, which lies beneath the apical tuft, and a second tubular evagination makes contact with the dorsal wall of the embryo. This latter tube opens to the outside forming the proboscis pore. The canal leading to the proboscis pore is clearly mesodermal in origin (see Fig. 8C).

The mode of formation of the four remaining coelomic cavities, the paired collar and trunk coeloms, varies from species to species. Van der Horst (1932–1939) summarizes some five different patterns of coelomogenesis. In *Saccoglossus pusillus*, posterolateral extensions of the protocoel give rise by dual constrictions to the collar and trunk coeloms. In *Saccoglossus kowalevskii* each coelomic compartment arises as a separate evagination of the archenteron wall (Bateson, 1884). A similar circumstance obtains in a tornaria from New England waters studied by Morgan (1891). Here, each collar and trunk sac arises independently but is first apparent as a cellular thickening of the archenteron wall which later develops a cavity. In *Balanoglossus clavigerus* the collar and trunk coeloms arise as single paired archenteral evaginations which later constrict to give an anterior collar, and a posterior trunk capsule on each side. Finally, in another paper, Morgan (1894) dealt with a large, tentaculate tornaria wherein the collar and trunk coeloms develop within masses of wandering mesenchymal cells which aggregate against the epidermis in the presumptive coelomic positions. Interestingly, not one modern paper on enteropneust development has dealt with the origin of coeloms.

From this point forward, differences in morphology most evident in external view separate quite clearly the directly developing forms from those with a tornaria larva.

7.2.3.3 ORGANOGENESIS

Since the time when various organs appear varies enormously from species to species within the enteropneusta, no discussion could adequately describe the staging of development for all species. For that reason, the following description of organogenesis will take a more classical approach by outlining the derivatives of each of the primary tissues thus far discussed: ectoderm (larval epidermis), endoderm (larval gut), and mesoderm (the walls of the coelomic vesicles and their proliferated mesenchyme).

Ectoderm gives rise not only to adult epidermis, but also the nervous system and perhaps the heart vesicle. There is present, in nearly all larval forms, an apical thickened ectodermal plate made up of tall columnar cells, underlain by nervous fibers. In many forms this area originally bears the apical ciliary tuft. In tornariae, after the tuft is lost, a small apical sensory plate with short cilia is present with a pair of eye cups on either side. These eye cups may or may not contain a lens (Stiasny, 1914a,b; Spengel, 1893; Morgan, 1891) or bristlelike projections from the cup cells. Pigment is present either in the cup cells or immediately beneath them. The apical plate becomes attached to the protocoel by mesenchymally derived muscle cells from the latter. The dorsal and ventral nerve cords arise as thickenings of ectoderm, *in situ*. The dorsal hollow nerve cord becomes separated from the epidermis in the collar either by a vertebratelike inrolling (most species) or by delamination (*Saccoglossus kowalevskii*, according to Bateson, 1885).

The heart vesicle may arise as a dorsal ectodermal invagination near the proboscis pore (Spengel, 1893; Stiasny, 1914a,b). Other authors (Bateson, 1885; Morgan, 1891), however, describe a mesodermal origin, either from the mesenchyme or from the protocoel wall, for the heart. In any case, the vesicle always comes to lie tightly against the protocoel and eventually within peritoneal folds derived from it. The glandular glomerulus, probably an excretory structure, develops from the peritoneal wall in contact with the heart vesicle.

Other ectodermal structures are, of course, the transitory larval ciliation and various glandular elaborations which consist mostly of mucus-secreting cells.

Endoderm, consisting of the larval gut which was directly derived from the archenteron, gives rise to all adult digestive structures. The larval foregut becomes the adult buccal cavity and pharyngeal region. The larval stomach elongates to contribute most of the definitive intestine and the hindgut probably persists as the terminal portion of the intestine.

The buccal diverticulum or "stomochord" arises as an outpocketing of the anteriodorsal buccal wall which penetrates the proboscis. The process of evagination varies in detail from species to species.

The gill slits arise as regional, paired, outpocketings of the gut wall in the presumptive pharyngeal region. The first gill slits to develop are the most anterior ones and successive ones are always added posterior to the last pair formed. When the endodermal outpocketings make firm contact with the ectoderm they perforate and small oval to elongate holes appear. Each hole is invaded by the downgrowth of a tongue bar from the dorsal roof of the gill slit, giving it the definitive **U**-shape. The skeletal rods which support the gill slits arise as thickenings of the basement membrane of the endodermally derived pharyngeal epithelium.

Mesoderm originally consists of the walls of the five coelomic vesicles described above and loose mesenchymal cells derived mainly from the walls of the protocoel in early development. The coelomic cavities enlarge very early to entirely eliminate the blastocoelic spaces. As well as forming the peritoneum and mesenteries, the mesoderm gives rise to the glomerulus (see above discussion of heart vesicle), the muscular and connective tissue, probably the gonads (see Section 7.2.2.2), delimits the circulatory passages, and contributes, with the epidermis, to the formation of the proboscis skeleton.

7.2.3.4 LARVAE AND METAMORPHOSIS

Direct Development. As previously stated, direct development is typical of such forms as *Saccoglossus* which have large eggs. Direct development has been best studied in S. *kowalevskii* (Bateson, 1884, 1885, 1886; Colwin and Colwin, 1953), and S. *horsti* (Burdon-Jones, 1952), and to a lesser extent in S. *pusillus* (Ritter and Davis, 1904; Davis, 1908). One would predict, on the basis of egg size, a direct development for species of *Harrimania* and *Protobalanus;* unfortunately, developmental studies are lacking for them. If *Xenopleura vivipara* is indeed viviparous as indicated by Gilchrist (1925), who reported finding embryos in a body cavity, this species must exhibit the most direct development of all.

In *Saccoglossus pusillus* and S. *horsti* hatching occurs 1 to 2 days after fertilization. The stages which escape from the egg membranes are rounded embryos with pointed anterior ends bearing an apical tuft of cilia, with a slight constriction denoting the proboscis-collar groove, and with a broad telotrochal ciliary band. Such larvae swim for only a brief period of time, during which further elongation of the larva occurs, particularly in the proboscis region. *Saccoglossus kowalevskii,* while hatching with both apical tuft and telotroch, apparently does not swim but immediately takes up a benthic existence. The juveniles of this species do not hatch until the seventh day after fertilization. *Saccoglossus otagoensis,* at hatching, lacks both apical tuft and telotroch and apparently has no planktonic stage (Kirk, 1938). Planktonic stages of all enteropneusts rely largely on the elongate telotrochal cilia for locomotion.

There is no evidence that the brief planktonic stages of directly developing enteropneusts utilize this period in feeding. It is more likely that the planktonic stage functions only in dispersal of the larvae.

An interesting aspect of the later development of *Saccoglossus kowalevskii* and S. *horsti* is the development in late larval stages of a ventral, posterior, postanal, taillike structure. This tail, or sucker, arises as a posteroventral derangement of the telotroch epidermis. It becomes quite muscular and is heavily supplied with mucous cells. Burdon-Jones (1952) reports that the tail is quite useful as, ". . . an anchor for the larva when feeding, exploring the surface of the substratum, or progressing through the burrow." The tail disappears only after considerable differentiation and elongation of the juvenile has occurred. Bateson (1885) noted that the tail is completely degenerated in S. *kowalevskii* by the time the juvenile has seven or eight pairs of gill pores.

The presence of a postanal tail in enteropneust juveniles has been seen by several authors as a point of homology with the stalk of adult pterobranchs (Hyman, 1959).

It is difficult to speak of a metamorphosis in the species discussed here. It should consist of nothing more than the disappearance of the apical tuft and telotroch. The juvenile takes the adult form simply by elongation of the entire body and *in situ* differentiation of definitive adult structures.

Indirect Development. As previously stated, the development of Ptychoderidae and Spengelidae proceeds via a typical pelagic, planktotrophic larva, the tornaria. The hatching stage of these species is generally a late, elongate gastrula which is uniformly covered with short cilia. The protocoel has already formed. A ciliary tuft develops at the anterior pole of the larva. Soon after hatching the larva begins to swell so that the blastocoel enlarges greatly and the epidermis becomes stretched quite thin (personal observation of *Ptychodera flava;* see Fig. 9). Connection of a ventral flexure of the anterior tip of the archenteron with the body wall gives rise to the mouth, and the anus opens at the point of blastopore closure (Fig. 9A). Ciliation becomes increasingly concentrated into a single elongate band which loops above the mouth, forming a preoral loop, connecting at the apical plate. The band runs posteriorly on each side and meets across the ventral surface between mouth and anus; this forms the ventral loop. A telotroch develops, encircling the anus. Such a larva is now called a "tornaria" and is achieved only after 3 to 4 weeks in *Balanoglossus clavigerus,* but in only 3 or 4 days in *Ptychodera.* Tornariae are characterized by their great transparency and the ease with which can be seen the gut, protocoel, musculatue, etc.

Development from zygote to early tornaria has been followed in *Balanoglossus clavigerus* (Heider, 1909; Stiasny, 1913, 1914a,b), *Ptychodera bahamensis* (Payne, 1937), and *P. flava* (Rao, 1954b). Only *B. clavigerus,* however, has been reared through to metamorphosis. Knowledge of later development and metamorphosis of the tornaria has been gained mainly from following these processes in larvae taken from the plankton (Morgan, 1891, 1894, for example.) This has injected a great confusion into subsequent literature since the specific identification of planktonic tornariae until recently has been difficult or impossible.

Changes with age in the tornaria include anterior inflation, circumferential constriction at the level of the mouth, and widening in the plane of the telotroch. On the inflated anterior end the ciliary bands undergo great elaboration, becoming thrown into long sinuous loops and "saddles" (Fig. 10). A series of tornaria stages has been described in the literature, with each stage bearing the name of its first describer. Table V is translated from van der Horst (1932–1939) and is included here because it seems not to have previously appeared in English. It was adapted by van der Horst (1932–1939) from the massive summaries of tornariae by Stiasny-Wijnhoff and Stiasny (1926, 1927).

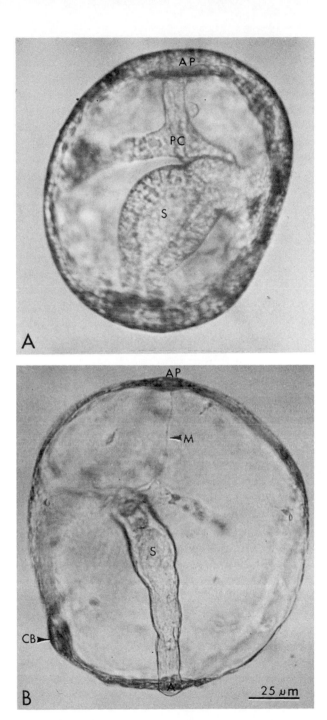

Fig. 9. Early development of *Ptychodera flava* (continued). (A) Posthatching tornaria; (B) tornaria 24 hours after hatching; note large increase in volume of blastocoel and thinning of all cell layers. A, anus; AP, apical plate; CB, tangential view of ciliated strip; M, muscle strand; PC, proboscis coelom; S, stomach.

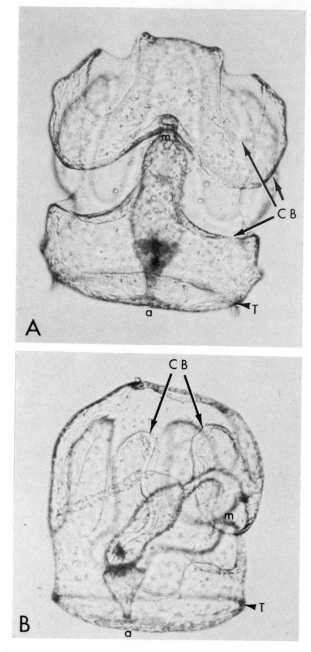

Fig. 10. Tornaria of *Ptychodera flava*, 4 months old, raised in laboratory. (A) Oral view; (B) seen from the right side. a, anus; CB, ciliated band; m, mouth; T, telotroch (about 1 mm high).

TABLE V

SYNOPTIC TABLE OF THE SUCCESSIVE DEVELOPMENTAL STAGES OF TORNARIA[a]

		Stage
1. Larva just recognizable as a tornaria, longitudinal ciliary bands formed, telotroch not yet present (Fig. 9B).	*Progressive*	Muller
2. Longitudinal ciliary bands without development of lobes and saddles; telotroch just formed.		Heider
3. Primary lobes and saddles in formation or formed. Trunk coeloms not yet or having just appeared (Fig. 10).		Metschnikoff
4. Secondary lobes and saddles or tentacles formed or forming. Trunk coelom usually present. Collar coelom not yet present or just first appearing. Large size. The high point of larval development (Fig. 11).		Krohn
5. Secondary lobes and saddles in regression or regressed. Collar and trunk coeloms present. Smaller than the previous stage, club shaped. Circular constriction about the middle of the body. Opaque.	*Regressive*	Spengel
6. Secondary lobes and saddles gone. Collar and trunk coeloms well developed. Regionalization indistinct. Entire body more elongate than in previous stage. Anal field a conical bulge. Longitudinal ciliary bands shifted toward the apical plate, the gut shifted analward. Protocoel very large.		Agassiz
7. Metamorphosis begins at the end of larval life. Proboscis, collar, and trunk regions delineated. Ciliary rows or tentacles in atrophy. Appearance of gill slits and buccal diverticulum (Fig. 12).		Metamorphosis

[a] Adapted from van der Horst (1932–1939).

The ciliary loops of the Krohn-stage in *Ptychodera* species develop small freely projecting ciliated lappets. Such larvae are referred to as *tentaculate tornariae* (Fig. 11). Richard Strathmann cultured larvae of *P. flava,* raised through early embryology in my laboratory, for more than 5 months before such a stage was achieved. This is further evidence for the apparently long planktonic life of such tornariae.

Tornariae are known from all seas and, while most do not exceed a millimeter in length, Spengel (1893) described one from Cape Verde Island which was 5–9 mm long. Bjornberg (1959) also mentioned finding "tornariae *chierchiae* I or *morgani*" in the Spengel-stage which measured 4.4–9 mm. The largest known tornariae appear to be tentaculate and thus probably are of species of *Ptychodera* (Hyman, 1959). Indeed, with the nearly complete Indo-Pacific tropical distribution of *P. flava* one might well predict that its larvae must be widely dispersed during

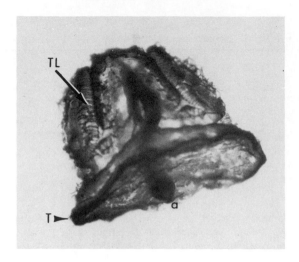

Fig. 11. Tentaculate tornaria, probably of *Ptychodera flava*, taken from Hawaiian plankton (preserved). a, anus; T, telotroch; TL, "tentacles" on ciliated bands (larva is 4 mm wide).

a long planktonic period. Bjornberg (1959), in fact, suggests that *P. flava* of the Indo-Pacific and *P. bahamensis* of the tropical Atlantic may be conspecific. In his massive study of planktonic tornariae, Bjornberg noted that Krohn- and Spengel-stage larvae are usually smaller in size in samples taken near shore than at some distance at sea. He remarks, "Large larvae captured at a greater distance from the coast seem to result from an ability of the species to prolong its larval existence, when unable to find a proper settling medium. It is probably a phenomenon of delayed metamorphosis." A long tenure in the plankton could well account for the great sizes attained by these planktotrophic larvae.

Far more species of tornariae are described in the literature than can be accounted for by presently described species of Ptychoderidae and Spengelidae. These "species" of tornariae all bear names such as "*Tornaria morgani*," which refer to their collectors. Bjornberg (1959) assayed a great number of planktonic tornariae, synonomizing many and relating them to adult worms where possible.

Metamorphosis of tornariae is not a sudden phenomenon. As can be seen in Table V, metamorphosis begins during the planktonic phase and consists, first of all, of the gradual reshaping of the preoral part of the tornaria into a proboscis. The anterior ciliation declines and disappears as the proboscis-collar groove becomes distinct. The larva is

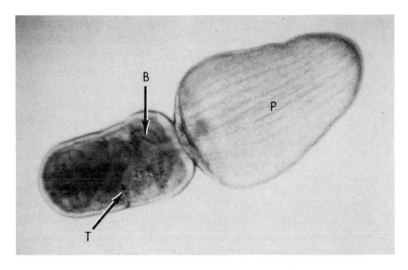

FIG. 12. Juvenile, probably of *Ptychodera flava*, achieved by metamorphosis from a tornaria collected from Hawaiian plankton. The pigmented band (T) marks the site formerly occupied by the telotroch. B, First pair of branchial slits; P, proboscis (about 2 mm long).

still motile due to the heavy ciliation of the telotroch. At last the collar-trunk region of the larva begins to elongate and the larva settles to the bottom. Larvae taken from near shore Hawaiian plankton which have a proboscis but still retain a large telotroch will complete metamorphosis within 24 hours; the telotroch disappears, the posterior collar groove appears, and the trunk elongates behind the zone of the telotroch. The line occupied by the telotroch remains obvious for at least 3 weeks after metamorphosis due to a line of darkly pigmented cells (see Fig. 12). Young worms possess all definitive adult regions. A postanal tail seems never to be present in benthic stages derived from tornariae.

Planctosphaera. Certain large and unusual planktonic larvae were described and given the name *Planctosphaera pelagica* by Spengel (1932). Van der Horst (1936) erected a class, within the Hemichordata, to embrace these larvae and called it Planctosphaeroidea. This class was accepted by Hyman (1959) and others.

Planctosphaera is known only from the tropical Atlantic (Scheltema, 1970) where its numbers are not great. These larvae are very large, often reaching a centimeter in diameter and ranging up to 22 mm. They are inflated, spherical, and mostly transparent. *Planctosphaera* possess a complete gut. Over the surface of the larva courses a very complex series of ciliated loops. The recently (1961) published, posthumous treatise on *Planctosphaera* and tornariae by D. Damas and G. Stiasny serves to clarify the proper position of the species. It is undoubtedly an enteropneust and *Planctosphaera* is its gigantic tornaria larva. Damas and Stiasny give very convincing point-for-point anatomical homologies between *Planctosphaera* and tornariae, even in the elaborate looping of the ciliary bands. The adults are unknown, but the purely oceanic distribution of the larva opens the possibility that it belongs to one of the few truly abyssal species of Enteropneusta. *Glandiceps abyssicola,* Spengel (1893), for instance was taken by the Challenger at 4500 m depth in the eastern tropical North Atlantic.

7.3 Pterobranchia

7.3.1 Asexual Reproduction

Colony formation by vegetative growth is a well-established phenomenon in two of the three recognized pterobranch genera, *Cephalodiscus* and *Rhabdopleura. Atubaria heterolopha,* the only species in the remaining genus, occurs only as naked solitary individuals and there is no evidence that this species does, or does not, proliferate asexually (Sato, 1936).

Cephalodiscus. The individual mature zooids within a coenecium of *Cephalodiscus* are not physically attached to one another. However, the evidence is good that all individuals in a single colony are asexually produced progeny of a single, sexually produced individual. The fact that both sexes may be found within a single colony seems to indicate only that sexual differentiation does not arise strictly through sexual reproduction. Ridewood (1907b) found both male and female buds arising from single individuals of a *Cephalodiscus.*

The method of asexual reproduction in *Cephalodiscus* is budding and the site is invariably near the bulbous end of the stalk or stolon which elongates posteriorly from the metasome of each adult. The number of buds which may form at any one time varies in different species, but the range seems fixed at 2 to 15 buds. The buds seen attached to the stalk of a single individual may represent several stages of development and it remains to be shown if budding is a continuous process or a short experience in the life of each zooid. Harmer (1905) has reported that, on completion of a budding phase, the adults of *C. gracilis* and *C. sibogae* may disintegrate.

Some details of budding have been reported for at least ten different species of *Cephalodiscus* (Harmer, 1905; Ridewood, 1907a,b, 1918a,b;

Schepotieff, 1908, 1909; John, 1932; and Dawydoff, 1948). In general, budding proceeds as follows:

1. From the parent stalk a bulge arises which is basically an epidermally covered evagination of the metacoel in the stalk with its associated connective and muscular tissues. The mesentery dividing the metacoel into right and left halves also extends into the bud.

2. The bud elongates and its distal end becomes swollen; the bulbous end soon takes the form of the typical pterobranch cephalic shield.

3. The coelom of the bud, derived from that of the parent, differentiates to form the tripartite condition known for all adult hemichordates: an unpaired anterior protocoel (proboscis coelom), paired mesocels (collar coeloms), and paired metacoels (trunk coeloms). Schepotieff (1908) felt that the anterior coeloms of C. dodecalophus buds were paired originally and that the main protocoel represents the left cavity, the pericardial sac the right. This interesting suggestion, also made for Rhabdopleura (see below), seems to have received no general support.

4. The gut may arise in all forms as an ectodermal derivative. Here is an exceptionally interesting point; most authors state that the gut of a budding Cephalodsicus forms as a stomodeal invagination, pushes inward within the meso- and metacoelic central mesentery, grows along a U-shaped course, and finally perforates an anus when contact is made with the epidermis in the definitive location. Thus a basically diploblastic bud becomes triploblastic by deriving endoderm from oral or stomodeal ectoderm. The endoderm of sexually derived embryos probably arises by a typical gastrular invagination (see Section 7.3.3.2).

Disagreement with the above description is found in John's (1932) discussion of budding in C. nigrescens. John maintains that a gut anlage, ". . . . a rod-shaped primitive gut," is present within the mesodermal mass in the bud. He gives no evidence concerning the source of this primordial gut tissue. A similar phenomenon has been reported to occur in the budding of Rhabdopleura.

5. The arms arise in pairs as projections of epidermis and coelom from the dorsal surface, behind the cephalic shield. This area where the arms arise eventually develops constrictions which set it off as the collar region.

6. When the above features are well developed, the bud becomes detached from the parent stalk and assumes the life style of the parent, crawling freely within the coenecium.

7. The remaining structures all develop in a manner identical to that known for normal embryology. The "stomochord" or buccal diverticulum is simply an anterior evagination of the buccal cavity into the cephalic shield. The gill slits form from evaginations of the pharyngeal wall which

contact the epidermis and perforate. The coelomoducts connecting the protocoel and mesocoels to the outside form from epidermal invaginations. The gonads are derived from the coelomic wall of the metacoels.

There is no information on factors which stimulate or mediate asexual propagation in *Cephalidiscus.*

Rhabdopleura. According to most modern authors only three valid species of recent *Rhabdopleura* have been described: *R. normani, R. annulata,* and *R. striata* (Dawydoff, 1948; Hyman, 1959). However, Stebbing (1968, 1970a,b) has given evidence for the reestablishment of *R. compacta* Hincks. This species and *R. normani* are the only two for which good information exists concerning asexual (or for that matter, sexual) reproduction and colony formation. A long series of papers discussing the biology of *R. normani* is climaxed by that of Burdon-Jones (1954). From this paper and the last of Stebbing (1970b) we take most of the following information.

Rhabdopleura, like *Cephalodiscus,* is colonial and the colonies are formed by the asexual propagation of a single, sexually produced primary zooid. Unlike *Cephalodiscus,* the tiny (average 0.5 mm long) *Rhabdopleura* zooids remain interconnected within their coenecium throughout their lives via an elongate stolon. This stolon provides the tissue which proliferates buds. The stolon is a simple tube with an outer epidermal cover and an inner coelomic cavity surrounded by a mesodermally derived peritoneum and transected longitudinally by a mesentery which is continuous with the median mesentery of the terminal zooid's metacoel.

Buds arise by simple evaginations of the stolon so that each bud originally is made up only of the components just described for the stolon. As the bud elongates, the peritoneal cells proliferate to form two cross septa, a distal one which separates the protocoel from the mesocoels and a more proximal one separating mesocoels from metacoels. According to Schepotieff (1907b), the protocoels are originally paired in *Rhabdopleura normani,* but through differential growth and morphogenetic movements, the left protocoel comes to dominate as the coelom of the cephalic shield while the right protocoel persists as the pericardial sac.

The nature and source of new buds in a colony of *Rhabdopleura* are unclear and the existing literature is sufficiently contradictory to leave these problems in a confused state. Schepotieff (1907b) notes that buds may arise in two distinct places: near the tip of a growing stolon, where buds always arise in unequal pairs, and from the "schwarzstolos," where buds arise singly. Schepotieff's written description and his figures clearly show that the production of paired buds at or near the tip of the stolon is always accompanied by the production of a new stolon bud. Thus the colony advances by what appears to be single growths of zooids at the stolon tips. The smaller zooid of the terminal pair only becomes obvious after a considerable length of time and its simultaneous appearance with the larger "terminal zooid" cannot be detected externally.

The interpretations of Schepotieff's work by Dawydoff (1948) and Hyman (1959) are at odds. Dawydoff calls the terminal zooid a "blasto-zoide inachevé" and suggests that it is arrested in its development. One assumes, from Dawydoff's brief discussion, that the terminal zooid serves no other function than the prolongation of the stolon from which true, maturing buds may arise. Hyman, on the other hand, does seem to acknowledge Schepotieff's actual description when she notes that buds arise from growing stolon ends and that ". . . such ends arise by out-growth from the base of the preceding young zooid . . ." (p. 188).

Stebbing (1970b), working with *Rhabdopleura compacta*, shows fairly clearly that new buds arise from the attachment stalk of previous zooids, become encapsulated and bud, in turn, from their stalks to produce additional zooids. He does not distinguish between terminal budding and budding along the length of the stolon. Any zooid of *R. compacta* may bud a new zooid from its base so that the colonies enlarge in all directions and such colonies are indeed "compact" with the tube of each individual lying closely applied to those around it. Lengthy stolons thus do not occur. It would be most interesting to learn, in this regard, if there are basic differences in the budding patterns of *R. compacta* and *R. normani*. Present evidence would certainly indicate that this is true; *R. compacta* buds only from the bases of preexisting zooids while *R. normani* may bud at numerous points on elongate, freely growing stolons.

Once a bud is formed in *Rhabdopleura* it secretes walls ahead of and behind itself creating an individual chamber within the growing tube. At a late stage of development the bud perforates the wall of the tube and secretes an erect open tube from which the zooid will project during feeding.

The origins of the internal structures in *Rhabdopleura* buds are apparently similar to that of *Cephalodiscus*. However, Schepotieff (1907b) describes a quite different origin for the gut. Where in *Cephalodiscus* the gut is probably entirely formed from stomodeal and proctodeal invaginations of the epidermis, in *Rhabdopleura* the gut arises as a cellular block from mesoderm in the mesenteries. Thus the mouth and oral cavity are of stomodeal origin, the intestine of proctodaeal origin, but the stomach and buccal diverticulum are from peritoneal, hence meso-dermal, tissue. Fowler (1904) said that the mid- and hindgut of buds were derived from a thin-walled tube which was contained within the parental stolonic mesentery and he supposed this tube to be of endodermal origin. This very interesting point is certainly deserving of further investigation.

In *Rhabdopleura normani* and *R. compacta* some buds do not immediately complete their differentiation but remain in a more or less dormant state. These buds may be rich in yolky material and capable of regression during times when the colony is subjected to starvation. Schepotieff (1907b) assumed that these buds never develop into normal zooids, but Stebbing (1970b) has shown that they do eventually give rise to zooids.

Stebbing (1970b) notes that few zooids in any given colony of *R. compacta* are ever sexually mature, but that both sexes may occur in a single colony.

Stebbing (1970b) is the only author to present information on seasonal variation in the state and composition of *Rhabdopleura* colonies. He notes that in November, as opposed to February and April, the number of degenerating zooids is greater, the number of empty tubes is greater, the number of dormant buds is lower, and the number of *empty* dormant bud capsules is vastly greater (50 in November, 4 in February, 5 in April). These data strongly suggest that cyclic reproductive activity does occur in the colonies of *R. compacta* and that this activity is environmentally mediated.

7.3.2 Sexual Reproduction

7.3.2.1 Sexual Dimorphism and Hermaphroditism

In *Cephalodiscus* and *Rhabdopleura* sexuality does not appear to be strongly determinate. Hermaphrodites and neuters are not uncommon and asexual propagation by budding frequently produces colonies containing both males and females. In fact, simultaneous buds from a single adults may be of either sex as demonstrated by Ridewood (1907b) in *C. nigrescens*. As one would suspect from this, sexual dimorphism is rare. In most species of *Cephalodiscus*, the sex of an individual zooid may be determined only by microscopic examination of the gonad. However, John (1931) reported that in *C. hodgsoni* the males were brown and had 10 to 11 arms and the females were crimson-brown and had 12 arms. In *C. sibogae* only males and neuters have been found (Harmer, 1905). The males are reduced, have only two arms which lack tentacles, and are without a gut. In *C. dodecalophus* only females and neuters have been seen (John, 1931). The oviduct wall contains red pigment.

In *Cephalodiscus,* when hermaphrodites occur, one gonad is a testis, the other an ovary (Johnston and Muirhead, 1951).

In *Rhabdopleura* hermaphroditism of the sort described above is rendered impossible by the presence of only a single gonad. Thus the sexes are separate, though both may be contained within a single clony (Stebbing, 1970b). Vaney and Conte (1906) suggested that successive hermaphroditism would explain the striking imbalance of numbers of males and females seen in colonies of *R. normani,* but this is disputed by Schepotieff (1907a).

7.3.2.2 ANATOMY OF THE REPRODUCTIVE SYSTEM AND ORIGIN OF GERM CELLS

The gonads of pterobranchs are situated in the anterodorsal regions of the metacoels, lying between the rectum and the pharynx. In *Cephalodiscus* and *Atubaria* the gonads are paired; in *Rhabdopleura* there is but one gonad and it lies in the right metacoel. The gonads open externally via short ducts and the gonopores are located on dorsal protuberances between the anus and the collar groove. The gonads of *Cephalodiscus* are supported by mesenteries connected to the medial dorsal metacoelic mesentery (Harmer, 1905; Schepotieff, 1907c).

Each gonad is a spherical or vase-shaped structure containing gametes. The embryological origin of the gonads has yet to be traced, but it appears likely that the situation is similar to that seen in enteropneusts. The pterobranch gonads are clearly in the same retroperitoneal position, relative to the trunk coelom, as that seen in the enteropneusts.

The structure of the gonads, too, is similar to that of enteropneusts. The sacs are clothed by coelomic epithelium under which is a vascularized basement membrane. Internally the gonad is lined by germinal epithelium which gives rise to gametes and follicle cells.

John (1932), judging from the manner of pigmentation of the gonoduct and its origin in budding, concludes that the gonoduct is mostly of ectodermal origin but with a short internal portion derived from the same tissue as the gonad. He refers to the latter germinal tissue as mesoderm. The relative length of the gonoduct varies in different species.

In *Rhabdopleura* the ovary is usually a rounded mass while the testis is elongate. In mature males of *R. normani* (Schepotieff, 1907a) the testis elongates even more and shows a median constriction. The portion closest to the gonoduct functions in sperm accumulation and is thus labeled a "seminal vesicle." The innermost portion of the testis continues to function in spermatogenesis.

Stebbing (1970b) describes an accessory structure in the trunk of mature female *Rhabdopleura compacta.* This consists of a "tapered, ring-shaped structure bearing a cap," and is located at the lower end of the metasome. Stebbing suggests that this structure functions in release of ripe, possibly already fertilized, eggs from the metacoel. Stebbing and others (Masterman, 1898; Harmer, 1905; Gilchrist, 1917) suggest that the gonoduct of pterobranchs is too narrow to allow passage of fully formed eggs and thus that release of eggs may occur through dehiscence of part of the ovary through the body wall or even that the adult zooid might entirely disintegrate to release the eggs. If any or all of these suggestions is correct, one must conclude that the "oviduct" itself can function only for sperm entry. Evidence for internal fertilization is, however, limited to a very few pterobranch species.

Usually only a few oocytes mature; the remainder disintegrate or disappear. They may thus serve as "nurse eggs" in the growth of the other oocytes. Certainly, yolk cells of the sort seen in many enteropneusts are not found in the pterobranchs. The growing oocytes are generally found at the point in the ovary farthest from the oviduct. Since these oocytes may achieve such size as to nearly occlude the metacoel, passage past the small elements of the ovary and through the oviduct seems nearly impossible.

7.3.2.3 Cytodifferentiation of the Gametes

Spermatogenesis. There is little detailed information on sperm formation in pterobranchs. In *Cephalodiscus,* where gonads are paired, the testes are large sacs lined with germinal epithelium. Extending from the germinal epithelium toward the lumen are successive stages of spermatogenesis; the mature sperm fill the lumen.

Mature sperm have been described by Ridewood for *Cephalodiscus gilchristi* (1906) and *C. densus* (1918b). The sperm has a pointed and sharply tapering head about 5 μm in length and a slender tail about 12 μm long. Ridewood (1918b) also saw sperm from *C. evansi* which he describes as resembling, but slightly smaller than, those of *C. densus.*

The two testes of *Cephalodiscus* are apparently equal in size and in state of maturity at any given time.

The single testis of *Rhabdopleura* is much more differentiated than those of *Cephalodiscus.* Here there is functional differentiation along the length of the sac. Only the innermost region of the male organ is given over to spermatogenic activity and this region is, in some species, separated into a distinct lobe by a tight constriction.

The germinal epithelium of male *Rhabdopleura normani* is a thick and infolded layer wherein the gonial cells are closely applied to the basement membrane (Schepotieff, 1907a). Successive stages of sperm development lie toward the lumen and are disposed in synchronous groups. Schepotieff noted that such groups were allied with single poorly defined nurse cells. Ripe spermatozoa lie in the lumen of the testis and are probably in the same size range as those of *Cephalodiscus*. With advancing sexual ripeness, the distal part of the male organ, the seminal receptacle, becomes distended with mature sperm.

Males of *Atubaria heterolopha* have never been reported.

Oogenesis. The ovaries of pterobranchs are simple, saclike structures hanging into the trunk coelom. They are paired in *Cephalodiscus* and *Atubaria*, but singular and lying in the right trunk coelom in *Rhabdopleura*. Ridewood (1906) noted that one ovary was always much larger than the other in mature females of *C. gilchristi*.

The ovaries are not uniformly lined by germinal epithelium. Rather, oocytes are proliferated in a rather narrow region at the distal end of the gonad, near the oviduct. This oogenic zone may run around the circumference of the ovary or be restricted to the lateral wall. The eggs are pushed posteriorly as they enlarge and may, in some species of *Cephalodiscus*, assume an elongate, tear-drop shape with the bulk of the egg lying free in the end of the ovary and the narrow end remaining attached near the oviduct.

Follicular elements are present in some, if not all, species of *Cephalodiscus* (van der Horst, 1932–1939). Andersson (1903, 1907) declared that there were basic differences between two groups of *Cephalodiscus* species, corresponding to the subgenera *Demiothecia* and *Orthoecus*. In the former, Andersson said, the eggs lie outside the ovarian epithelium, in a blood space between the germinal epithelium and the peritoneum. In *Orthoecus* species the eggs lie internal to the germinal eipthelium and are surrounded by follicle cells. Van der Horst (1932–1939) disputed this and suggested that there is probably no major difference between the two forms; the confusion arises only from the appearance of .the segregated cytoplasm of yolk-filled follicle cells in *Demiothecia* species. Van der Horst's idea is very likely correct, but a reinvestigation of this problem is needed.

Neither follicle nor yolk cells have been reported to occur in the ovaries of Rhabdopleura.

John (1932) noted that immature ovaries of *Cephalodiscus fumosus* contain five or six eggs of about equal size, while ovaries of mature animals contain but one or two. From this he assumed that some of the eggs seen in the immature ovaries must be used as food by the developing eggs. If this is so, it constitutes the only known instance of nurse eggs in the Hemichordata. One wonders if John could have mistaken yolk cells, such as are known from many Enteropneusta and from *C.*

hodgsoni, for oocytes. Alternatively, there may be a set season for reproduction in *C. fumosus* so that the animals John found with only one
or two eggs were those which had spawned the remainder before
collection.

In *Rhabdopleura* the blind end of the ovarian sac never encloses
more than one large yolky egg at a time. I have not been able to find
actual measurements of such eggs in the literature, but an extrapolation
of Dilly's (1973) scaled photographs of ovarian sections of *R. compacta* suggests that fully-grown eggs would be about 300 μm in diameter.

The number of growing oocytes to be found in the ovary at any
one time is apparently a species characteristics in *Cephalodiscus* (van
der Horst, 1932–1939). Those species with large eggs produce fewer
eggs at one time than those with smaller eggs. The eggs of *C. hodgsoni*
are 250 μm in diameter, those of *C. solidus* 450 μm in diameter. Van der
Horst quotes Ridewood as finding large encapsulated eggs (600 × 500
μm to 1000 × 600 μm) within the coenecium of *C. nigrescens* and smaller
ones (400 × 330 μm to 500 × 400 μm) in *C. gilchristi,* but it is not known
if these dimensions include the capsular walls; it would seem that they
must.

Komai (1949) illustrates two very large eggs in an ovary of *Atubaria.*

7.3.2.4 GAMETOGENIC CYCLE

The studies of Stebbing (1970b) on *Rhabdopleura compacta* provide
the only data on seasonal reproductive changes in any pterobranch species. Stebbing made field collections of *R. compacta* near Plymouth,
England in April, November, February, and April of two consecutive
years. In all of the last three collections he found embryos being brooded
within the parent tubes, but the number was somewhat greater in April.
Stebbing concludes, ". . . it appears that *R. compacta* is capable of successful sexual reproduction throughout the year."

Van der Horst and Helmcke (1956) suggest that in *Cephalodiscus*
sexual reproduction is limited to a distinct season. This is an assumption
based only on the low number of eggs produced per female at one
time, the abundant occurrence, in preserved expedition collections, of
nonreproductive animals, and the massive occurrence of active budding.

There are no data on seasonal changes in the gonads of pterobranchs.
This gap in our knowledge of this group can probably only be filled
by studies of *Rhabdopleura* from Europe where such species do occur
in sufficiently shallow depths to allow regular collection by divers ((Laubier, 1964; Stebbing, 1968). *Cephalodiscus gracilis* was originally taken
from an intertidal coral reef in Borneo (Harmer, 1905) and may offer

a good opportunity for reproductive studies in the genus when this geographical area becomes better studied.

7.3.2.5 FACTORS INFLUENCING GAMETOGENESIS

We have only Stebbing's (1970b) observations on *Rhabdopleura compacta* as an indication of the role of nutrition in reproductive cycles in the Pterobranchia. He reports that dormant buds apparently serve for nutrient storage in this species and that stored material is utilized during the winter months. Thus the occurrence in April of brooded larvae noted by Stebbing may be related to a vernal increase in food availability.

7.3.2.6 REPRODUCTIVE BEHAVIOR

No aspects of reproductive behavior have ever been observed in a pterobranch.

7.3.2.7 SPAWNING

Mechanisms. While spawning has never been observed in any pterobranch, an interesting controversy has continued in the literature for nearly 70 years. It began when Masterman (1898) observed that eggs of *Cephalodiscus dodecalophus* were so large that it is difficult to imagine them being able to pass through the tiny neck of the oviduct. Indeed, in illustration after illustration of pterobranchs in the literature, one must marvel at the immense size of full-grown eggs relative to that of the tiny zooids containing them. Masterman suggested then that the oviduct must serve only as a passage for entering sperm and that release of eggs must be accomplished by the death and disintegration of the parent. Gilchrist (1917) thought that the eggs of *C. gilchristi* might be released through the body wall by a reparable breakage. Ridewood (1918a) and van der Horst (1932–1939), however, argued against such a mechanism, pointing out that, in contradiction to Masterman, ovaries contain numerous eggs in all stages of development which probably ripen sequentially. Moreover, *Cephalodiscus* embryos are frequently encapsulated and attached by stalks to the inner wall of the parent coenecium. Such encapsulating can be seen by van der Horst as being accomplished only by the oviduct. In fact, van der Horst and Helmcke (1956) admit no other possibility.

Stebbing (1970b) adds fuel to the fire with his noteworthy observation that developing embryos of *Rhabdopleura compacta* lie in the blind ends of tubes beneath parent zooids; the parent may still be in an oogenic state. Stebbing suggests that a peculiar "cap and ring" structure seen at the posterior end of the metasome of ripe animals may serve as a trap door for the release of eggs.

If eggs of *Cephalodiscus* are released in any way other than through the oviduct, then a mechanism for encapsulation and attachment to the coenecium must still be found. For a lengthy discussion of the possible role of the oviduct and its red pigment, see van der Horst (1932–1939).

It is noteworthy that early embryonic development may proceed in two different locations in different pterobranchs. Van der Horst and Helmcke (1956) note that in *Cephalodiscus* species belonging to the subgenus *Demiothecia* (*C. dodecalophus, C. gracilis, C. sibogae, C. hodgsoni,* and *C. aequatus*) eggs are invested with a capsule and attached to the parent coenecium by a stalk; the stalk is composed of the same material as the coenecium. The eggs of the remaining *Cephalodiscus* species, like those of *Rhabdopleura* (Stebbing, 1970b) are found unstalked, lying beneath the parent in the blind end of a coenecial tube.

There is no information available on synchronization of spawning or on reproductive periods of any pterobranch other than that cited above for *Rhabdopleura compacta* (Stebbing, 1970b), which apparently reproduces year around but may have an increase in egg production in the spring.

7.3.3 Development

7.3.3.1 FERTILIZATION

Dilly's (1973) recent observation that the first cleavage stages of *Rhabdopleura compacta* are found within the maternal ovary indicates that fertilization is internal in this species. The embryo is between the 8-cell and the blastula stage when discharged. The case is not so clear for *Cephalodiscus*. Andersson (1907) saw sperm inside the ovary of *C. densus.* Masterman (1898) has considered internal fertilization to be a necessity in *C. dodecalophus,* but his reasons for assuming so were probably faulty. The fact that laid eggs of *Cephalodiscus,* found in the coenecium, are always encapsulated would seem to necessitate internal fertilization if, as most authors assume, the capsule is laid down by the oviduct.

No method of sperm transfer has been deduced for any pterobranch, or has the stage of development of the female sex cells (oocyte or mature ovum) at fertilization been determined.

7.3.3.2 CLEAVAGE AND EARLY EMBRYOLOGY

Cleaving eggs of several *Cephalodiscus* species have been seen. Although the eggs are very yolky, cleavage seems to be total and equal at least in the early stages. There is insufficient evidence to allow a clear statement on the nature of the cleavage pattern, but it is most likely radial. Gastrulation has variously been ascribed to invagination (Andersson, 1907), unipolar proliferation (Gilchrist, 1917), and delamination (Harmer, 1905). These patterns may correlate with presence or absence of a cavity in the blastula. Harmer (1905) said that the blastula of *C. gracilis* was solid, the internal cells become endoderm, the external ones ectoderm. Yolk disappears from the external cell layer and Harmer figures the presumptive endoderm only as a "yolk mass." The unipolar proliferation of endoderm described by Gilchrist (1917) for *C. gilchristi* proceeds from the posterior end of the embryo.

Schepotieff's (1909) illustrations of *Cephalodiscus indicus* show that these embryos soon become covered with cilia and possess an anterior sensory tuft of elongate cilia. The ectoderm of *Cephalodiscus* and *Rhabdopleura* gastrulae contains rodlike pigment bodies which are frequently more concentrated on one surface. Also, a ventral patch on some *Cephalodiscus* embryos thickens and becomes highly glandularized. Harmer (1905) suggested that this glandular patch is the anlage of the thickened glandular epidermis of the adult cephalic shield.

Schepotieff (1907a), Stebbing (1970b), and Dilly (1973) have described various aspects of early development in *Rhabdopleura* species. Schepotieff found a single egg, covered by a membrane and stalked to the arm of an adult *R. normani*. Stebbing observed living larvae of *R. compacta*, but provides little information on their development or internal structure. The recent publication of Dilly (1973) considerably expands our knowledge of development of *Rhabdopleura*. Dilly notes that early cleavages in *R. compacta* are equal and occur while the embryo is still within the maternal ovary. During early cleavage a primary cavity is formed which quickly becomes filed with yolk which is apparently excluded from the blastomeres during cleavage. Primary mesenchyme cells, arising from the ectoderm,. come to form a more or less continuous layer between the basement membrane of the outer cell layer and the acellular yolk mass within.

The epithelium of the *Rhabdopleura* blastula contains cells with a variety of pigment types (Dilly, 1973). The pigmented cells are excluded from several areas of the embryo, one of which is the anteroventral site of gastrular invagination. Gastrulation occurs in a manner comparable to most deuterostomes, except for the anteroventral rather than posterior location of the invagination site. The archenteron pushes posteriorly in the blastocoel, carrying with it most of the internal yolk mass which comes to lie between the tip of the archenteron and the posterior end of the embryo. Dilly suggests that the definitive mesoderm cells arise from endoderm at the tip of the archenteron. Neither the fate of these mesoderm cells nor the origin of the coelomic cavities has been followed.

The method of formation of coelomic cavities in pterobranchs remains in some doubt. The protocoel very likely arises in much the same manner as has been described in the enteropneusts: as an anteriorly constricted segment of the archenteron. The figures of Harmer (1905) strongly indicate that the mesocoels and metacoels of *Cephalodiscus levinseni* and *C. gracilis* are derived from posterior extensions of the protocoel. These extensions are solid in early stages, as is also the case in early stages of coelomogenesis of some enteropneust species. Schepotieff (1909) figures a larva of *C. indicus* in which the middle and posterior cavities seem to be arising by a posterior proliferation from the archenteron (a phenomenon also noted in some enteropneust species), and John (1932) illustrates an embryo of *C. nigrescens* in which, almost without doubt, the meso- and metacoels have arisen jointly from an anterolateral position on the archenteron.

7.3.3.3 ORGANOGENESIS

We know very little about the formation of specific internal organs in *Cephalodiscus* and nothing of the process in *Rhabdopleura*. In *Cephalodiscus* the coelomic capsules retain their primitive positions and provide the five body cavities of which the paired metacoels are by far the largest. The protocoel is the space within the cephalic shield which is the homolog of the enteropneust proboscis. According to Andersson (1907) the heart vesicle is derived from this coelomic compartment. Such an origin is in agreement with the manner in which the heart is formed in budding (see Section 7.3.1). The collar coelom exists mainly as tubular spaces in the arms of the adults. The mesentery formed from the inner walls of the trunk coeloms suspends the gut.

The archenteron apparently produces all of the adult gut. Dilly's (1973) account of the fate of the blastopore and the origin of the mouth in *Rhabdopleura compacta* leaves these matters in confusion. His figures clearly show the blastopore in an *anterior* ventral position. Further, the advancement of the growing archenteron is toward the posterior end of the embryo. Dilly states that the mouth arises in a dorsal position (Dilly, 1973; p. 83), but this seems clearly not to be the case as judged from his illustrations.

An anterior buccal diverticulum develops from an evagination of the roof of the buccal cavity and penetrates the cephalic shield in *Cephalodiscus*. Also in this genus a pair of gill pores arises, probably in identical manner to that in the Enteropneusta. The origins of nervous system, gonads, and circulatory system have not been traced, but it may be safely assumed that they arise as they do in the enteropneusts.

7.3.3.4 LARVAL STAGES AND METAMORPHOSIS

In 1909, Schepotieff described a larva which he found in the tubes of *Cephalodiscus indicus* and which he assumed to belong to that species. The larva, which greatly resembles the buguliform larva of certain ectoproct bryozoans, has been figured in every major text since the original description. Hyman (1959) notes that these larvae are so similar, ". . . that the author suspects some sort of mistake or mixup with adjacent ectoprocts." Most larvae of pterobranchs are elongate ovoid, ciliated embryos so much resembling the young of cnidarians that many early workers referred to them as "planulae."

Reviewers (van der Horst and Helmcke, 1956, for instance) have noted that larval structure is already well advanced toward that of the adult and since no marked metamorphosis occurs we could well speak of pterobranch development as direct.

In most *Cephalodiscus* species the embryos at hatching possess a posterior ectodermal invagination. This may be an eversible stalk primordium, as reported by Schepotieff (1909) for *C. indicus* and John (1932) for *C. nigrescens,* or it may be the anlage of the anus (deduced from Harmer, 1905). The older larvae of *C. nigrescens* seen by John (1932) were elongate with a straight gut, a differentiated head shield, and arm buds. The coelomic cavities were apparent and the pericardium was distinct. Metamorphosis then must consist mainly of the formation of the stalk and the bending of the trunk into its characteristic U-shape, carrying along with it the gut.

Stebbing (1970b) and Dilly (1973) have described the larva of *Rhabdopleura compacta.* It has an elongate ovoid form, is uniformly ciliated, lacks an apical ciliary tuft, and is pigmented except for some specialized areas: the ventral gastrular plate, two dorsal areas, and the posterior tip. Dilly suggests that the patch of clear cells at the posterior extremity of the larva is an adhesive plaque which functions in securing the larva to the substratum at settling. *Rhabdopleura compacta* larvae are lecithotrophic, demersal, and probably have a short free-swimming period. They attach at the posterior tip and gradually differentiate the structures of the adult. No sudden metamorphosis occurs.

7.4 Conclusions

The Enteropneusta provide locally very abundant and easily obtained material for research. The endeavors of A. L. and L. H. Colwin have already pointed the way toward a very sophisticated use of enteropneust gametes in the investigation of fertilization. The nature of the yolk cells and the method of nutrient transfer in the ovary are rich in problems deserving of modern treatment. There also awaits for some modern student a rewarding study of the internal organogenesis of enteropneust embryos, a problem untouched since the 1920's. The fascinating tornaria larvae have yet to reveal much information on the capabilities of oceanic larvae to disperse over very long distances and probably to delay their metamorphosis until suitable habitats are found.

The greatest gaps in our current knowledge of the reproduction of Hemichordata lie among the Pterobranchia. We know very little of the cytology of gametogenesis, fertilization, cleavage, and embryology of these forms. Such information would provide invaluable insights into the relationships of pterobranchs and enteropneusts, of hemichordates with other phyla, and into the very origins of the deuterostome-chordate line.

Acknowledgments

Numerous people have helped me immensely and in various ways in the completion of this chapter. Mrs. Paula Szilard of the Science and Technology Division of Hamilton Library at the University of Hawaii was incredibly diligent in helping me track down and obtain even the most obscure references. To her go my deepest thanks. Dale B. Bonar, Stephen Michael, and Michael P. Yunker have contributed many hours of their time collecting animals, preparing sections, and helping with manuscript preparation. None of the original contributions of this chapter would have been possible without their help. Some monies from a University of Hawaii Intramural Research Grant were utilized in various aspects of the *Ptychodera* project.

7.5 References

Andersson, K. A. (1903). Eine Wiederentdeckung von *Cephalodiscus*. *Zool. Anz.* 26, 368–369.

Andersson, K. A. (1907). Die Pterobranchier der Schwedischen Südpolar-Expedition 1901–1903. *Wiss. Ergeb. Schwedischen Südpolar-Exped.* 5, 1–122.

Assheton, R. (1908). A new species of *Dolichoglossus*. *Zool. Anz.* 33, 517–520.

Barrington, E. J. W. (1965). "The Biology of Hemichordata and Protochordata," 177 pp. Oliver and Boyd, London.

Bateson, W. (1884). The early stages in the development of *Balanoglossus*. *Quart. J. Microsc. Sci.* 24, 208–236.

Bateson, W. (1885). The later stages in the development of *Balanoglossus kowalevskii*, with a suggestion as to the affinities of the Enteropneusta. *Quart. J. Microsc. Sci.* 25, (supplement), 81–122.

Bateson, W. (1886). Continued account of the later stages in the development of *Balanoglossus kowalevskii*, and of the morphology of the Enteropneusta. *Quart. J. Microsc. Sci.* 26, 511–533.

Bathen, K. H. (1968). "A Descriptive Study of the Physical Oceanography of Kaneohe Bay, Oahu, Hawaii," Tech. Rept. No. 14, 353 pp. University of Hawaii, Institute of Marine Biology, Hawaii.

Beklemishev, W. N. (1969). "Principles of Comparative Anatomy of Invertebrates" Vol. 1, 490 pp., and Vol. II, 529 pp. Univ. Chicago Press, Chicago, Illinois. (English edition translated by J. M. MacLennan and edited by Z. Kabata.)

Bjornberg, T. K. S. (1959). On Enteropneusta from Brazil. Bol. Inst. Oceanogr. Sao Paulo, Brazil 10, 1–104.

Brambell, F. W. R., and Cole, H. A. (1939). Saccoglossus cambrensis sp. n., an enteropneust occurring in Wales. Proc. Zool. Soc. London B109, 211–236.

Burdon-Jones, C. (1951). Observations on the spawning behavior of Saccoglossus horsti Brambell and Goodhart, and of other Enteropneusta. J. Mar. Biol. Ass. U.K. 29, 625–638.

Burdon-Jones, C. (1952). Development and biology of the larva of Saccoglossus horsti. Phil. Trans. Roy. Soc. London B236, 553–590.

Burdon-Jones, C. (1954). The habitat and distribution of Rhabdopleura normani Allman. Univ. Bergen Årb. (Mat. Natur.) 11, 1–17.

Burdon-Jones, C., and Patil, A. M. (1960). A revision of the genus Saccoglossus in British waters. Proc. Zool. Soc. London 134, 635–645.

Caullery, M., and Mesnil, F. (1904). Contribution à l'étude des Entéropneustes (Protobalanus koehleri n. gen.). Zool. Jahrb. Abt. Anat. Ontog. Tiere 20, 227–256.

Colwin, A. L., and Colwin, L. H. (1953). The normal embryology of Saccoglossus kowalevskii. J. Morphol. 92, 401–432.

Colwin, A. L., and Colwin, L. H. (1962a). Fine structure of acrosome and early fertilization stages in Saccoglossus kowalevskii. Biol. Bull. 123, 492–493.

Colwin, L. H., and Colwin, A. L. (1962b). Induction of spawning in Saccoglossus kowalevskii at Woods Hole. Biol. Bull. 123, 493.

Colwin, A. L., and Colwin, L. H. (1963a). Role of the gamete membranes in fertilization in Saccoglossus kowalevskii. I. The acrosomal region and its changes in early stages of fertilization. J. Cell Biol. 19, 477–500.

Colwin, L. H., and Colwin, A. L. (1963b). Role of the gamete membranes in fertilization. II. Zygote formation by gamete membrane fusion. J. Cell Biol. 19, 501–518.

Colwin, L. H., and Colwin, A. L. (1967). Membrane fusion in relation to sperm-egg association. In "Fertilization" (C. B. Metz and A. Monroy, eds.), Vol. 1, pp. 295–367. Academic Press, New York.

Cori, C. I. (1902). Über das Vorkommen des Polygordius und Balanoglossus (Ptychodera) im Triester Golfe. Zool. Anz. 25, 361–365.

Costello, D. P., Davidson, M. E., Eggers, A., Fox, M. H., and Henley, C. (1957). "Methods for Obtaining and Handling Marine Eggs and Embryos," pp. 199–201. Marine Biological Laboratory, Woods Hole, Massachusetts.

Damas, D., and Stiasny, G. (1961). Les larves planctoniques d'entéropneustes. Mem. Acad. Roy. Belg. Sci. 15, 1–68.

Davis, B. M. (1908). The early life history of Dolichoglossus pusillus Ritter. Univ. Calif. Publ. Zool. 4, 187–226.

Dawydoff, C. (1902). Über die Regeneration der Eichel bei den Enteropneusten. Zool. Anz. 25, 551–556.

Dawydoff, C. (1907a). Sur la morphologie des formations cardio-péricardique des Entéropneustes. Zool. Anz. 31, 352–362.

Dawydoff, C. (1907b). Sur le développement du néphridium de la trompe chez les Entéropneustes. *Zool. Anz.* 31, 576–611.

Dawydoff, C. (1909). Beobachtungen über den Regeneration prozess bei den Enteropneusten. *Z. Wiss. Zool.* 93, 237–305.

Dawydoff, C. (1928). Quelques observations sur le développement des Entéropneustes. *C. R. Acad. Sci. Paris,* 186, 173–175.

Dawydoff, C. (1948). Embranchement des Stomocordés. In "Traité de Zoologie" (P. Grassé, ed.), Vol. XI, pp. 367–494. Masson et Cie, Paris.

De Jorge, F. B., and Petersen, J. A. (1968). Accumulation of "glycogens" in the genital region of *Balanoglossus gigas* Fr. Muller (Spengel, 1893) during reproduction. *Comp. Biochem. Physiol.* 26, 737–740.

De Jorge, F. B., and Petersen, J. A. (1970). Biochemical studies in *Glossobalanus crozierei. Comp. Biochem. Physiol.* 35, 245–249.

Devanesan, D. W., and Varadarajan, S. (1940). The occurrence of 'Tornaria larva' at Krusadai. *Curr. Sci.* 9, 375–377.

Dilly, P. N. (1973) The larva of *Rhabdopleura compacta. Mar. Biol.* 18, 69–86.

Endo, Y. (1961a). Changes in the cortical layer of sea urchin eggs at fertilization as studied with the electron microscope. I. *Clypeaster japonicus. Exp. Cell Res.* 25, 383–397.

Endo, Y. (1961b). The role of the cortical granules in the formation of the fertilization membrane in the eggs of sea urchins. II. *Exp. Cell Res.* 25, 518–528.

Fowler, G. H. (1904). Notes on *Rhabdopleura normani. Quart. J. Microsc. Sci.* 48, 23.

Gilchrist, J. (1917). On the development of the Cape *Cephalodiscus (C. gilchristi* Ridewood). *Quart. J. Microsc. Sci.* 62, 189–211.

Gilchrist, J. (1923). A form of dimorphism and asexual reproduction in *Ptychodera capensis. J. Linn. Soc. London* 35, 393–398.

Gilchrist, J. (1925). *Xenopleura vivipara,* g. et sp. n. (Enteropneusta). *Quart. J. Microsc. Sci.* 69, 555–570.

Harmer, S. F. (1905). The Pterobranchia of the Siboga Expedition, with an account of other species. *Siboga Exped. Monogr. No.* 26, 1–132.

Heider, K. (1909). Zur Entwicklung von *Balanoglossus clavigerus. Zool. Anz.* 34, 695–704.

Hill, J. P. (1894). On a new species of Enteropneusta (*Ptychodera australiensis*) from the coast of New South Wales. *Proc. Linn. Soc. N.S.W.* 10 (Ser. 2), 1–42.

Hyman, L. H. (1955). "The Invertebrates: Echinodermata," Vol. 4, 763 pp. McGraw-Hill, New York.

Hyman, L. H. (1959). "The Invertebrates: Smaller Coelomate Groups," Vol. 5, 783 pp. McGraw-Hill, New York.

Ikeda, I. (1908). On the swimming habit of a Japanese enteropneust, *Glandiceps hacksii* Marion. *Annot. Zool. Jap.* 6, 255–257.

John, C. C. (1931). Cephalodiscus. *Discovery Rept.* 3, 223–260.

John, C. C. (1932). On the development of Cephalodiscus. *Discovery Rept.* 6, 193–204.

Johnston, T., and Muirhead, N. (1951). Cephalodiscus. *Brit. Aust. N.Z. Antarct. Res. Exped. Rept., Ser.* B1 (pt. 3), 89–120.

Kirk, H. (1938). Notes on the breeding habits and early development of *Dolichoglossus otagoensis* Benham. *Trans. Proc. Roy. Soc. N.Z.* 68, 49–50.

Knight-Jones, E. W. (1954). Relations between metachronism and the direction of ciliary beat in Metazoa. *Quart. J. Microsc. Sci.* 95, 503–521.

Komai, T. (1949). Internal structure of the pterobranch *Atubaria heterolopha* with an appendix on the homology of the "notochord." *Proc. Jap. Acad.* **25**, 19–24.

Kuwano, U. (1902). On a new enteropneust from Misaki, *Balanoglossus misakiensis*. *Annot. Zool. Jap.* **4**, 77–84.

Laubier, L. (1964). Découverte de la classe Ptérobranches en Méditerranée. *C. R. Acad. Sci., Paris*, **258**, 4340–4342.

Masterman, A. F. (1898). On the further anatomy and the budding processes of *Cephalodiscus dodecalophus*. *Trans. Roy. Soc. Edinburgh* **39**, 507–527.

Morgan, T. H. (1891). The growth and development of tornaria. *J. Morphol.* **5**, 407–458.

Morgan, T. H. (1894). The development of *Balanoglossus*. *J. Morphol.* **9**, 1–86.

Packard, A. (1968). Asexual reproduction in *Balanoglossus*. *Proc. Roy. Soc. London* **B171**, 261–272.

Payne, F. (1937). Early development of *Ptychodera bahamensis*. *Pap. Tortugas Lab.* (Carnegie Inst. Wash.) **31**, 71–76.

Petersen, J. A., and Ditadi, A. S. F. (1971). Asexual reproduction in *Glossobalanus crozieri* (Ptychoderidae, Enteropneusta, Hemichordata). *Mar. Biol.* **9**, 78–85.

Rao, K. P. (1953). The development of *Glandiceps*. *J. Morphol.* **93**, 1–18.

Rao, K. P. (1954a). Bionomics of *Ptychodera flava*. *J. Madras Univ. Sec.* **B24**, 1–5.

Rao, K. P. (1954b). The early development of *Ptychodera flava*. *J. Zool. Soc. India* **6**, 145–152.

Rao, K. P. (1955a). Tornaria from Madras. *Hydrobiologia* **7**, 269–278.

Rao, K. P. (1955b). Morphogenesis during regeneration in an enteropneust. *J. Anim. Morphol. Physiol.* **1**, 1–7.

Ridewood, W. G. (1906). A new species of *Cephalodiscus* (*C. gilchristi*) from the Cape seas. *Mar. Invest. S. Afr.* **4**, 173–192.

Ridewood, W. G. (1907a). On the development of plumes in buds of *Cephalodiscus*. *Quart. J. Microsc. Sci.* **51**, 221–252.

Ridewood, W. C. (1907b). Pterobranchia; *Cephalodiscus*. *Nat. Antarct. Exped. 1901–1904 ("Discovery")* *Nat. Hist. Rept.* **2**, 1–67.

Ridewood, W. G. (1918a). Cephalodiscus. *Br. Antarct. ("Terra Nova") Exped. 1910, Nat. Hist. Rept. Zool.* **4**, 11–82.

Ridewood, W. G. (1918b). Pterobranchia. *Austr. Antarct. Exped. Sci. Rept.*, **C3**, 5–25.

Ritter, W., and Davis, B. M. (1904). Studies on the ecology, morphology, and speciology of the young of some Enteropneusta of Western North America. *Univ. California Publ. Zool.* **1**, 171–210.

Sato, T. (1936). Vorläufige Mitteilung über *Atubaria heterolopha* gen. nov., sp. nov., eines in freiem Zustand aufgefundenen Pterobranchier aus dem Stillem Ozean. *Zool. Anz.* **115**, 97–106.

Scheltema, R. S. (1970). Two new records of Planctosphaera larvae. *Mar. Biol.* **7**, 47–48.

Schepotieff, A. (1907a). Die Anatomie von *Rhabdopleura*. *Zool. Jahrb. Abt. Anat. Ontog. Tiere* **23**, 463–534.

Schepotieff, A. (1907b). Knospungsprozess und Gehäuse von *Rhabdopleura*. *Zool. Jahrb. Abt. Anat. Ontog. Tiere* **24**, 193–238.

Schepotieff, A. (1907c). Die Anatomie von *Cephalodiscus*. *Zool. Jahrb. Abt. Anat. Ontog. Tiere* **24**, 553–608.

Schepotieff, A. (1908). Knospungsprozess von *Cephalodiscus*. *Zool. Jahrb. Abt. Anat. Ontog. Tiere* **25**, 405–494.

Schepotieff, A. (1909). Die Pterobranchier des Indischen Ozeans. *Zool. Jahrb. Abt. System Okol. Geogr. Tiere* **28**, 429–448.

Spengel, J. W. (1893). Die Enteropneusten des Golfes von Neapel. *Fauna Flora Golfes von Neapel Monogr.* **18**, 758 pp.

Spengel, J. W. (1909). Pelagisches Vorkommen von Enteropneusten. *Zool. Anz.* **34**, 54–59.

Spengel, J. W. (1932). *Planctosphaera pelagica. Sci. Results "Michael Sars"* N. *Atlantic Deep-Sea Exped.* **5**, 1–27.

Stebbing, A. R. D. (1968). Discovery of *Rhabdopleura* at Plymouth. *Nature (London)* **217**, 1284.

Stebbing, A. R. D. (1970a). The status and ecology of *Rhabdopleura compacta* from Plymouth. *J. Mar. Biol. Ass. U.K.* **50**, 209–221.

Stebbing, A. R. D. (1970b). Aspects of the reproduction and life cycle of *Rhabdopleura compacta. Mar. Biol.* **5**, 205–212.

Stiasny, G. (1913). Studien über die Entwicklung von *Balanoglossus clavigerus. Zool. Anz.* **42**, 487–500.

Stiasny, G. (1914a). Studien über die Entwicklung von *Balanoglossus clavigerus*. I. Die Entwicklung der Tornaria. *Z. Wiss. Zool.* **110**, 36–74.

Stiasny, G. (1914b). Studien über die Entwicklung des *Balanoglossus clavigerus* Delle Chiaje. II. Darstellung der weiteren Entwicklung bis zur Metamorphose. *Mitt. Zool. Sta. Neapel* **22**, 255–290.

Stiasny-Wijnhoff, G., and Stiasny, G. (1926). Über Tornarian-Typen und ihre Beziehung zur Systematik der Enteropneusten. *Zool. Anz.* **68**, 159–165.

Stiasny-Wijnhoff, G., and Stiasny, G. (1927). Die Tornarian. Kritik der Beschreibungen und Vergleich sämlicher bekannter Enteropneustenlarven. *Ergeb. Fortschr. Zool.* **7**, 38–208.

Tweedel, K. S. (1961). Regeneration of the enteropneust, *Saccoglossus kowalevskii. Biol. Bull.* **120**, 118–127.

van der Horst, C. J. (1932–1939). Hemichordata. *In* "H. G. Bronn's Klassen und Ordnungen des Tier-Reichs." *4*(Abt. 4, Buch 2), 1–737.

van der Horst, C. J. (1936). Planctosphaera and tornaria. *Quart. J. Microsc. Sci.* **78**, 605–613.

van der Horst, C. J. and Helmcke, J. G. (1956). Cephalodiscidea. *In* "Handbuch der Zoologie" (J. G. Helmcke and H. v. Lengerken, eds.), Vol. 3, pp. 33–66. Gruyte, Berlin.

Vaney, A., and Conte, A. (1906). Recherches sur le *Rhabdopleura normani* Allman. *Rev. Suisse Zool.* **14**, 143–183.

Willey, A. (1899). Enteropneusta from the South Pacific, with notes on the West Indian species. "A. Willey's Zoological Results," Part III, pp. 223–234. Cambridge Univ. Press, London.

Chapter 8

CHORDATA: TUNICATA

N. J. Berrill

8.1 Introduction

Tunicates, fully qualified members of the phylum Chordata, exhibit virtually every conceivable form of sexual and asexual reproduction, a versatility matched only by the cnidarians. Both kinds of reproduction appear to have been fully exploited in relation to environmental circumstances. The habitats range from the intertidal zone to the oceanic abyss and the open sea (Table I). The immense variability seen in both sexual and asexual forms of development offers unlimited opportunity for experimental and comparative study, while the unmistakable chordate

241

TABLE I
Classification of the Tunicata[a]

Class Ascidiacea (Sedentary, marine, living free in sand or mud, or attached to rocks and other firm surfaces.)

    Order 1. Enterogona. Colony-forming or solitary ascidians. Body divided into thorax and abdomen, or undivided. Gonad unpaired and lying within or behind the loop of the intestine.

        Suborder Aplousobranchia Lahille (= Krikobranchia Seeliger)

          Families: Clavelinidae, Polyclinidae, Didemnidae

        Suborder Phlebobranchia Lahille (= Dictyobranchia Seeliger)

          Families: Cionidae, Diazonidae, Perophoridae, Corellidae, Ascidiidae

    Order 2. Pleurogona. Colony-forming or solitary ascidians. Body never divided into thorax and abdomen. Gonads in lateral mantle, usually on both sides.

        Suborder Stolidobranchia Lahille (= Ptychobranchia Seeliger)

          Families: Styelidae, Pyuridae, Molgulidae

Class Thaliacea (Pelagic, oceanic.)

    Order 1. Pyrosomida. Single genus *Pyrosoma*. Free-swimming hollow tubular colonies consisting of numerous individuals embedded in wall of gelatinous tube.

    Order 2. Doliolida. Barrel-shaped body with branchial and atrial apertures at opposite ends of body, with single row of gill slits on each side.

    Order 3. Salpida. Barrel-shaped body similar to that of doliolids but without gill slits.

Class Larvacea (Pelagic, oceanic.)

    Order 1. Copelata. Planktonic forms that become sexually mature as small tadpole-like organisms similar to the tadpole larvae of ascidians.

[a] From Berrill 1950c.

character of the tadpole larva, first recognized by Kowalesky a century ago, has made the ancestry of tunicates a highly controversial topic (Berrill, 1955; Barrington, 1965, Millar, 1966).

8.2 Asexual Reproduction

Asexual reproduction in animals appears to be an exploitation of a general capacity for regenerative growth that is common in some degree to most living organisms. As a rule the process consists of the setting apart of a small fragment of tissue, physically or physiologically separate from the parental body, capable of developing into a new whole. Such a procedure may be employed for the production of large numbers of individuals constituting a clonal population as in polychaete fragmentation, in the budding of hydras and medusae among the cnidarians, and in salps and doliolids among the tunicates. More often the process is associated with the establishment of colonies in which the individual zooids are in varying ways combined to form a true colonial organism, as in cnidarians, tunicates, and bryozoans. In tunicates the colonies may consist of numerous separate ascidiozooids more or less embedded in a common mass of tunicin, with or without actual tissue continuity between the constituents. In fact, in tunicates as a whole the various types of asexual reproduction are all primarily or primitively concerned with colony formation.

Ascidians as a class are sessile organisms, whether solitary or colonial, and colony formation and asexual reproduction are virtually synonymous. Even among the thaliaceans, all pelagic, budding is exclusively concerned with colony formation and growth in pyrosomids, while in both the salps and doliolids colonies of a sort are first formed, from which ascidiozooids are subsequently set free as independent individuals. In these last two orders there is a regular alternation between asexual and sexual generations: individual zooids produced by budding and eventually set free become sexually mature; the fertilized egg, which develops viviparously, becomes an individual with a budding stolon but lacks gonads of its own.

8.2.1 Ascidians

Ascidians are with little doubt the most primitive of the three classes of tunicates. Even though the asexual reproductive processes in this class lead only to colony formation or growth, they do represent true asexual reproduction so far as the individual constituent zooids are concerned. The subject is vast, however, both as it relates to colony formation as such and as it expresses various forms of asexual development. A few of the more basic processes are described here.

The process of budding in ascidians varies according to the general morphology of the organism. There are two main types: merosomatous forms in which the zooid is divided into an elongate thorax and a stalklike abdomen containing the looped intestine, heart, and gonads, although in some the body wall extends beyond the intestine as a postabdomen containing gonads and heart; and holosomatous forms in which the intestinal loop, heart, and gonads within the intestinal loop, lie at one side of the thorax.

In most merosomatous forms, a series of epidermal constrictions (strobilation) divide the abdomen or postabdomen into a series of short pieces capable of reconstituting a whole organism by a combination of reorganization and regeneration. The particular course taken depends

on the character of the tissues contained within a given epidermal envelope. In *Diazona, Eudistoma, Archidistoma,* and *Aplidium,* where there
is no postabdomen, strobiliation is confined to the abdomen between
the base of the thorax and the stomach (Fig. 1A). In most species
of *Polyclinum, Amaroucium, Morchellium, Sidnyum,* and *Synoicum*
strobilation affects only the postabdomen (Fig. 1B). In the first group,
therefore, a segment of intestine is included, in the second it is not.
In other species of *Polyclinum* and *Amaroucium,* strobilation includes
both the lower part of the abdomen and most of the postabdomen (Brien,
1925, 1937, 1939; Oka, 1942; Kott, 1952, 1969; Oka and Watanabe, 1961;
Nakauchi, 1966a,b,c, 1970; Freeman, 1971).

Commonly, in natural circumstances, budding of this kind is preceded
by a progressive regression of the anterior region of the body, usually
following the end of the sexual breeding season. The thorax becomes
shortened and slender, the gill structure disappears, while the general
reduction of anterior tissue is accompanied by the appearance of so-
called trophocytes, which are large phagocytic cells heavily laden with
reserves. They were originally thought to be formative cells themselves,
but Spek (1927) showed that they merely support the proliferation of
formative cells during the process of reconstitution. They migrate posteriorly. Where there is no postabdomen or a hypertrophied posterior vascular ongrowth, the migration is halted at the lower end of the abdomen
and the abdomen alone strobilates. If a postabdomen is present, the
trophocytes congest it, with reduction of the abdomen and with subsequent postabdominal strobilation. In *Clavelina,* which has no true postabdomen but has a hypertrophied posterior vascular process, the trophocytes migrate into this process and congest the ampullary branches.
In northern species such as *Clavelina lepadiformis* only such residual
masses survive the reduction process, and in early spring each congested
branch gives rise to a new zooid, initially as an epidermal evagination
enveloping a local outgrowth of the mesenchymatous septum of the vascular stolonic structure (Brien and Brien-Gavage, 1927; Berrill and Cohen,
1936). In species of warmer waters, e.g., *Clavelina picta* and *C.
phregaea,* regression may affect only the thoracic tissues and a colony
is reconstituted through seasonal regeneration rather than by a form
of total development (Berrill, 1935b). Trophocytes are not formative
cells as such but are primarily nutritive reserve cells, rich in glycopro-
teins, serving to sustain the extensive proliferation of lymphocytes and
other tissues involved in reconstitution of new zooids.

Anteroposterior regression and associated appearance and migration of
trophocytes is not, however, a necessary accompaniment of abdominal
or postabdominal strobilation. Strobilation has been observed in colonies

of *Eudistoma* and *Aplidium* without such regression, while brief exposure of colonies of *Diazona* and of oozooids of *Eudistoma* to mammalian gonadotropic hormones initiates abdominal strobilation (Levine, 1963). In *Clavelina*, pieces of the vascular stolon develop new zooids if separated from the parent zooid even in the case of juvenile and sexually breeding individuals. Gonadotropic hormones have been identified in the neural complex of ascidians and are, therefore, possibly involved (Carlisle, 1951).

Operative experiments on *Amaroucium* oozooids indicate some sort of thoracic control of the timing of the strobilating process (Freeman, 1971). Following metamorphosis the oozooid grows at a logarithmic rate for about 2 weeks, after which it ceases to grow and shortly afterward reproduces asexually by strobilating, the abdomen dividing into from five to twelve segments, each of which reconstitutes a new individual zooid. If the thorax is removed during the first week of growth, a new thorax regenerates, growth continues for about 2 weeks, and strobilation follows; but if the thorax is removed after the oozooid has grown for more than 7 days, strobilation occurs within 24 hours. Strobilation, accordingly, may be the result of stimulation from the thorax, perhaps from the neural complex, or possibly it may be triggered by the removal of an inhibiting agent, which may be the presence of the growing thorax itself.

In two families of ascidians, the Perophoridae and Styelidae (subfamily Botryllinae), belonging to different orders, the developing buds remain attached to the parental colonial system by means of stolonic connections (Fig. 1B and C). In the Botryllinae, which includes eight genera besides the botryllids proper (Berrill, 1950c), buds originate as lateral outgrowths of the mantle wall, known as palleal outgrowths, and consist primarily of an outer epidermal vesicle enclosing an internal vesicle of atrial epithelium, together with some cells of mesenchymatous origin in the space between the two epithelial tissues. The details of this type of budding in *Botryllus* and *Botrylloides* are given by Pizon (1893), Berrill (1941, 1947b), Watterson (1945), Sabbadin (1958), Izzard (1973) and in other genera, particularly *Distomus* and *Stolonica* by de Sélys-Longchamps (1917), and Berrill (1948c, 1950c); *Symplegma* was discussed by Berrill (1940), and *Metandrocarpa* by Abbott (1953) and Newberry (1965a). An additional means of budding, known as vascular budding, occurs in species of *Botrylloides* (Oka and Watanabe, 1959, 1960), and *Botryllus* (Oka and Watanabe, 1957). This form of budding serves to restore a colony when functional zooids and their palleal-type buds are no longer present and only a viable, colonial vascular system of epidermal vessels and ampullae remain. The origin of

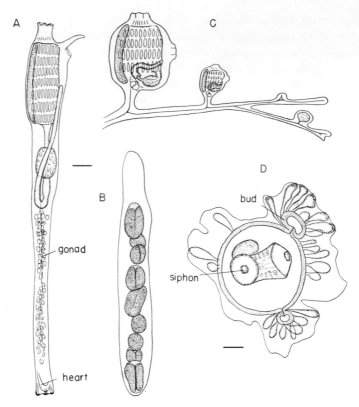

Fig. 1. Asexual reproduction in ascidians. (A) Zooid of a polyclinid colony showing gonad confined to the postabdomen which is an extension of the body between the abdomen and heart. (B) Strobilation of the postabdomen following resorption of thorax and abdomen and congestion of postabdomen by trophocytes. (C) Buds developing from stolons of *Perophora*. (D) Oozooid of a botryllid, *Distomus*, showing two buds growing as evaginations of the body wall, associated with locally intensive outgrowth of epidermal ampullae. (From Berrill, 1935b.) Scale in (A–C) is 1 mm and in (D) is 0.3 mm.

the cells constituting the inner vesicle of the vascular bud has yet to be finally determined although lymphocytes appear to be involved (Miyamoto and Freeman, 1970).

Budding in the Perophoridae, i.e., in species of *Perophora* and *Ectein-ascidia*, may be compared with the vascular budding of botryllids and also with the stolonic budding of *Clavelina*. Perophorids are holosomatous ascidians, i.e., with intestine and gonads on one side at the level of the lower part of the branchial sac. As in *Clavelina*, a hypertrophied branching vascular stolon extends from the posterior end of the body, and also as in *Clavelina* a mesenchymatous septum divides the interior of the epidermal stolon into an afferent and efferent circulatory channel. Buds arise successively at a certain distance from the tip of a growing stolon, without separation of the stolon from the parental zooid (Barth and Barth, 1966). Each bud arises as a local epidermal evagination which encloses a vesicular growth derived from the adjacent mesenchymatous septum (Brien and Brien-Gavage, 1927; Brien, 1948) and, according to Freeman (1964), lymphocytes from the circulating blood. Budding in the related genera *Euherdmania* and *Pycnoclavella* has been described by Trason (1957, 1963). For a general review, see Brien (1968).

8.2.2 Thaliaceans

Colony formation in ascidians is mainly associated with postmetamorphic growth of a multiple-zooid organism, leading to an overall state of sexual maturity and reproduction. The same holds for certain thaliacean tunicates as well, namely, the various species of *Pyrosoma*. In *Pyrosoma* each mature zooid carries a single large yolky egg, close to the cloacal cavity, which develops directly into a more or less abortive oozooid that gives rise precociously to a new, pelagic colony by strobilation of a stolon. The new organism is always a colony of ascidiozooids embedded in a common matrix and the whole so organized so as to function as a single, motile structure. The number of zooids produced through growth and strobilation of the primary stolon (i.e., of the oozooid) ranges from 4 in *Pyrosoma atlanticum*, to 40 in *P. spinosum*, to about 100 in *P. vitjasi* (Godeaux, 1963). Growth of each colony as a whole continues by stolonic budding by each new generation of constituent ascidiozooids (Berrill, 1950b, Godeaux, 1958).

If we assume that the stolonic budding characteristic of *Pyrosoma* is a derivation from an ascidian ancestry, as seems probable, then in the remaining thaliaceans, i.e., salps and doliolids, the original process of colony formation and growth by means of stolonic budding has been exploited as a truly asexual reproductive procedure giving rise to large populations of separate individuals. In both of these groups there is a regular alternation of nonsexual individuals that give rise to numerous sexual progeny by stolonic strobilation. The budding stolon in all thaliaceans forms as an epidermal (ectodermal) outgrowth from near the posterior end of the endostyle, as in perophorids. In all three thaliacean orders, represented by *Pyrosoma,* the stolon contains an endodermal tube continuous with the posterior end of the endostyle, and a genital strand derived from mesenchymal tissue; the stolon of *Salpa* is similar

(Berrill, 1950b,c; Godeaux, 1958). In *Doliolum* the stolon also receives a pair of epicardial strands, derived from two parental epicardial saclike outgrowths, and a pericardial strand; the parental atrial (or cloacal) walls contribute a pair of atrial ribbons (Godeaux, 1958). Thus the outgrowing stolon of *Doliolum* consists of seven parallel rudiments from the start, whereas in *Pyrosoma* and *Salpa* the stolon is initially more simple and the atrial components derive from the endodermal tube during stolonic extension.

In all three types the stolon grows in length primarily as a result of growth at its base, which, in effect, pushes the older stolonic tissue progressively farther from its point of origin. As it grows, a process of epidermal segmentation or strobilation occurs, whereby the stolonic tube subdivides into prospective but definitive individuals. Whereas in *Pyrosoma* these separate to become constituent zooids of a colony, in *Salpa* they separate from the end of the chain of developing zooids as free sexual individuals, each of which produces a single egg that develops within the maternal organism by means of placentation until it literally bursts the parental body wall (Fig. 2). The sexually produced individual already possesses a budding stolon, and the cycle repeats (Brien, 1928; Berrill 1950a). Strobilation and bud development in *Doliolum* is comparable to that of *Salpa,* although the buds thus formed are not immediately set free but migrate to take up a position on the dorsal or blastophore process of the parental individual, thus forming a true colony. Those buds that settle as a double row along the side of the growing blastophore remain permanently attached and serve as nutritive zooids for the colony. Wandering buds also settle on the median-dorsal line of the blastophore. These remain attached for a short time after they become functional but soon break away as free individuals, the phorozooids, each carrying a wandering bud that had settled on its stalk. This in turn elongates as a probud and gives rise to a series of small buds that remain attached to the stalk. These final, definitive buds, however, develop into the gonozooid and are eventually set free as hermaphroditic sexual individuals to start the cycle over again (Neumann, 1906; Brien, 1948; Berrill, 1950c; Godeaux, 1958).

8.3 Sexual Reproduction

8.3.1 Hermaphroditism and Sex Determination

Tunicates are typically hermaphrodites, the few exceptions being certain members of the colonial ascidian family Polycitoridae, namely, some

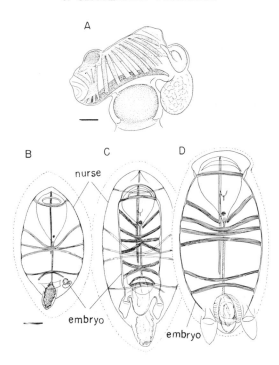

Fig. 2. The life cycle of *Salpa*. (A) Embryo with placental organ attaching it to parent, and eleoblast at right containing reserve material. (B) Sexual individual (nurse) with hermaphroditic gonad and embryo at rear end. (C) Nurse with half-grown embryo occupying most of the internal space. (D) Embryo (asexual individual) freed from parent by rupture of parental body, with curved, strobilating stolon at rear end but no gonad. Dotted lines indicate enveloping tunic. (From Berrill, 1935a.) Scale in (A) is 0.2 mm and in (B–D) is 0.5 mm.

species of *Sycozoa, Distaplia,* and *Colella,* which may be exclusively male or female. The mechanism of sex determination in these species is unknown. Self-fertility is common in hermaphroditic species even though cross-fertilization is usually effected by a variety of means.

8.3.2 Anatomy of the Reproductive System

Primitively, in ascidians and thaliaceans alike, a single hermaphroditic gonad lies within the loop of the intestine or as close to this situation as available space permits. Ascidian families with this disposition comprise the Enterogona. Dislocation of the gonad into a postabdomen in certain of these ascidians and to a thoracic location alongside the endostyle in doliolid thaliaceans is evidently due to spatial circumstances.

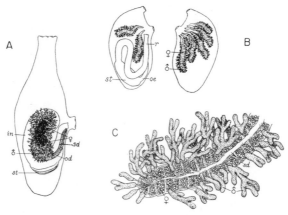

Fig. 3. Location and nature of sexual reproductive organs in enterogonid and pleurogonid ascidians. (A) Left side of *Ascidia*, an enterogonid, showing hermaphrodite gonad and ducts lying within and about the primary loop of the intestine. (B) Left and right sides of *Styela*, a pleurogonid, with gonads on both sides as multiple hermaphrodite units. (C) Enlargement of part of a hermaphrodite unit of *Styela*. in, Intestine; od, oviduct; oe, oesophagus; r, rectum; sd, sperm duct; st, stomach; ♀, ovary; ♂, testis. (After Van Name, 1945.) A and B are about one-half natural size.

Among the Pleurogona, which represent the structurally most advanced types of ascidians, the gonads are associated with the mantle wall on both sides of the body.

The gonad of most enterogonid ascidians, e.g., of *Ciona, Ascidia, Perophora*, and *Diazona*, consists of a lobulated ovary, with oogenesis proceeding within each lobule, partly surrounded by a mass of branching testicular diverticula (Fig. 3A). The number of ovarian lobes and testicular masses varies greatly, in direct proportion to the size of the parental organism. An oviduct and sperm duct pass from the ovary and testes, respectively, and extend alongside one another close to the intestine to open into the atrial chamber near the exhalant siphon.

Among the Enterogona the gonad remains essentially a single unit, if for no reason than that it lies in most forms some distance from the atrial cavity and exhalant siphon, and may require a single conducting system for the transport of its products. In the Pleurogona, one end of the gonadal system, on both sides of the body, is inevitably close to the exhalant siphon and a single conducting system no longer has any particular merit (Fig. 3B). Accordingly, there is a tendency for numerous secondary ducts, especially of the sperm duct, to open independently into the atrial cavity, which is adjacent to the gonads as a whole, allowing the massive gonad of each side to break up into smaller units more suitable for incorporation in the relatively thin but broad mantle wall. The gonad remains single on both sides of the Molgulidae, but in both the Styelidae and Pyuridae, especially on the right side, the gonad is commonly present as several elongate units of considerable size, each with its own ducts (Fig. 3C). In the styelid genus *Polycarpa* the process of subdivision is carried further and numerous small units known as polycarps, each consisting of a small ovary and peripheral testicular lobes, line the mantle wall. In the styelid subfamily Botryllinae this form of specialization reaches the extreme.

In oviparous forms both the oviduct and sperm duct extend close to the exhalant siphon, so that eggs and sperm are carried out by the strong exhalant current. Shortening of the oviduct results in delivery near the base of the atrial chamber, which may then serve as the incubator, as in species of *Clavelina* and *Dendrodoa*. With few exceptions, however, viviparity is accompanied by increase in egg size, and where individual zooid size is small the anterior part of the oviduct itself acts as a uterus and may even bulge out into a pediculate brood pouch as in *Distaplia* and *Sycozoa*. In the didemnid ascidians, which have the smallest zooids, the eggs burst directly from the ovary into the adjacent matrix of the colony, where development takes place. Only the so-called solitary ascidians, which do not reproduce asexually, grow to comparatively large individual size, and, with one exception, only these are oviparous breeders that shed eggs freely, to be fertilized and undergo development external to the parent. Of the colonial ascidians only the relatively massive *Diazona*, with correspondingly large constituent zooids, is oviparous. All others are viviparous and so are certain solitary types. Viviparity is readily attained in tunicates since the atrial chamber provides protection and other developmental conditions necessary for egg development. In many cases the chamber is used to house the later stages of development preceding the functional larval stage, while the earlier phases of development are completed within a dilated and necessarily shortened oviduct.

8.3.3 Origin of Germ Cells and Gonads

In all tunicates, the gonads originate as a single rudiment, the ovotestis, consisting of a small mass of undifferentiated cells. In *Clavelina* (Van Beneden and Julin, 1886), in *Ecteinascidia* (Simkins, 1924), and in *Corella* (Hŭus 1937) this cell mass derives from the posterior termination of the dorsal cord, which is a cellular extension from the subneural gland. In other forms, where multiple gonads are present, comparable cell masses develop in association with the mantle tissue (Tucker, 1942;

Imai, 1969). In any case there is no germ line traceable back to the egg and the gonadal rudiment may derive from dorsal cord, atrial epithelium, mesenchyme, or lymphocytic tissue according to the nature of the reproductive process. However this may be, the initial cell mass gives rise to an ovarian vesicle and a testicular vesicle. Oocytes differentiate from the ovarian epithelium and each becomes lodged within a follicle also derived from the epithelium. The lining of the testicular vesicle persists as a polystratified layer which becomes subdivided into lobules attached to the central cavity by a stalk. Each testicular lobe is limited by a flat epithelium enveloping polystratified strands of spermatogonia.

8.3.4 Gametes

The ascidian egg typically consists of the oocyte itself and a chorion, or egg membrane, surmounted by a layer of follicle cells. The perivitelline space between the chorion and the egg surface contains so-called "test cells," together with some colloidal organic material. Egg size ranges from about 100 to 180 μm diameter for eggs of oviparous species and up to about 750 μm for viviparous species. Spermatozoa, both in ascidians and thaliaceans, characteristically have an elongated nucleus and a long tail filament but appear to have a poorly developed acrosome and are seemingly without a middle piece (Franzén, 1958). Spermatogenesis has been studied in *Styela* by Tuzet and Harant (1930).

Gametogenesis in tunicates has been studied almost entirely with regard to the role of the follicle cells, with particular reference to growth of the oocyte. According to earlier workers, both the oocytes and the primary follicle cells arise from the unspecialized germinal epithelium, with the inner, test cells differentiating from primary follicle cells (Hŭus, 1937). Different conclusions had been reached by Spek (1927) and Knaben (1936), Spek employing vital staining methods in *Clavelina*, Knaben exploiting the exceptional clarity of the *Corella* egg and its accessories. They concluded that following the separation of the test cells from the primary follicle, the latter differentiates into inner and outer follicle epithelia, the outer epithelium remaining continuous with the germinal epithelium and not carried over into the ripe egg. The chorion is deposited between the test cells and the inner follicle cells as a basement membrane to which the inner follicle cells adhere.

Contrary to Hŭus, however, both Spek and Knaben derive the follicle cells and chorion from the migration and activity of ameboid mesenchymal cells which apply themselves to the surface of the egg. Tucker (1942) investigated *Styela* with this difference of opinion in mind and confirmed the earlier account. (Fig. 4). Considering the differences both

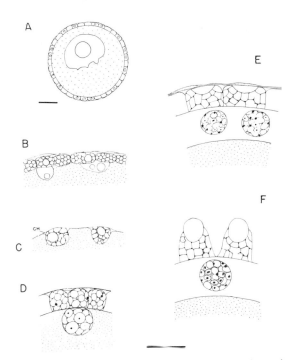

Fɪɢ. 4. Follicular cells in ascidian oogenesis. (A) Oocyte of *Styela* showing inner and outer follicle epithelia prior to test cell formation, the peripheral line representing the outer follicle epithelium which was continuous with the germinal epithelium. (B) Periphery of a full-grown ovarian egg with outer follicle epithelium intact and two test cells indenting the peripheral cytoplasm of the oocyte. (C) Two test cells indenting oocyte, showing presence of chorion (CH). (D) Test cell within oocyte cytoplasm, with follicle cells firmly attached to chorion. (E) Ripe egg at time of ovulation, with chorion and follicle cells elevated from oocyte surface, and test cells extruded into perivitelline space. (F) Oocyte following ovulation, showing elongation and vacuolation of inner follicle cells and the loss of the outer follicle epithelium. (After Tucker, 1942.) Scale in (A) is 20 μm and in (B–F) is 10 μm.

in methods of investigation and in the taxonomic position of the three genera studied, it is possible that there is some variation in the original derivation of the follicle cells. More recently, Kessel and Kemp (1962) working with *Molgula*, support the conclusion that the primary follicle cells derive from the germinal epithelium. However, since the existence of a germ line has not been demonstrated in the development of ascidian or other tunicate eggs, and since mesenchymal and other cell types in tunicates are generally totipotent, the issue seems to lack significance.

Quite apart from the question of initial derivation, follicle and test cells raise problems of their own with regard to function. In their final state the test cells lie free in the perivitelline space (e.g., *Corella*), or are held closely to the surface of the egg by a fine membrane probably of their own production (e.g., *Ascidia*). According to Mancuso (1965) many problems concerning the origin, differentiation, and function of these cells remain unsolved. At first the test cells are more or less distant from one another. Then, at the initiation of yolk deposition in oocyte cytoplasm, they form a more or less continuous layer around the oocyte. The increase in numbers appears to be the result of mitotic cell division (Mansueto, 1964).

Test cells at the surface of the oocyte have a variable form, seem to be ameboid, and have a diameter, in *Ciona*, of about 5 μm, with nuclear diameter about 4 μm. When they have penetrated the oocyte, to lie in the superficial layer of the cytoplasm, the nuclei have a rounded form. During vitellogenesis the test cells become greatly changed. Free ribosomes of the endoplasmic reticulum disappear, large vesicles appear, and test cell granules accumulate. During the later stage of oocyte growth the test cells form small groups separated from one another by lobular masses of oocyte cytoplasm which extend to the chorion. These lobes often exhibit microvilli which penetrate the chorion, and they contain numerous mitochondria and small yolk granules which lack maturation centers, while numerous pinocytotic vesicles appear immediately beneath the cytoplasmic membrane.

The formation of yolk granules in *Ciona* has at least two distinct phases. The ground substance in the earliest yolk granules is derived by the transformation of multivesicular bodies; yet in a later stage the ground substance for new granules seems to be derived by pinocytosis. Test cell granules, probably lipids (Harvey, 1927), are apparently transferred into the oocyte and used for the maturation of the yolk granules. At the same time, the presence of pinocytotic vesicles beneath the plasma membrane of the oocyte during vitellogenesis indicates a passage of material from the follicle cells to the oocyte through the chorion, as reported for *Molgula* by Kessel and Kemp (1962) and Kessel (1966a). Test cells eventually are extruded from the surface of the oocyte, to form an irregular layer of free cells between oocyte and chorion. Meanwhile the follicle cells undergo vacuolization. In *Ciona* and *Molgula* (Kessel, 1967) they at first contain secretion granules, embedded in a less electron-opaque matrix, synthesized by both Golgi elements and endoplasmic reticulum. Vacuolization of the follicle cells before oocyte release into oviduct or seawater results from the progressive dissolution of their secretory products (Fig. 5).

Following the release of eggs into seawater, irrespective of fertilization, the perivitelline space expands osmotically (Berrill, 1929). This expansion

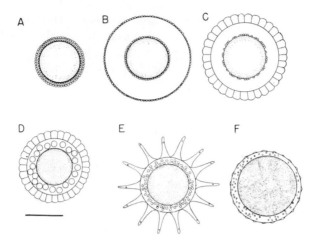

FIG. 5. Liberated eggs of various ascidians. (A) Egg of *Ascidiella scabra* at moment of liberation. (B) Same egg 30 minutes later showing great increase in perivitelline space, and investiture of ovum by layer of test cells. (C) Egg of *Ascidiella aspersa*, with large follicle cells which enable it to float. (D) Egg of *Corella parallelogramma* with large follicle and test cells. (E) Egg of *Ciona intestinalis* with greatly extended follicle cells containing refringent granule. (F) egg of ovoviviparous species, *Molgula companata*, showing stretching out of the limited number of follicle cells over the expanded chorion. All are drawn to the same scale; scale is 0.2 mm. (From Berrill, 1929.)

together with the vacuolization of the follicle cells attached to the chorion results in a greater degree of buoyancy of the egg compare with that of a naked oocyte. Accordingly, the eggs of oviparous ascidians do not settle except in completely still water, while in certan species, e.g., *Ascidiella aspersa, Corella parallelogramma,* and *Ciona intestinalis,* eggs complete with their accessory structures tend to float (Berrill, 1929; Knaben, 1936).

8.3.5 Maturation

In *Ciona* the eggs leave the ovary as soon as they are full grown and accumulate for 24 hours or longer in the oviduct (Morgan 1942; Peres, 1952, 1953). The germinal vesicle disappears and the polar spindle comes to lie at the pole of the egg, maturation being arrested at the metaphase of the first maturation division. Polar bodies are not given off until the eggs have been shed and a spermatozoon has entered the egg. As a rule, spawning takes place every 24 hours (Whittingham, 1967) or it may be delayed for several days, in which case eggs may accumulate in the oviduct in very large numbers. The same holds for species of *Ascidia, Ascidiella,* and *Phallusia* during the height of the breeding season. Sperms also tend to accumulate. Except in extreme cases (Berrill, 1929), delayed spawning is compatible with normal development.

8.3.6 Spawning

A number of ascidians are known to spawn in response to light following darkness, although this may not always be the case. *Styela partita,* in aquariums, spawns in the late afternoon (Conklin, 1905; Rose, 1939). Rose found that ripe individuals of this species spawn immediately after subjection to 12 hours of darkness followed by 11–12 hours of light. In *Ciona intestinalis* and *Molgula manhattensis,* ripe individuals usually spawn at dawn, but if kept in darkness will do so at any time shortly after exposure to light (Conklin, 1905; Berrill, 1947a). *Corella parallelogramma* also normally spawns in early morning, but dark-adapted individuals can be induced to spawn at any hour following exposure to the light of a 60-candle bulb at a distance of 25 cm for 2 minutes (Hǔus, 1939, 1941). A so-called dormant period is observed between stimulation and spawning, 11 minutes at 24°C and 17 minutes at 10.5°C, i.e., at the upper and lower temperature limits at which spawning occurs. According to Hǔus the temperature dependency of the length of the dormant period may indicate that light stimulates the production of a hormone which in turn triggers the spawning response. Light also triggers the release of larvae in viviparous species such as *Distaplia occidentalis* and *Metandrocarpa taylori* (Watanabe and Lambert, 1973).

Lambert and Brandt (1967) have investigated the light effect in *Ciona intestinalis,* and Whittingham (1967) in *Molgula* and *Ciona.* Continuous illumination prevents spawning but dark-adapted individuals can be induced to spawn within ½ hour, following a 2-minute exposure to light of sufficient intensity and return to darkness. The action spectrum for spawning suggests cytochrome *c* as a chromophore. Reese (1967) has demonstrated that the gonoducts themselves respond to light (see also Watanabe and Lambert, 1973).

Carlisle (1951) reported that ingestion of gametes, particularly through the hypophyseal duct into the lumen of the neural gland, or injection of chorionic gonadotropin provoked spawning in *Ciona intestinalis* and *Phallusia mammillata,* although only after 20 hours following treatment. Commenting on these experiments, Lambert and Brandt (1967) suggest that the effect of light and dark had not been considered and that Carlisle may have been inducing ovulation as distinct from spawning. However, inasmuch as in *Ciona* and *Phallusia* mature individuals are generally obtained with both oviduct and sperm duct fully packed with gametes, spawning must have been induced either by direct stimulation or by augmentation of ovulation thereby overdistending the oviduct and sperm duct. In other forms, however, *Corella parallelogramma* and species of *Mologula*, the ducts may be nearly empty prior to spawning and, therefore, ovulation and spawning occur closer together.

8.3.7 Breeding Season

The breeding season of ascidians depends mainly on temperature, although the location of a particular population within the latitudinal range of the species is also influential. In most recorded cases there is a typical breeding season limited to August and September, the warmest months, or to a season extending from May to October. Sabbadin (1957) records that in the Venice lagoon the breeding season for *Botryllus schlosseri*, *Ciona intestinalis*, *Molgula manhattensis*, and *Styela plicata* extends from early spring to late autumn when the lagoon temperature exceeds 10°–11°C, and that several reproductive cycles occur during one such season. Similarly, Millar (1952, 1954b, 1958) found that in *Dendrodoa grossularia*, a more northern species, gonads ripen when the sea temperature is 7°–7.5°C, and breeding begins when it is 8°–9°C, although the whole process is reduced when the temperature exceeds 15°C, and is suppressed at 20°C. However, presumably depending on the annual temperature range of the regional environment, certain species may breed throughout the year. In the English Channel, for example, *Molgula manhattensis* breeds throughout the year, as do *Ascidiella aspersa Ascidia mentula*, *Phallusia mammillata*, and probably others, depending on the size of the individuals (Berrill, 1957); also a north Pacific species, *Corella willmeriana*, which is a small viviparous relative of the oviparous *C. parallelogramma*, breeds all year round (Lambert, 1968), although individuals commence breeding when about 3 months old and have a life span of only 5 months. Continuous breeding, in populations, is reported to prevail in tropical waters (Goodbody, 1961, 1963). Breeding in colonial ascidians relates both to the individual zooid and to the colony as a whole. Each constituent zooid typically reproduces sexually but once, although individual production of eggs may extend through weeks or months where zooids are large, as in *Diazona*, *Ecteinascidia*, and *Clavelina* species. In any case it is limited to the life span of the zooid. A colony, however, may live for 1 or 2 years, and form sexually mature zooids in season (Kott, 1952, 1969; Haven, 1971).

8.4 Development

8.4.1 Fertilization

Tunicates, as hermaphroditic organisms, exhibit various degrees of self-sterility or self-fertility. From a general study of the genetic and physiological problems of self-sterility in *Ciona intestinalis*, T. H. Morgan (1942, 1945) concluded that the block to self-fertilizaton in this species lies in the egg membrane (chorion) and that spermatozoa fail to penetrate the membrane of eggs from the same individual. The chorion with its associated follicle cells may be removed from eggs of *Ciona*, and even more readily from those of species of *Ascidia, Ascidiella, Phallusia,* and *Corella,* by means of trypsin or by diluted crustacean stomach juice in seawater, thus yielding naked, fertilizable eggs for experimental studies (Berrill, 1932).

Successful interspecific fertilization between *Ciona intestinalis* eggs and *Phallusia mammillata* sperms, after removal of the egg membranes (Reverberi, 1937) support the conclusion that in *Ciona* only the membranes oppose self-fertilization. Minganti (1948), employing needles rather than enzymes, so as not to modify the egg cortex chemically, found that interspecific crosses between species of *Ascidia, Ascidiella, Phallusia,* and *Ciona* resulted in development in a high percentage of cases which varied with the particular combination, in the absence of membranes. In no case did interspecific fertilization succeed with eggs still inside their membranes, in spite of the fact that in certain species (*Phallusia mammillata, Ascidia mentula, Ascidiella aspersa*), unlike *Ciona,* self-fertilization is a normal procedure. All the species mentioned above are oviparous and shed large numbers of small eggs which are fertilized outside the body.

Among colonial ascidians, all of which with the exception of *Diazona* species are viviparous and produce small numbers of relatively large yolky eggs, only *Botryllus schlosseri* has been investigated with regard to fertilization. Both self- and cross-fertilization occur (Sabbadin, 1955, 1958, 1971), resulting in great genetic heterogeneity among individual colonies, particularly with regard to the striking pigmentation patterns (Watterson, 1945).

Individual zooids within a *Botryllus* colony are typically hermaphroditic, but, because ripening of the sperms is somewhat delayed relative to ripening of the eggs during each generation of developing buds, self-fertilization does not occur in the presence of spermatozoa from other colonies (Sabbadin and Graziana, 1967). In this genus, zooids of one generation are resorbed and replaced by those of the next in fairly well synchronized sexual and budding cycles. Eggs leave the egg pouches on the sides of the new zooids and enter the atrial cavity. A few hours later the germinal vesicle ruptures and only then is fertilization possible. Isolation of a colony shortly before the new siphons open usually prevents cross-fertilization from other colonies and allows self-fertilization to occur about a couple of days later when the new testes have become mature. However, eggs of *Botryllus* have been removed and fertilized externally by Milkman and Borgmann (1963). Mature eggs are placed in a syracuse dish of seawater and testes are added; eggs are swirled to the center of the dish and the testes crushed, so that the eggs lie at the center of a thick layer of sperms. Such eggs and early embryos may be raised simply by placing them on a piece of filter paper in a vessel such as a finger bowl containing either Instant Ocean or filtered seawater (Milkman, 1967). Other ascidians, such as species of *Ciona, Molgula, Phallusia,* and *Pyura,* require far fewer sperms, usually only enough to cause a faint milkiness in the water (Berrill, 1950c; Millar, 1951; Costello *et al.,* 1957). In all cases, all surplus spermatozoa must be eliminated from the system for normal development to take place, since any subsequent bacterial contamination causes abnormal development.

8 4.2 Embryonic Development

The development of the ascidian egg has long been noted for its determinative (mosaic) character (Fig. 6A). Conklin (1905) gave a complete and accurate account of embryonic development in *Styela* and *Ciona* in a classical monograph that laid the foundation not only for the nature of ascidian development but for cell lineage studies in general. This early account has been substantially confirmed by Tung (1934) employing vital staining procedures and following the development of living material, and more recently in a series of papers by Reverberi (1961), Minganti (1948), Mancuso (1964, 1965), and Ortolani (1958).

In *Styela partita* Conklin was able to follow ooplamic rearrangement in living eggs, following fertilization, by virtue of natural coloration of lipid inclusions (mitochondria). With the rupture of the germinal vesicle, the ensuing mixture of nuclear sap and cytoplasm flows across the equator of the egg, and at the time of polar body formation a yellow crescent appears in the equatorial zone and extends halfway around the egg. In addition, two other materials were recognized, a slate-gray substance which goes into the endoderm of the embryo, and a transparent cytoplasm which becomes the ectoderm. The yellow crescent material later becomes segregated within the muscle tissue of

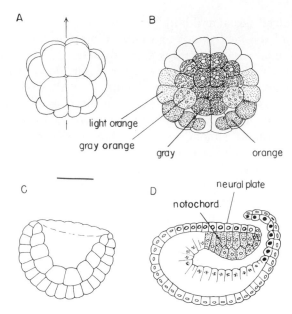

FIG. 6. Development of the ascidian egg. (A) 16-cell stage, showing bilateral symmetry of cleavage pattern, with arrow indicating the future embryonic axis. (B) Mid gastrular stage of *Boltenia echinata,* from blastoporal side, showing natural coloration of blastomeres, the orange crescent representing the future tail muscle tissue, the light orange the prospective mesenchyme of the trunk, the gray, yolky cells being prospective notochord and endoderm, and the gray orange being of uncertain destiny. (C) Side view of typical gastrula. (D) Optical longitudinal section of neurula, with neuropore and temporary neurenteric canal. (From Berrill, 1929.) Scale is 50 μm.

the larval tail, as it does also in *Styela coriacea* (Millar, 1963). A similar but deep orange-colored crescent (Fig. 6B) forms the prospective mesoderm in *Boltenia echinata* (Berrill, 1929, 1948d) and in *Pyura squamulosa* (Millar, 1951). The ultrastructure of such ooplasms has been studied by Kessel (1966b).

The precocious segregation of the egg cytoplasm is reflected in the precise cleavage patterns which follow and which are determinative in the extreme. In all ascidians studied, the first cleavage divides the egg into prospective right and left half embryos, i.e., the plane of cleavage coincides with the future embryonic axis. Gastrulation commences in every case after the sixth cleavage and is completed before the seventh, when about 74 cells are present (Fig. 6C). In *Styela,* according to Conklin, at the 64-cell stage, the animal (ventral) half consists of 26 ectoderm and 6 neural plate cells; the vegetal (dorsal) half consist of 4 neural plate, 4 chordal, 10 mesenchymal, 4 muscle, and 10 endodermal cells.

In later development the cellular constitution is equally rigid in that the notochord consists finally of about 42 cells; tail muscle, 38—40 cells; otolith, one cell; and the ocellus with three unicellular lenses and a single pigment cell. the neural plate invaginates to form a neural tube as in all other chordates (Fig. 6D). The presumptive tissues or organ-forming regions, therefore, relate primarily to the organization and cellular constitution of the tadpole larva, rather than to the postmetamorphic functioning juvenile ascidian. Accordingly, the segregative process in the ascidian egg has lent itself to extensive investigation cytologically and experimentally (Cowden, 1961).

Employing the glass-clear eggs of *Phallusia mammillata*, which nevertheless reveal a faint yellow crescent after fertilization, Reverberi (1961) followed the fate of the mitochondria by means of vital staining with Janus green and found that they become segregated first into a typical crescent and subsequently through cleavage and gastrulation become mainly confined to the differentiating muscle tissue of the developing tail of the tadpole larva.

The early view of the ascidian egg as a strict mosaic has undergone considerable modification. Twin larvae have been obtained from eggs of *Ascidia malaca* that were divided before fertilization into equal halves by equatorial, meridional, or oblique sections (Reverberi and Ortolani, 1962). On the other hand, Von Ubisch (1938) obtained single tadpoles from pairs of embryos fused at the 2-cell stage, if the morphogenetic movements were congruent. Further, if two ascidian eggs at the 8-cell stage are pressed together so as to bring their homologous territories into contact to form a topographical continuum, the combination can give rise to a single gigantic, well-proportioned tadpole, although only if there has been a single gastrulation (Reverberi and Gorgone, 1962).

At the 8-cell stage it is possible to isolate blastomeres so that whole presumptive territories are segregated. As a basis for such isolation experiments, Ortolani (1958) reinvestigated the presumptive territory map by marking the egg surface with colored chalk granules, confirming in the main that isolated blastomeres (with one exception) differentiate as they would if they had been left in place. Correlated histochemical studies and fragmentation experiments (Materazzi and Ortolani, 1969) support this conclusion and offer promising leads into molecular analysis of cytodifferentiation in general.

Comparative studies on the effects of protein synthesis inhibitors during the development of regulative (sea urchin) and determinate (ascidian) eggs, made by Markert and Cowden (1965), demonstrate greatly different temporal patterns. Actinomycin D, puromycin, and analogs of amino acids and nucleic acid constituents were tested in cultures of the two types of embryos. Whereas synthesis of ribosomes and proteins occur gradually and at relatively early stages in sea urchin (*Lytechinus*) embros, the principal synthesis of both ribosomes and proteins occurs abruptly at the onset of metamorphosis in *Ascidia nigra*. This is in keeping with the conclusion based on centrifugation, isolation, and recombination experiments with ascidian blastomeres, namely, that determination is mainly complete by the 4- to 8-cell stage, and that this precocious determination is probably mediated by selective segregation of cytoplasmic constituents. The course of development of the tadpole larva is accordingly set by properties of the egg immediately or shortly following fertilization, and the upsurge of ribosomal and protein synthesis at the onset of metamorphosis is associated with the acceleration of development of the permanent organization and the correlated destruction of the precocious larval organization.

Gastrulation, by invagination, is begun and completed in ascidians in the interval between the sixth and seventh cleavages. Preceding this event the mesodermal crescent material, whether yellow, orange, or colorless, exists as a crescent of cells symmetrically placed around the future posterior side of the larval organism. The cells which comprise the chordaneural material are also symmetrically placed about the embryonic axis, but on the anterior side. During gastrulation the two arms of the chordal cells are brought together in the middle line, and fit as a short column, four cells across and four cells in length, between the two arms of the posterior mesodermal crescent. The central core of chordal cells gives rise to the notochord; the two arms of the mesodermal crescent give rise to the lateral bands of tail muscle tissue. The cells constituting the chordal group at the time of gastrulation undergo no more than two or three divisions, so that a notochord is formed which consists of 40 to 44 cells. The process of determination in this respect appears to be rigid, for in all tadpole larvae of ascidians, and even in the embryo of the thaliacean *Doliolum*, the cellular constitution of the notochord remains the same irrespective of egg size or yolk content. In the larvacean tunicate *Oikopleura*, and without doubt in other genera, the determinative process as a whole is relatively precocious compared to cleavage. Gastrulation, by invagination, begins and ends one cleavage sooner, and there is one division less in the presumptive regions giving rise to notochord and tail muscle, i.e., the notochord consists of about 20 cells and the lateral muscle on each side consists of only 9 cells. In this type of tunicate the tail is retained throughout life, the tail cells become giant but do not change in number, and the tadpole stage as such becomes sexually mature.

The one presumptive territory of the ascidian tadpole not yet determined at an early stage is that the nervous system. Rose (1939) showed that the differentiation of the presumptive neural territory requires the induction action of the anterior vegetal blastomeres. Isolation and recombination experiments by Ortolani (1958) demonstrate that neural differentiation actually depends on induction, and not on the critical size of the differentiating territory, subsequent to the 64-cell stage. The presumptive neural ectoderm is converted into the neural system only under the inductive capacity of the immediately underlying chordal and endodermic cells, as a consequence of contact between the correlated territories. According to Reverberi (1961) the formation of the neural system in ascidians is strictly directed in the same way as in amphibians, the main difference being that in ascidians the inductor is chorda-endoderm rather than chorda-mesoderm.

As already stated, gastrulation in ascidians begins and ends during a precise, short period during early cleavage, and is typically invaginative even in the case of relatively large eggs, except where eggs a densely packed with yolk, when epiboly is evident to some degree. The gastrulative movement brings previously separated presumptive territories into juxtaposition, as originally described by Conklin. Following gastrulation the neurula forms by typical invagination of the neural plate. In fact, the living neurula stage of many ascidians, such as *Phallusia, Corella,* or *Clavelina,* remarkably resembles, in miniature, that of an amphibian.

The inclusive process of tail formation consists of: (1) localized histogenetic ooplasmic regions of the egg prior to the onset of cleavage. (2) spatial convergence during gastrulation of the previously separated presumptive muscle and notochordal territories, (3) rapid histodifferentiation of muscle and notochordal regions, with consequent cessation of cell division preceding final differentiation, so that these tissues accordingly consist of small numbers of relatively large cells, (4) induction of the neural plate, followed by typical neurulation to form a neural tube with an anterior sensory vesicle, (5) outgrowth of the tail consequent upon the activity of the notochordal cells, and (6) final differentiation of tail muscle bands.

The actual force which causes the outgrowth of the tail derives from the swelling of the individual chordal cells. This commences after the completion of gastrulation and only after the full complement of chordal cells are present, and is the result of vacuolation; a single fluid-containing vacuole appears in each cell shortly after the final division and progressively enlarges until the cell is several times its original size. The enlarging chordal cells progressively interdigitate, sliding among themselves, so that a single row of forty or so replace the original mass. The extension of the tail is as much the result of this rearrangement as it is of individual cell expansion directly, although the latter is in great part the cause of the other (Scott, 1945; Berrill, 1955).

At the close of gastrulation a neural groove appears and the lateral ectodermal margins of the neural plate approach the middorsal line and fuse in typical manner to form the neural tube. An anterior neuropore forms, which shifts progressively anteriorly until the wider anterior region of the neural plate becomes completely closed. During the stage in which the tail is extending rapidly, the neural tube consists of a wide vesicle anteriorly and an extremely narrow tube overlying the notochord and extending the length of the tail. An otolith and an ocellus differentiate from the wall of the anterior part of the vesicle, while the posterior part of the vesicle becomes the operative ganglion of the tadpole larva although destined later to give rise to the neural complex (neural ganglion, subneural gland, and dorsal cord) of the adult organism. Finally, three adhesive papillae form from the anterior ectoderm of the larval trunk, and are connected to the larval cerebral ganglion by fine nerves (Grave, 1921; Kasas, 1940; Hirai, 1941, 1969; Scott, 1945; Sebastian, 1958).

The sensory organs of the ascidian tadpole have been described in detail for *Amaroucium, Molgula, Perophora, Styela, Ascidia,* and *Botryllus* by Grave (1921, 1926, 1932, 1944), for *Ciona* with electron micrographs by Dilly (1962), and for *Styela* and *Ciona* by Whittaker (1966, 1973), using phase contrast and various biochemical experiments concerning the development of the pigment granules associated with the otolith and ocellus. In all forms, the otolith consists of a single cell attached to the wall of the sensory vesicle by a fine protoplasmic connection. The ocellus, typically, as in *Ciona* and most other ascidians, consists of three clear lens bodies which are probably single cells, a number of neurosensory (retinal) cells, and a single large unicellular pigmented structure with the shape of a flat cup (Fig. 7A). In *Styela,* however, the ocellus is reduced to a single small melanic granule surrounded by a thin layer of optically clear material (Grave, 1944), located in the brain wall. In all tadpole larvae reported for species of *Molgula* and other members of the Molgulidae, an otolith alone is present. Since this family is highly specialized, with no evidence of primitive characters, the ocellus has presumably been lost. Among the Styelidae, the tadpole larvae of species of *Polycarpa* and *Dendrodoa* (*Styelopsis*) also possess an otolith but no ocellus, but in the styelid subfamily Botryllinae, which includes all the colony-forming styelids, the pigmented mass of the otolith develops an invagination into which project neurosensory cells from the larval brain at the rear of the sensory vesicle, to form a "photolith"

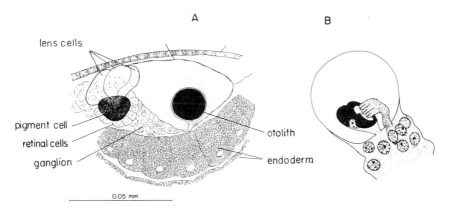

FIG. 7. Sensory vesicles of ascidian larvae. (A) Typical structure (of *Boltenia echinata*), with unicellular otolith and an ocellus consisting of three lens cells, a cup-shaped pigment cell, and a number of neurosensory, or retinal, cells adjoining the neutral ganglion. (From Berrill, 1948d.) (B) The photolith, characteristic of all botryllids, with unicellular otolith hollowed out to receive terminals of several neurosensory cells. (After Grave, 1932.)

(Fig. 7B) sensitive to both light and gravity (Grave and Riley, 1935; Berrill, 1949; Abbott, 1955).

8.4.3 Hatching and Tadpole Activity

In oviparous species, typically with many non-yolky small eggs and tadpole larvae, hatching is effected by digestion of the chorion and consequent collapse of the follicular envelope by a proteolytic enzyme, apparently produced by the embryo. In *Phallusia mammillata*, at least, unfertilized eggs mixed in a dense culture of developing embryos become hatched at the same time as the tadpole larvae, and such eggs will do so in water collected from a culture (Berrill, 1929; Wallace, 1961). Crab stomach juice diluted in sea water is also very effective (Berrill, 1932). Nothing is known concerning the specific source of the hatching enzyme. In this type of development, hatching coincides with the attainment of functional maturity by the tadpole larva and, as a rule, the last stages of hatching are effected by the final straightening of the tail as the chordal cells acquire their full volume. When fully developed the tadpole tail is also equipped with a noncellular, cuticular fin extending the length of the tail on two sides and ending as a blade. In all oviparous species, all of which have small eggs (diameter 100–150 μm), the fin is vertical, in the same plane as the vertical longitudinal axis of the trunk. This is made possible by the existence of considerable perivitelline space and the coiling of the tail, during late development, around the trunk more or less at right angles to its original direction of outgrowth. In viviparous species of the Styelidae and other pleurogonids the same condition prevails, e.g., in *Dendrodoa, Stolonica, Distomus, Metandrocarpa, Symplegma, Botryllus,* all with relatively large, yolky eggs, with diameters from 420 to 720 μm (Berrill, 1935a, 1950c); the tail develops as before, partly curled around the trunk in such a way that the fin is vertical.

In all the viviparous enterogonid species, i.e., all colonial forms except *Diazona,* the developing tail remains in its original axis and encircles the trunk in the meridion. As the tail fin forms, especially in the virtual absence of perivitelline space, it lies flat against the embryonic trunk and to do so the tail itself is inevitably twisted through 90 degrees at its base. On hatching, this twist is retained and the tadpole larvae of the Clavelinidae, Polyclinidae, Didemnidae, and Perophoridae have tail fins and an oscillating contraction that is horizontal rather than vertical and from side to side.

All ascidians, solitary or colonial, are sessile. Oviparous development ensures extensive dispersal, as in most solitary species. Tadpole larvae, depending on the type, may or may not significantly contribute further to dispersal and are primarily instruments for site selection for settlement. They can be grouped in three categories: those developing from the small eggs of oviparous species (Fig. 8A), that have developed during one to several days according to the temperature, drifting in the sea, with a subsequent free-swimming period of 12–36 hours; those developing from much larger eggs of pleurogonid viviparous species but with a comparably long free-swimming period; and those that develop from larger and often very yolky eggs of viviparous enterogonid species, together with those of *Botryllus,* all of which have a much abbreviated free- swimming period lasting from a few minutes to no more than 2–3 hours. These differences are correlated with habitat. In the first category the various solitary species, together with the colonial *Diazona,* are extremely widespread on the floor of shallow seas, yet they require a hard substrate such as rock, embedded stones, or heavy shells as a base for settlement and growth. Away from the intertidal or subtidal zone such sites are at a premium.

8.4.4 Settling of Larvae

Whatever the duration of the free-swimming period of the tadpole larvae, the terminal reaction is the vital one. A tadpole, descending

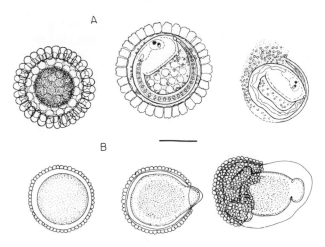

Fɪɢ. 8. Development and hatching. (A) Egg of *Ascidia aspersa* with follicle epithelium and chorion, fully developed tadpole within chorion and follicular layer, and tadpole hatching by means of proteolytic digestion of chorion (B) Egg of *Molgula retortiformis,* a species with direct development, showing egg at same scale as *A. aspersa,* and embryos hatching by rupture of chorion by protrusion of enlarging tunic and ampulla; no sign of tadpole development is evident. (From Berrill, 1931.) Scale is 100 μm.

from the upper zone of greater light intensity, turns toward any dark or shaded surface it may be passing, such a surface being more likely to be firm, clean, and perhaps more sheltered compared with horizontal surfaces. During the first part of this period the tadpole is positively phototactic and negatively geotactic but during the later part it becomes positively geotactic and negatively phototactic. The net result of the locomotory activity, the sensory reactions to light and gravity, and the presence of the three adhesive organs at the anterior end of the tadpole is the attachment of the tadpole to what may well be a suitable spot for permanent settlement and postlarval growth (Grave, 1920, 1926, 1932, 1944; Grave and Woodbridge, 1924; Grave and Nicoll, 1939). It must be emphasized that the tadpole larva has dispersal and site-selection functions only. In no case is the permanent organization so advanced as to enable it to feed, even in such complex larvae as those of *Ecteinascidia* (Plough and Jones, 1939).

An argument has been presented elesewhere (Berrill, 1955) that the ascidian tadpole is an ascidian evolutionary creation and that it evolved in connection with the selection of suitable sites for settlement. There is no doubt that all of the structure and behavior of the tadpole is involved in the procedure of site selection. Where site selection has become less rigorous there is often some degree of degeneration of purely larval structure and activity. Millar (1966) has suggested that the specialization of the ascidian (or tunicate) epidermis as a tunicin or cuticle-secreting tissue and the consequent loss of all external ciliation may have led to the evolution of a muscular mechanism for locomotion.

The importance of the typical tadpole larva, with full sensory equipment, as a vital instrument for settlement on wide-spread but local hard surfaces is shown by the apparent ease with which the tadpole organization is suppressed when such site selection becomes unnecessary. This is shown particularly in the various species of the family Molgulidae. Ancestral molgulids presumably were originally attached to relatively firm substrates, and produced fully equipped tadpole larvae. The majority of existing species, however, now inhabit sand, shell-gravel, or muddy bottoms at various depths from low tide down to more than 200 m. There is little doubt that the larvae of many ascidians settle on these extensive areas of the continental shelf but fail to survive the shifting instability of the immediate environment. Molgulids (species of *Molgula* and *Eugyra*) and also many species of the styelid genus *Polycarpa* are able to survive and flourish under such circumstances because of two features: their more or less globular shape, and the extension of the outer tunic, or test, into innumerable fine filamentous processes capable of anchoring the animal to a zone of sand grains or other particles. Not only does this ability to survive and grow in such a habitat open up huge territories for colonization, but individuals occur in enormous numbers matted together to form a closely woven carpet.

In species of *Molgula* and *Polycarpa* the tadpole larvae, when present at all, possess an otolith but no ocellus, and are responsive only to gravity. The most striking phenomenon in molgulid development is the suppression of the tadpole stage (Fig. 8B) altogether in a number of species which cannot be brought together by any means of classification based on adult structure. There is, however, a close but not complete correlation of such anural development with the sand-flat habitat. Thus of 9 sand-flat species, 8 have anural development and only one, *Molgula oculata,* develops tadpoles. Of 9 attached species, 7 have tadpoles and only two are anural (Berrill, 1931). The early cleavage pattern remains unchanged in anural development, but prospective notochord cells fail to enlarge and in consequence no tail outgrowth occurs; nor is there any development of sensory vesicle or sense organs. All attached species of *Polycarpa* have tadpoles with otolith alone, but the only recorded development of a sand-inhabiting species is anural (Millar, 1962b). In another styelid, *Pelenaia corrugata,* which is an unattached sand-living oviparous species, with a deeply embedded cylindrical body, relatively large eggs are shed which develop without forming a tadpole larva (Millar, 1954b).

In the case of ascidians living in the shallow littoral or the intertidial zone, attached to rocks, weeds, corals, mangrove roots, and various other firm surfaces, the opportunities for settlement are obviously various and specialized. In most cases a low profile is necessary and individuals, whether solitary or the constituents of colonies, are generally small. Many are truly dwarfed, and virtually all are characteristically viviparous and produce comparatively large eggs. Even solitary species that are small and viviparous restrict dispersal in favor of maintaining a local population in their special habitat. Small species such as *Molgula citrina* and *M. complanata* (3 mm at maturity) are not only viviparous but the free-swimming period of the tadpole larvae is no more than 1 or 2 hours; in the European variety of *M. citrina* the tadpole larva breaks out of the egg membrane only after the activity period has passed (Berrill, 1951). Similarly in certain interstitial species of *Psammostyela* (Styelidae) and *Heterostigma*, (Pyuridae) living freely in sand and maturing at a remarkably small size of 1 to 2 mm, tadpole larvae are also short lived (Monniot, 1965). The same is true for the vast assembly of colonial, or compound, ascidians with the exception of the oviparous genus *Diazona* and the open-site botryllids, i.e., excluding *Botryllus* and *Botrylloides*. The tadpole larvae of these last two genera, together with those of all colonial enterogonid families other than the diazonids, have a free-swimming period of not more than 2–5 hours and often merely a few minutes, which, with viviparity, limits the range for settlement.

Viviparous ascidians typically have larger eggs than have oviparous species. Whatever the causal relation between these two features may be, one important consequence of larger eggs is the increased size of the tadpole larvae, for swimming speed increases directly with tadpole length and results in greater independence from local water turbulence during the critical period of settling.

A second correlation relates to the obvious duality of ascidian egg development. This has already been evident in the apparent ease with which the tadpole stage may be eliminated from development without otherwise altering the course of developmental events. In most, although not all cases of development of large eggs, by the time the tadpole stage attains functional maturity and the tadpole larvae are released, permanent organization of the ascidian has reached a comparatively advanced state of development. This condition is definitely correlated with the relatively brief duration of the free-swimming period of the tadpole larva. Where comparable increased egg size exists but tadpole larvae still have the much longer free-swimming period characteristic of oviparous, small egg types, there is no such apparent telescoping of development of larval and adult type organization. Analysis of the temporal course in these several types, however, shows that the course of development of the permanent structure and organs remains essentially unaffected by increase in egg size, but that where the apparent telescoping effect is evident, it is due to a relative retardation of the development of the tadpole structure rather than a relative acceleration of adult structure development (Berrill, 1935a). The immediate effect, however, is the production of tadpole larvae which not only have a more or less abbreviated free-swimming period and are relatively large, but are generally highly complex compared with the small, standard type. This complexity is expressed to some extent in the tadpole structure as such, as in greater cellular construction of the anterior adhesive papillae and the lateral muscle bands of the tail. Sensory organs and notochord remain unaffected apart from increase in constituent cell size. The more striking difference is the presence of a branchial sac with gill slits already formed, a heart that may already be functioning, a well differentiated although not functional digestive tract, and ampullary outgrowths or vesicles associated with either permanent anchoring or tunic maintenance (Fig. 9). In some cases asexual budding leading to colony formation is already underway.

Whatever the general character of the tadpole larva, however, whether small or large, elaborate or otherwise, the process of settling and metamorphosis is essentially the same in all. Initial attachment to the substrate is by secretions from the three adhesive papillae. These are conical groups of secretory cells or else cups with a conical structure protruding from the center. Once attachment is made, regression of the tail and associated larval sensory equipment begins, accompanied by continuing and often accelerated development of the permanent organization. Both the triggering mechanism for the onset of metamorphosis and the tissue mechanics involved in tail resorption have been in debate for many years.

8.4.5 Metamorphosis

Metamorphosis may be said to begin with the initial attachment of the tadpole larva to a substrate by means of the three anterior adhesive papillae. Following this there is a progressive resorption of the tail (Fig. 10) commencing at the tail tip in some species (Berrill, 1947c) but at the base in others (Cloney, 1961), together with dissolution of other components of the larval action system; in *Amaroucium* cuticular molting

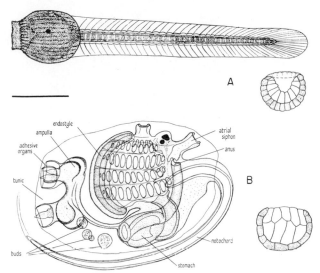

Fig. 9. Types of tadpole larvae. (A) Tadpole and gastrula of the viviparous species *Dendrodoa* (*Styelopsis*) *grossularia*, which develop within the large atrial chamber of the parent. (From Berrill, 1929.) (B) Tadpole and gastrula of *Distaplia magnilarva*, which develop in a brood sac formed by an evagination of the parental body wall containing a hypertrophied loop of the oviduct. Scale is 0.5 mm. (From Berrill, 1948a.) The *Dendrodoa* larva possesses an otolith alone, and most of the 40 notochord cells are readily discernible, together with a large number of epidermal ampullae at the anterior end; the development of the permanent ascidian organ systems is still at a rudimentary stage. The *Distaplia* larva has developed for a longer period as such and exhibits elaborate adhesive organs, virtually all permanent ascidian structure (except gonad), and even the first generation of asexually produced buds. The former has a long free-swimming period and a long postmetamorphoic phase of development before a functional ascidiozooid is produced. The latter has a very short free-swimming period and is ready to function almost immediately after absorption of the tail.

occurs at this time (Oka, 1960). Coinciding with this destructive process, epidermal stolonic or ampullary outgrowths from the trunk take over the function of attachment of the body to the substrate, while development of the ascidiozooid continues from whatever stage may have been attained in the tadpole larva. Investigations of the metamorphic phenomenon, as in other animals, relate to the initiating or control mechanism and to the transformative process itself.

Many chemical agents have been found to be effective in triggering metamorphosis, particularly trace amounts of heavy metals in the seawater, such as copper (Grave, 1935; Glaser and Anslow, 1949). Many different agents, however, have been found either to induce or inhibit

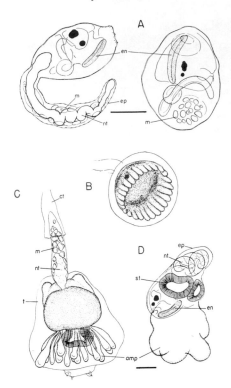

FIG. 10. Metamorphosis of ascidian tadpole larvae. (A) Tail resorption in *Phallusia mammillata*. (B) *Dendrodoa grossularia* following attachment, with tail fully resorbed (otolith still visible in trunk) and a ring of about thirty ampullae which serve in permanent attachment of ascidian. (C) Tail resorption and ampullary outgrowth in *Distomus variolosus* at time of settling (the three adhesive papillae visible and extruded). (D) Tail resorption in *Clavelina huntsmani*. amp, Ampulla; ct, cuticle of tail; en, endostyle; ep, epidermis; m, muscle cells of tail; nt, notochord; st, stomach; t, tunic. (D, after Bradway, 1936; A, B, C, from Berrill, 1947c.) Scale for (A) is 0.1 mm, for (B and C) is 0.4 mm, and for (D) is 0.2 mm.

metamorphosis (reviewed by Lynch, 1961); while Whittaker (1964) found that chelation of the copper, in the case of *Styela* tadpoles, causes no delay in the onset of metamorphosis. As a rule metamorphosis does not commence until attachment has been made, although *Amaroucium* tadpoles may undergo metamorphosis without attachment when removed from seawater containing potassium cyanide (Lynch, 1961), admittedly a most abnormal circumstance. Enzymes are clearly involved. Cloney (1961) suggests that tail resorption occurs through the interaction of a hydrolytic enzyme and the contraction of the epidermis, and has shown that epidermal contraction is effected by oriented contractile microfilaments present in the epidermal cells (Cloney, 1966, 1969).

Whatever the biochemical and histological mechanism of resorption of the larval action system may be, the process as a whole is undoubtedly initiated immediately following attachment and may proceed very rapidly indeed (Scott, 1954). It is significant that the larval action system includes peripheral nerves that pass forward from the larval visceral ganglion to anterior ganglia and nerves associated with the adhesive papillae (Grave, 1921, 1934) in *Amaroucium* and *Botryllus*. Working with tadpole larvae of *Halocynthia roretzi*, Hirai (1969) has shown that an explosive discharge of secretion occurs when the adhesive papillae first make contact, as in the case of *Amaroucium* (Scott, 1952), and presents evidence that indicates a slow passage of a stimulating agent from the anterior end of the tadpole in a posterior direction. He suggests that an endocrine agent in the papillae undergoes tissue transmission and triggers tail resorption. Tail resorption, however, although the most striking event in normal metamorphosis, is not essential to the constructive aspects of metamorphosis. If the anterior end of a tadpole of *Polycitor* is cut off, together with the papillae, involution of the tail is suppressed but progressive metamorphosis proceeds to the completion of a young ascidian. When the tail is cut off precociously, progressive metamorphosis proceeds at the same tempo as in the normal case (Oka, 1943). A comprehensive review of ascidian metamorphosis is given by Barrington (1968).

8.4.6 Postlarval Growth

Postlarval growth in ascidians is directed along one of two pathways. In solitary species the oozooid persists as the prospective sexually reproducing adult, with continual elaboration of the branchial structure to maintain efficiency of the water-filtration apparatus as body size increases (Van Beneden and Julin, 1886; de Selys-Longchamps and Damas, 1900; Julin, 1904; Berrill, 1947a), and with control of body size at sexual maturity in accordance with nature of habitat.

In colonial species the oozooid serves primarily as a means for establishing a young growing colony and only after a large number of asexually produced ascidiozooids have become established as a colonial system do these become sexually mature. Only rarely, as in *Botryllus* and *Botrylloides*, budding and sexual reproduction continue side by side. Where strobilation of the abdomen or postabdomen is the means for zooid multiplication, the processes of sexual and asexual reproduction are clearly incompatible. In certain related forms, however, budding has become a highly specialized process, as in the Didemnidae and the subfamily Holozoinae. In the didemnid genus *Diplosoma,* the free-swimming tadpole larva has not only a near-functional oozooid but a blastozooid ascidiozooid as well. In the holozoan *Hypsistozoa fasmeriana,* however, the tadpole larva itself produces, through a process of stolonic strobilation, from 9 to 14 buds, which become arranged about a common cloacal aperture, and a small colony functions immediately following metamorphosis (Brewin, 1956, 1959).

The viviparous state discussed so far in the case of ascidians is, strictly, ovoviviparity. True viviparity involving extraembryonic nutrition is rare, being known only in *Botrylloides, Hypsistozoa,* and salps. In all three it is associated with reduction in size of eggs. Thus in *Botrylloides,* which is a genus very closely related to *Botryllus,* the egg diameter, in the species *Botrylloides leachi* (Berrill, 1935a, 1947b), is only 260 μm, compared with 450 μm for *Botryllus schlosseri.* The embryo grows throughout the period of embryonic development, nourished by the secretory lining of the brood pouch, and becomes a tadpole larva similar in body size and duration of free-swimming period to that of *Botryllus* and, as in *Botryllus,* metamorphoses to form an oozooid with a single developing blastozooid. In *Botrylloides digense,* similar tadpole larvae are produced during the summer season, but during winter months giant tadpole larvae develop, five times as large as the summer season larvae (G. Freeman, personal communication).

In *Hypsistozoa* the eggs are reduced to what may well be the minimum size possible in the animal kingdom, being only 25 μm in diameter, in comparison to 420 μm diameter in the case of its nearest relatives such as *Distaplia* species. Yet the tadpole larvae attain a size twice as large as the maximum recorded for *Distaplia.* The egg develops within an oviducal brood pouch and acquires endodermal tubes with a placental function, remaining attached to the parent for the remarkably long period of 5½ months before being liberated as a tadpole with its freight of ready-to-function blastozooids (Brewin, 1956).

The most striking departure from the precise and intriguing development of the tadpole larva of tunicates is that of salps. The egg is small, comparatively yolkless, and is retained within an ovarian follicular envelope. During early cleavage the blastomeres become separated by follicle cells and continue division to form separate islands of cells. The disaggregated units eventually come together to form an embryo the scale of which has already been determined, at least in part, by the growth and form of the associated follicular envelope. A true placentation is established between embryo and parent and development and growth proceed until the parent becomes destroyed (Brien, 1928; Berrill, 1950a;

Sutton, 1960). The assembly of islands of cells to form an embryo may be compared with the integration of dissociated fragments of oozoids of *Amaroucium* to form functioning individuals (Scott, 1961, 1962).

8.5 Bibliography

Abbott, D. P. (1953). Asexual reproduction in the colonial ascidian *Metandrocarpa taylori* Huntsman. *Univ. Calif. Publ. Zool.* **61**, 1–78.

Abbott, D. P. (1955). Larval structure and activity in the ascidian *Metandrocarpa taylori*. *J. Morphol.* **97**, 569–594.

Barrington, E. J. W. (1965). "The Biology of Hemichordata and Protochordata," 176 pp. W. H. Freeman, San Francisco, Calif.

Barrington, E. J. W. (1968). Metamorphosis in Lower Chordates. *In* "Metamorphosis" (W. Etkin and L. E. Gilbert, eds.), pp. 223–270. Appleton-Century Crofts, New York.

Barth, L. G. and Barth, L. C. (1966). A study of regression and budding in *Perophora viridis*. *J. Morphol.* **118**, 451–460.

Berrill, N. J. (1929). Studies in tunicate development. I. General physiology of development of simple ascidians. *Phil. Trans. Roy. Soc.* **B218**, 37–78.

Berrill, N. J. (1931). Studies in tunicate development, II. Abbreviation of development in the Molgulidae. *Phil. Trans. Roy. Soc.* **B219**, 281–346.

Berrill, N. J. (1932). Ascidians of the Bermudas. *Biol. Bull.* **52**, 77–88.

Berrill, N. J. (1935a). Studies in tunicate development, III. Differential retardation and acceleration. *Phil. Trans. Roy. Soc.* **B225**, 255–326.

Berrill, N. J. (1935b). Studies in tunicate development, IV. Asexual reproduction. *Phil Trans Roy Soc.* **B225**, 327–379

Berrill, N. J. (1940). The development of a colonial organism: *Symplegma viride*. *Biol. Bull.* **79**, 271–281.

Berrill, N. J. (1941). Spatial and temporal growth patterns in colonial organisms. *Growth Symp.* **3**, 89–111.

Berrill, N. J. (1947a). The development and growth of *Ciona*. *J. Mar. Biol. Ass. U.K.* **26**, 616–625.

Berrill, N. J. (1947b). The developmental cycle of *Botrylloides*. *Quart. J. Microsc. Sci.* **88**, 393–407.

Berrill, N. J. (1947c). Metamorphosis in ascidians. *J. Morphol.* **81**, 249–268.

Berrill, N. J. (1947d). The structure, development and budding of the ascidian, *Eudistoma. J. Morphol.* **81**, 269–281.

Berrill, N. J. (1948a). Budding and the reproductive cycle of *Distaplia*. *Quart. J. Microsc. Sci.* **89**, 253–289.

Berrill, N. J. (1948b). Structure, tadpole and bud formation in the ascidian *Archidistoma. J. Mar. Biol. Ass. U.K.* **28**, 380–388.

Berrill, N. J. (1948c). The gonads, larvae, and budding of the polystyelid ascidians *Stolonica* and *Distomus*. *J. Mar. Biol. Ass. U.K.* **27**, 633–650.

Berrill, N. J. (1948d). The nature of the ascidian tadpole, with reference to *Boltenia echinata. J. Morphol.* **82**, 269–285.

Berrill, N. J. (1948e). The development, morphology and budding of the ascidian *Diazona. J. Mar. Biol. Ass. U.K.* **27**, 389–399.

Berrill, N. J. (1948f). Tadpole larvae of the ascidians *Polycitor, Euherdmania* and *Polysyncraton. J. Morphol.* **82**, 355–363.

Berrill, N. J. (1949). The gonads, larvae and budding of the polystyelid ascidians *Stolonica* and *Distomus. J. Mar. Biol. Ass. U.K.* **28**, 633–650.

Berrill, N. J. (1950a). Budding and development in *Salpa. J. Morphol.* **87**, 553–606.

Berrill, N. J. (1950b). Budding in *Pyrosoma. J. Morphol.* **87**, 537–552.

Berrill, N. J. (1950c). "The Tunicata," Vol. 133, 354 pp. Roy. Society, London.

Berrill, N. J. (1951). Regeneration and budding in tunicates. *Biol. Rev.* **26**, 456–475.

Berrill, N. J. (1955). "The Origin of Vertebrates," 257 pp. Oxford Univ. Press, Oxford.

Berrill, N. J. (1957). Tunicata in "The Plymouth Fauna List," pp. 371–382. Mar. Biol. Assoc. U.K., Plymouth.

Berrill, N. J. and Cohen, A. (1936). Regeneration in *Clavelina lepodiformis. J. Exp. Biol.* **13**, 352–362.

Bradway, W. (1936). The experimental alteration of the rate of metamorphosis in the tunicate, *Clavelina huntsmani* (Van Name). *J. Exp. Zool.* **72**, 213–224.

Brewin, B. I. (1956). The growth and development of a viviparous compound ascidian *Hypsistozoa fasmeriana. Quart. J. Microsc. Sci.* **97**, 434–454.

Brewin, B. I. (1959). An account of larval budding in the compound ascidian, *Hypsistozoa fasmeriana. Quart. J. Microsc. Sci.* **100**, 575–589.

Brien, P. (1925). Contribution à la blastogénèse des Tuniciers I. Bourgeonnement chez *Aplidium zostericola* (Giard). *Arch. Biol.* **35**, 155–205.

Brien, P. (1928). Contribution à l'étude de l'embryogénèse et de la blastogénèse des Salpes. *Rec. Inst. Zool. Torley-Rousseau* **2**, 5–116.

Brien, P. (1937). Formation des coenobies chez les Polyclinidae. *Ann. Soc. Roy. Zool. Belg.* **67**, 63–73.

Brien, P. (1939). Contribution à l'étude du bourgeonnement et de l'organogénése du blastozooide des Distomidae (Polycitoridea). *Distaplia. Ann. Soc Roy. Zool. Belg.* **70**, 101–152.

Brien, P. (1948). Tuniciers. Morphologie et Reproduction. *In* "Traité de Zoologie" (P. P. Grassé, ed.), Vol. 9, pp. 553–1040. Masson, Paris.

Brien, P. (1968). Blastogenesis and Morphogenesis. *Advan. Morphog.* **7**, 151–204.

Brien, P. and Brien-Gavage, E. (1927). Contribution à l'étude de la blastogénése des Tuniciers. *Rec. Inst Zool. Torley-Rousseau.* **1**, 31–81, 123–152.

Carlisle, D. B. (1951). On the hormonal and neural control of the release of gametes in ascidians. *J. Exp. Zool.* **28**, 463–472.

Cifuentes, G. F. A. Cea. (1969–1970). Estados primarios del desarrollo y metamorfosis de *Pyura chilensis molina* (Tunicata, Ascidiacea, Pyuridae). *Bol. Soc. Biol. Concepcion* **42**, 317–331.

Cloney, R. A. (1961). Observations on the mechanism of tail resorption in ascidians. *Amer. Zool.* **1**, 67–88.

Cloney, R. A. (1966). Cytoplasmic filaments and cell movements: epidermal cells during ascidian metamorphosis. *J. Ultrastruct. Res.* **14**, 300–328.

Cloney, R. A. (1969). Cytoplasmic filaments and morphogenesis: the role of the notochord in ascidian metamorphosis. *Z. Zellforsch.* **100**, 31–53.

Conklin, E. G. (1905). The organization and cell-lineage of the ascidian egg. *J. Acad. Nat. Sci. Phila.* **13**, 1–119.

Conklin, E. J. (1931). The development of centrifuged eggs of ascidians. *J. Exp. Zool.* **60**, 1–119.

Costello, D. P., Davidson, M. E., Eggers, A., Fox, M. H., and Henley, C. (1957). "Methods for Obtaining and Handling Marine Eggs and Embryos," 247 pp. Marine Biological Laboratory, Woods Hole, Mass.

Cowden, R. (1961). A comparative study of oocyte growth and development in two species of ascidians. *Acta Embryol. Morphol. Exp.* **4**, 123–141.

Cowden, R. R. and Markert, C. L. (1965). A cytochemical study of the development of *Ascidia nigra. Acta Embryol. Morphol. Exp.* **4**, 142–160.

Dalcq, A. (1932). Etudes des localisations germinales dans l'oeuf vierge d'Ascidie par dex expériences de merogonie. *Arch. Anat. Microsc. Morphol. Exp.* **28**, 223–233.

de Sélys-Longchamps, M. (1917). Sur le bourgeonnement des Polystyelinés *Stolonica* et *Heterocarpa. Bull. Sci. Fr. Belg.* **50**, 170–276.

de Sélys-Longchamps, M. and Damas, D. (1900). Récherches sur le développement post-embryonaire et l'anatomie définitive de *Molgula ampulloides. Arch. Biol.* **17**, 385–488.

Dilly, P. N. (1962). Studies on the receptors in the cerebral vesicle of the ascidian tadpole. *Quart. J. Microsc. Sci.* **103**, 393–396.

Franzén, A. (1958). On sperm morphology and acrosome filament formation in some Annelida, Echiuroidea, and Tunicata. *Zool. Bidr. Uppsala* **33**, 1–28.

Freeman, G. (1964). The role of blood cells in the process of asexual reproduction in the tunicate *Perophora viridis. J. Exp. Zool.* **156**, 157–184.

Freeman, G. (1969). The control of the initiation of asexual reproduction in the tunicate *Amaroecium constellatum. Biol. Bull.* **137**, 399–400.

Freeman, G. (1971). A study of the intrinsic factors which control the initiation of asexual reproduction in the tunicate *Amaroecium constellatum. J. Exp. Zool.* **178**, 433–456.

Fukumoto, M. (1971). Experimental control of budding and stolon elongation in *Perophora orientalis*, a compound ascidian. *Develop Growth Differentiation* **13**, 73–88.

Glaser, O. and Anslow, G. A. (1949). Copper and ascidian metamorphosis. *J. Exp. Zool.* **111**, 117–140.

Godeaux, J. (1958). Contribution à la Connaissance des Thaliaces. *Ann. Soc. Roy. Zool. Belg.* **88**, 1–285.

Godeaux, J. (1963). Tuniciers: Embryologie, Histologie et Morphologie (1958–1961). *Fortschr. Zool.* **16**, 333–394.

Goodbody, I. (1961). Continuous budding in three species of tropical ascidians. *Proc. Zool. Soc. London* **136**, 403–409.

Goodbody, I. (1963). The biology of *Ascidia nigra* (Savigny). II. The development and survival of young ascidians. *Biol. Bull.* **124**, 31–44.

Grave, C. (1920). *Amaroecium constellatum.* I. The activities and reactions of the tadpole larvae. *J. Exp. Zool.* **30**, 239–259.

Grave, C. (1921). *Amaroecium constellatum.* II. The structure and organization of the tadpole larva. *J. Morphol.* **36**, 71–101.

Grave, C. (1926). *Molgula citrina* (Alder and Hancock)—activities and structure of the free-swimming larva. *J. Morphol.* **42**, 453–471.

Grave, C. (1934). The botryllus type of ascidian larva. *Carnegie Inst. Wash. Publ. No.* **435**, 143–156.

Grave, C. (1935). Metamorphosis of Asciadian larvae. *Carnegie Inst. Wash. Publ. No.* **452**, 209–292.

Grave, C. (1944). The larva of *Styela* (*Cynthia*) *partita*. Structure, activities and duration of life. *J. Morphol.* **75,** 173–188.

Grave, C. and Nicoll, P. A. (1939). Studies of larval life and metamorphosis in *Ascidia nigra* and species of *Polyandrocarpa*. *Carnegie Inst. Wash. Publ. No.* **452,** 209–292.

Grave, C. and Riley, G. (1935). Development of the sense organs of the larva of *Botryllus schlosseri*. *J. Morphol.* **57,** 185–211.

Grave, C. and Woodbridge, H. (1924). *Botryllus schlosseri* (Pallas): The behavior and morpology of the free-swimming larva. *J. Morphol.* **39,** 207–247.

Harvey, L. A. (1927). The history of cytoplasmic inclusions of the egg of *Ciona intestinalis* (L.) during oogenesis and fertilization. *Proc. Roy. Soc. London* **B101,** 136–162.

Haven, N. D. (1971). Temporal patterns of sexual and asexual reproduction in the colonial ascidian *Metandrocarpa taylori* Huntsman *Biol. Bull.* **140,** 400–415.

Hirai, E. (1941). An outline of the develoment of *Cynthia roretzi* Drasche. *Sci. Rep. Tohoku Univ.* (*Ser. 4*) **16,** 257–261.

Hirai, E. (1969). On the early stages of metamorphosis of the ascidian, *Halocynthia roretzi*. (v. Drasche). *Sci. Rep. Tohuku Univ.* (*Ser. V*) **33,** 349–358.

Hǔus, J. (1937). Tunicata: Ascidiaceae. *Handb. Zool. Kukenthal Krumbach* **5** (part 2), 545–672.

Hǔus, J. (1939). The effect of light on the spawning of ascidians. *Avh. Nor. Vidensk. Akad. Oslo Mat. Naturvidensk. Kl. No.* **4,** 5–49.

Hǔus, J. (1941). Effects of physical factors on the spawning in ascidians. *Avh. Nor. Vidensk. Akad. Oslo Mat. Naturvidensk. Kl. No.* **8,** 1–13; **9,** 2–12.

Imai, T. (1969). Microscopic studies on the gonad of the ascidian *Cynthia roretzi* von Drasche. *Tohoku J. Agr. Res.* **19,** 63–73.

Izzard, C. S. (1973). Development of polarity and bilateral asymmetry in the palleal bud of *Botryllus schlosseri* (Pallas). *J. Morph.* **139,** 1–26.

Julin, C. (1904). Recherches sur la phylogénèse des Tuniciers développement de l'appareil branchial. *Z. Wiss. Zool.* **74,** 544–64.

Kasas, O. M. (1940). Structure of the larvae of ascidians *Dendrodoa grossularia* and their metamorphosis. *Bull. Acad. Sci. U.S.S.R. Biol.*, pp. 862–883.

Kessel, R. G. (1966a). Electron microscope studies on the origin and maturation of yolk in oocytes of the tunicate, *Ciona intestinalis*. *Z. Zellforsch.* **71,** 525–544.

Kessel, R. G. (1966b). Ultrastructure and relationships of ooplasmic components in tunicates. *Acta Embryol. Morphol. Exp.* **9,** 1–24.

Kessel, R. G. (1967). The origin and fate of secretion in the follicle cells of tunicates. *Z. Zellforsch.* **76,** 21–30.

Kessel, R. G. and Kemp, N. E. (1962). An electron microscope study on the oocyte, test cells and follicular envelope on the tunicate. *Molgula manhattensis*. *J. Ultrastruct. Res.* **6,** 57–76.

Knaben, N. (1936). Uber entwicklung und funktion der testazellen bei *Corella parallelogramma*. *Bergens Mus. Arb. No.* **1,** 1–33.

Kott, P. (1952). Observations on compound ascidians of the Plymouth area, with descriptions of two new species. *J. Mar. Biol. Ass. U.K.* **31,** 65–83.

Kott, P. (1969). Antarctic ascidiacea. *Antarct. Res. Ser.* **13,** 1–239. (N.A.S.—N.R.C. Publ. in 1725.)

Lambert, C. C. and Brandt, C. L. (1967). The effect of light on the spawning of *Ciona*. *Biol. Bull.* **132,** 222–228.

Lambert, G. (1968). The general ecology and growth of a solitary ascidian, *Corella willmeriana*. *Biol. Bull.* **135**, 296–307.

Levine, E. P. (1963). XVI. Intern. Congr. Zool. (oral presentation).

Lynch, W. (1961). Extrinsic factors influencing metamorphosis in Bryozoan and Ascidian larva. *Amer. Zool.* **1**, 59–66.

Mancuso, V. (1964). Ultrastructural changes in the cytoplasm of *Ciona intestinalis* oocytes. *Acta Embryol. Morphol. Exp.* **7**, 269–295.

Mancuso, V. (1965). An electron microscope study of the test cells and follicle cells of *Ciona intestinalis* during oogenesis. *Acta Embryol Morphol. Exp.* **8**, 239–266.

Mansueto, C. (1964). Sulla riproduzione per divisione mitotica delle cellule testati delle Ascidie. *Accad. Naz. Lincei Rome* **36**, 683–689.

Markert, C. L. and Cowden, R. R. (1965). Comparative responses of regulative and determinate embryos to metabolic inhibitors. *J. Exp. Zool.* **160**, 37–45.

Materazzi, G. and Ortolani, G. (1969). A study of the origin of the cells containing sulfated mucopolysaccharides in the cephalic portion of the larvae of *Phallusia mamillata* and *Ascidia malaca*. *Develop. Biol.* **20**, 378–385.

Milkman, R. (1967). Genetic and developmental studies on *Botryllus schlosseri*. *Biol. Bull.* **132**, 229–243.

Milkman, R. and Borgmann, M. (1963). External fertilization in *Botryllus schlosseri* eggs. *Biol. Bull.* **125**, 383.

Millar, R. H. (1951). The development and early stages of the ascidian *Pyura squamulosa* (Alder). *J. Mar. Biol. Ass. U.K.* **30**, 27–31.

Millar, R. H. (1952). The annual growth and reproductive cycle in four ascidians. *J. Mar. Biol. Ass. U.K.* **31**, 41–61.

Millar, R. H. (1954a). The breeding and development of the ascidian *Pelonaia corrugate* Forbes and Goodsir. *J. Mar. Biol. Ass. U.K.* **33**, 681–687.

Millar, R. H. (1954b). The annual growth and reproductive cycle of the ascidian *Dendrodoa grossularia* (Van Beneden). *J. Mar. Biol. Ass. U.K.* **33**, 33–48.

Millar, R. H. (1954c). The annual growth and reproductive cycle of *Dendrodoa grossularia* (van Beneden). *J. Mar. Biol. Ass. U.K.* **33**, 33–48.

Millar, R. H. (1958). The breeding season of some littoral ascidians in Scottish waters. *J. Mar. Biol. Assoc. U.K.* **37**, 649–652.

Millar, R. H. (1962a). Budding in the ascidian *Aplidium petrense* Michaelsen. *Ann. Mag. Nat. Hist.* (*Ser. 13*) **5**, 337–340.

Millar, R. H. (1962b). The breeding and development of the ascidian *Polycarpa tinctor*. *Quart. J. Microsc. Sci.* **103**, 399–403.

Millar, R. H. (1963). The development and larva of *Styela coriacea J. Mar. Biol. Ass. U.K.* **43**, 71–74.

Millar, R. H. (1966). Evolution in Ascidians. In "Some Contemporary Studies in Marine Science" (H. Barnes, ed.), pp. 519–534. George Allen & Unwin, Ltd., London.

Minganti, A. (1948). Interspecific fertilization in ascidians. *Nature* (*London*) **161**, 643–644.

Miyamoto, D. and Freeman, G. (1970). The origin of the cells which form zooids during vascular budding in the ascidian *Botrylloides digense*. *Amer. Zool.* **10**, 533–534.

Monniot, F. (1965). Ascidies interstitielles des cotes d'Europe. Thesis. Univ. Paris.

Morgan, T. H. (1942). Cross-and self-fertilization in the ascidian *Molgula manhattensis*. *Biol. Bull.* **82**, 172–177.

Morgan, T. H. (1945). The conditions that lead to normal or abnormal development of *Ciona*. *Biol. Bull.* **88**, 50–62.

Nakauchi, M. (1966a). Regeneration in the zooid of *Polycitor mutabilis*. *Sci. Rep. Tokyo Kyoiku Daigaku* **B12**, 151–189.

Nakauchi, M. (1966b). Budding and colony formation in the ascidian *Amaroucium multiplication*. *Jap. J. Zool.* **15**, 151–172.

Nakauchi, M. (1966c). Budding and growth in the ascidian, *Archidistoma aggregation*. *Rep. Usa Mar. Biol. Stat.* **13**, 1–10.

Nakauchi, M. (1970). Asexual reproduction in *Amaroucium yamazii* (a colonial ascidian). *Publ. Seto Mar. Biol. Lab.* **17**, 309–328.

Neumann, G. (1906). *Doliolum. Ergbr. Deut. Tiefsee-Exped.* **12**(2), 93–243.

Newberry, A. T. (1965a). The structure of the circulatory apparatus of the test and its role in budding in the polystyelid ascidian *Metandrocarpa taylori* Huntsman. *Mem. Acad. Roy. Belg.* **16**, 1–57.

Newberry, A. T. (1965b). Formation et évolution des gonades chez *Distomus variolosus* (Ascidiace, Stolidobranche, Polystyelidae). *C. R. Acad. Sci.* (*Paris*) **260**, 6685–6688.

Oka, H. (1942). On a new species of *Polycitor* from Japan with some remarks on its mode of budding. *Annot. Zool. Jap.* **21**, 155–162.

Oka, H. (1943). Metamorphosis of *Polycitor mutabilis*. *Annot. Zoo. Jap.* **22**, 54–58.

Oka, H. (1960). Moulting at metamorphosis in ascidians. *Bull. Mar. Biol. Stat. Asamushi Tohoku Univ.* **10**, 177–180.

Oka, H. and Watanabe, H. (1957). Vascular budding, a new type of budding in *Botryllus*. *Biol. Bull.* **112**, 225–240.

Oka, H. and Watanabe, H. (1959). Vascular budding in *Botrylloides*. *Biol. Bull.* **117**, 340–346.

Oka, H. and Watanabe, H. (1960). Problems of colony-specificity in compound ascidians. *Bull. Mar. Biol. Stat. Asamushi Tohoku Univ.* **10**, 153–155.

Oka, H. and Watanabe, H. (1961). Kunstliche Auslosung der Strobilation bei den Synascidien. *Embryologia* **6**, 135–150.

Ortolani, G. (1958). Cleavage and development of egg fragments in ascidians. *Acta Embryol. Morphol. Exp.* **1**, 247–272.

Peres, J. M. (1952). Recherches sur le cycle sexuel de *Ciona intestinalis*. *Arch. Anat. Microsc. Morphol. Exp.* **41**, 153–183.

Peres, J. M. (1953). Considerations sur le fonctionnement ovarien chez *Ciona intestinalis* (L.). *Arch. Anat. Microsc. Morphol. Exp.* **43**, 58–78.

Pizon, A. (1893). Histoire de la blastogénèse chez les Botryllides. *Ann. Sci. Nat.* (*Ser.* 7) **14**, 1–386.

Plough, H. H. and Jones, N. (1939). *Ecteinascidia tortugenesis*, sp. nova. with a review of the Perophoridae (Ascidiacea) of the Tortugas. *Carnegie. Inst. Wash. Publ. No.* **517**, 47–60.

Reese, J. P. (1967). Photoreceptive regulation of spawning in *Ciona intestinalis*. Master's thesis, 81 pp. San Diego State College, San Diego, California.

Reverberi, G. (1937). Richerche sperimentali sulla structura dell'vovo fecondata delle Ascidie. *Comm. Pont. Acad. Sci.* **1**, 135–172.

Reverberi, G. (1961). The embryology of ascidians. *Advan. Morphog.* **1**, 55–101.

Reverberi, G. and Gorgone, I. (1962). Gigantic tadpoles from ascidian eggs fused at the 8-cell stage. *Acta Embryol. Morphol. Exp.* **5**, 104–112.

Reverberi, G. and Ortolani, G. (1962). Twin larvae from halves of the same egg in ascidians. *Develop. Biol.* **5**, 84–100.

Rose, S. M. (1939). Embryonic induction in the ascidia. *Biol. Bull.* **78**, 216–232.

Sabbadin, A. (1955). Il ciclo biologico di *Botryllus schlosseri* (Pallas) nella Laguna di Venezia. *Arch. Oceanogr. Limnol.* 10, 217–230.

Sabbadin, A. (1957). Il ciclo biologico di *Ciona intestinalis* (L), *Molgula manhattensis* (De Kay) e *Styela plicata* (Lesueur) nella Laguna Veneta. *Arch. Oceanogr. Limnol.* 17, 1–27.

Sabbadin, A. (1958). Analisi sperimentale della sviluppo delle colonie di *Botryllus schlosseri*. *Arch. Ital. Anat. Embriol.* 63, 178–221.

Sabbadin, A. (1971). Self- and cross-fertilization in the compound ascidian, *Botryllus schlosseri*. *Develop. Biol.* 24, 379–391.

Sabbadin, A. and Graziani, G. (1967). New data on the inheritance of pigmentation patterns in the colonial ascidian *Botryllus schlosseri* (Pallas). *Riv. Biol.* (*Peruvia*) 60, 581–598.

Scott, F. M. (1945). Developmental history of *Amaroecium constellatum*. I. Early embryonic development. *Biol. Bull.* 88, 126–138.

Scott, F. M. (1946). The Developmental history of *Amaroecium constellatum*. II. Organogenesis of the larva action system. *Biol. Bull.* 91, 66–80.

Scott, F. M. (1952). Developmental history of *Amaroecium constellatum*. III. Metamorphosis. *Biol. Bull.* 103, 226–241.

Scott, F. M. (1954). Metamorphic differentiation in *Amaroecium constellatum* treated with nitrogen mustard. *J. Exp. Zool.* 127, 331–366

Scott, F. M. (1961). Tissue affinity in *Amaroecium*. I. Aggregation of dissociated fragments and their integration into one organism. *Acta Embryol. Morphol. Exp.* 2, 209–226.

Scott, F. M. (1962). Tissue affinity in *Amaroecium*. II. Reaggregation of three partial zooids into functioning siamese twins. *Biol. Bull.* 122, 396–416.

Sebastian, V. O. (1958). On the anatomy and larval organisation of *Polyclinum* sp. *J. Madras Univ.* 14, 251–276.

Simkins, C. S. (1924). Origin of the germ cells in *Ectoinascoidia*. *J. Morphol.* 39, 295–321.

Spek, J. (1927). Uber die Winterknospenentwicklung, Regeneration und Reduktion bei *Clavellina lepadiformis* und die Bedeutung besonderst 'omnipotenter' Zellemente für diese Vorgange. *Arch. Entwicklungsmech. organismen* 111, 119–172.

Sutton, M. (1960). The sexual development of *Salpa fusiformis*. Part I. *Embryol. Exp. Morphol.* 8, 268–290.

Trason, W. B. (1957). Larval structure and development of the oozooid in the ascidian *Euherdmania claviformis*. *J. Morphol.* 100, 509–546.

Trason, W. B. (1963). The life cycle and affinities of the colonial ascidian *Pycnoclavella stanleyi*. *Univ. Calif. Publ. Zool.* 65, 81–113.

Tucker, G. H. (1942). The histology of the gonads and development of the egg envelopes of an ascidian (*Styela plicata* Lesueur) *J. Morphol.* 70, 81–113.

Tung, T. C. (1934). Recherches sur les potentialités des blastomères chez *Ascidiella scabra*. *Arch Anat. Microsc.* 30, 381–410.

Tuzet, O. and Harant, H. (1930). Sur la spermatogénèse d'une ascidie *Styela partita* (Stimpson). *C. R. Soc. Biol.* (*Paris*) 105, 292–301.

Van Beneden, E. and Julin, C. (1886). Recherches sur la morphologie des Tuniciers. *Arch. Biol.* 6, 237–476.

Van Name, W. G. (1945). The North and South American ascidians. *Bull. Amer. Mus. Natur. Hist.* 84, 1–462.

Von Ubisch, L. (1938). Uber Keimverschmelzungen an *Ascidiella aspersa*. *Arch. Entwicklungsmach. Organismen* **138**, 18–36.

Wallace, H. (1961). The breeding and development of *Styela mammiculata* Carlisle. *J. Mar. Biol. Ass. U.K.* **41**, 187–190.

Watanabe, H. and Lambert, C. C. (1973). Larva release in response to light by the compound ascidians *Distaplia occidentalis* and *Metandrocarpa taylori*. *Biol. Bull.* **144**, 556–566.

Watterson, R. L. (1945). Asexual reproduction in the colonial tunicate *Botryllus schlosseri* (Pallas) Savigny, with special reference to the developmental history of intersiphonal bands of pigment cells. *Biol. Bull.* **88**, 71–103.

Whittaker, J. R. (1964). Copper as a factor in the onset of ascidian metamorphosis. *Nature (London)* **202**, 1024–1026.

Whittaker, J. R. (1966). An analysis of melanogenesis in differentiating pigment cells of ascidian embryos. *Develop. Biol.* **14**, 1–39.

Whittaker, J. R. (1973). Segregation during ascidian embryogenesis of egg cytoplasmic information for tissue-specific enzyme development. *Proc. Nat. Acad. Sci.* **70**, 2096–2100.

Whittingham, D. G. (1967). Light-induction of shedding of gametes in *Ciona intestinalis* and *Mogula manhattensis*. *Biol. Bull.* **132**, 292–298.

Chapter 9

CHORDATA: ACRANIA (CEPHALOCHORDATA)

John H. Wickstead

9.1 Introduction

The phylum Chordata usually is considered to consist of four subphyla (1) Hemichordata (= Adelochordata), (2) Urochordata (= Tunicata), (3) Acrania (= Cephalochordata), and (4) Craniata (= Vertebrata). The first three categories are often referred to collectively as the Protochordata. More recently (see Barrington, 1965) the Hemichordata have been excluded from the Chordata, being raised to phylum level. It is the Acrania, commonly called amphioxus or lancelets, with which we are concerned here.

The Acrania is a small, distinctive group of closely related animals consisting of two families, the Branchiostomidae and the Asymmetrontidae, the former with a single genus, *Branchiostoma,* the latter with the single genus, *Asymmetron,* and possibly a second, *Epigonichthys.* There are two main distinguishing features between the

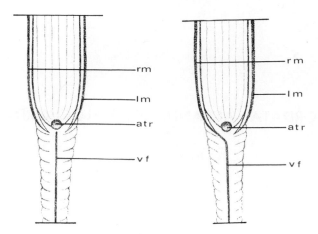

FIG. 1. Diagram showing the relationship between the metapleural folds and ventral fin in *Branchiostoma* (left) and *Asymmetron* (right). atr, Atriopore; lm, left metapleure rm, right metapleure; vf, ventral fin.

families: (1) adult Branchiostomidae develop a double row of gonads, while adult Asymmetrontidae develop a single row of gonads along the right side only. (2) In the Branchiostomidae the right and left metapleures are symmetrical, terminating at the atrioporal region, with the ventral fin medially disposed, and unconnected with either right or left metapleure (Fig. 1). In the Asymmetrontidae, the right metapleure is continued to join the ventral fin.

The Acrania are essentially shallow-water tropical and subtropical in their distribution, some species extending to temperate areas. *Branchiostoma lanceolatum* (Pallas) from the Norwegian coast are perhaps farthest removed from the equator (see Tambs-Lyche, 1967). The greatest depth from which an acraniate has been taken appears to be 100 fm (Zietz, 1908). The best known species is *Branchiostoma lanceolatum*, the original specimen being taken from the south Cornwall coast of England and described by Pallas in 1774. The classical studies of Willey (1891, 1894b) were on this same species from the Messina area of the Mediterranean. The most widely distributed species appears to be *Asymmetron lucayanum* Andrews, originally described from the Bahamas, subsequently found circumequatorially in warm, shallow waters.

9.2 Asexual Reproduction

As far as is known, asexual reproduction does not occur in the Acrania either as part of the natural life cycle or as regeneration of severed

parts of the body. It appears that powers of regeneration in this group
are very limited indeed. Chin (1941) commented: "If any part of the
body of amphioxus is injured it does not regenerate due to the presence
of numerous ciliates." I have seen numerous ciliates when keeping
damaged larvae. Of all the individuals examined by Chin (the species
here is *Branchiostoma belcheri* Gray) he found only one which showed
signs of regeneration, in this case, the tail. Probst (1930) examined
regeneration in *B. lanceolatum* by amputation of the postanal section
of individuals in the 6–9 mm size range; he was unsuccessful. Biberhofer
(1906) experimented, rather superficially, on regeneration of the anterior
end, with the same lack of success. Andrews (1893) observed the beginnings of regeneration in *A. lucayanum,* but did not continue long
enough to record complete regeneration.

Thus, as far as is known, asexual reproduction in the Acrania is unknown, and powers of regeneration virtually nonexistent.

9.3 Sexual Reproduction

9.3.1 Sexual Dimorphism

In some, and probably all, of the Acrania, sexes can be differentiated
by superficial examination provided the gonads are mature. However,
true sexual dimorphism has not been positively recorded in any species.

Having made this statement it must be noted that *Asymmetron cultellus* (Peters) can be variable in form. Two forms can be recognized,
one in which the dorsal fin rises steeply from the rostrum, the other
in which the dorsal fin rises gradually (see Franz, 1922, p. 420). The
latter form is the norm in Acrania. It has been noted that this is not
a geographical variation in that both types have been taken from the
same locality (Wickstead, 1964a).

These different types might be a sexual dimorphism, a size and/or
maturity feature, or something else again; there is, as yet, no positive
evidence in any direction. It is a point which should be relatively easy
to resolve, however, and it is to be hoped that a worker in the western
Pacific area will collect sufficient fresh material for this purpose.

9.3.2 Sex Determination and Hermaphroditism

Normally the sexes are separate in the acraniates. X and Y chromosomes have been recognized (Nogusa, 1957), and it is assumed that
sex determination conforms to the usual X and Y mechanism. Howell

and Boschung (1971), however, could not identify X and Y chromosomes in *Branchiostoma floridae* Hubbs, nor could Colombera (1974) in *B. lanceolatum*.

Hermaphrodite individuals in acraniates are well authenticated. Goodrich (1912) found a specimen of *B. lanceolatum* with 25 gonads (all testes) on both sides, with the exception, that on the left side, ". . . a single ovary containing numerous large ova, which could be distinguished even in the living animal . . . No trace of ova can be seen in the other gonads; and no trace of spermatozoa can be seen in the ovary."

Orton (1914) noted a hermaphroditic specimen from Plymouth. This specimen had 43 gonads, 22 testes along the right side, two of which were seen to discharge spermatozoa, and 21 gonads along the left side. Of these 21, 20 were testes, but one, the fifth from the anterior, was typically female, and full of eggs.

Both of the above instances are very similar. In both, one single gonad toward the anterior of the left series is completely female, the remaining gonads being completely male.

A different form of hermaphroditism in *B. lanceolatum* was reported by Langerhans and by Riddell. In the first case, Langerhans (1875) found the tails of spermatozoa among the young oocytes of amphioxus. Riddell (1922), when examining prepared slides of an amphioxus, noted that individual gonads contained both spermatozoa and eggs.

Chen (1931) described a specimen of *B. belcheri* having equal numbers of male and female gonads. Chin (1941) also found a specimen of *B. belcheri* with equal numbers of testes and ovaries, and goes on to say that "the hermaphrodite specimen from Amoy is an extraordinarily beautiful find, due to the arrangement of the creamy-white testes and the lemon-yellow ovaries." This appears to be the first observation of color differences between testes and ovaries.

9.3.3 Anatomy of the Reproductive System

As has been said above the sexes are normally separate in the Acrania and, apart from the reproductive organs, there is no known difference between the male and female.

The gonads are hollow sacs developed metamerically in association with the myotomes. They are associated with that part of the body between the mouth and atriopore. In the Branchiostomidae, a row of is developed on either side of the body (Fig. 2); in the Asymmetrontidae, gonads are developed along the right side only. In the Branchiostomidae the gonads on the right side usually number one or

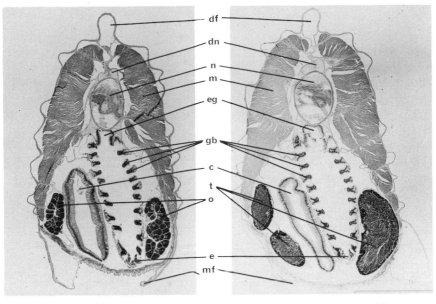

FIG. 2. Transverse sections of *Branchiostoma caribaeum;* female, left; male, right. df, Dorsal fin support; dn, dorsal nerve cord; n, notochord; m, myotome, eg, epipharyngeal groove; gd, gill bars; c, caecum; o, ovaries; t, testes; e, endostyle; mf, metapleural fold.

two more than on the left. The number of gonads is usually in the order of about 28 on the right side and 27 on the left. Species with a large number of myotomes, e.g., *Branchiostoma elongatum* (Sundevall), which has approximately 79 myotomes, can have 38 gonads along one side. Reference should be made to Franz (1922) for numbers of gonads usually found in different species.

Each gonad is a separate sac, with no common duct. As the gonads develop, they project into the atrial cavity. Surrounding the gonad is an epithelium of coelomic origin; thus each gonad is surrounded by a coelomic sac.

Release of the mature gametes is by a rupture of the atrial wall of the gonads, the gametes thus being released into the atrium. According to Young (1962, p. 41), this rupture subsequently closes and the gonads redevelop. This presupposes that amphioxus is capable of breeding more than once. Probably this is true in species living in relatively cool areas, e.g., *B. lanceolatum* and *B. belcheri*, but it is thought that *Branchiostoma nigeriense* Webb spawns but once, the act of spawning being followed by death (Webb, 1958).

9.3.4 Origin of Germ Cells and Gonads

Much of the original research on the origin of the gonads in amphioxus is to be found in the works of Boveri (1892a,b), and relates to *B. lanceolatum*.

From the arms of the mesodermal crescent of the gastrula originates a band of small, rapidly dividing mesodermal cells which forms the mesodermal bands (see Conklin, 1932). Constrictions of these mesodermal bands show the boundaries of the somites. The somites are later pinched off to form the myotomes. Thus each myotome is bounded by mesodermal epithelium, and each consecutive pair of myotomes is separated by two layers of mesodermal epithelium between which persists some of the myocoel (see de Beer, 1928).

It is said that the development of the gonads begins when the young amphioxus is about 5 mm long (see Willey, 1894b, p. 151). I cannot find any specific reference to the relationship between the onset of the development of the gonads and metamorphosis of the larva but, almost certainly, development of the gonads is associated with metamorphosis rather than with a given size. Off Messina the larvae metamorphosed at approximately 3.5 mm (Willey, 1894b). Off Plymouth, where the water is appreciably cooler, the larvae are usually more than 5 mm long before the onset of metamorphosis (Wickstead, 1967). I have not seen any evidence of developing gonads in these larvae. Thus, rather than give a particular size at which the gonads begin to develop, it is probably more accurate to say that the development of gonads begins after metamorphosis (see, however, p. 313).

The sexual cells are initiated by differentiation of some of the epithelial cells lining the myocoel, these cells being at the base of the myotome (Fig. 3A–H). As the cells develop, they protrude into the myocoel anterior to them, to the extent that the gonad becomes shut off from its original myocoel and hangs freely into the next. As this primitive gonad enlarges, so does the basal wall of the myocoel in which it now lies evaginate. The cavity of the sacs thus formed, the gonadial pouches, constitutes the perigonadal coelom and, at the time of maturity, is filled completely with the sexual elements. These developments can be followed in Fig. 3A–H, which are taken from Boveri (1892a). The developing mass of sexual cells, at first solid, develop a cavity, the gonadal sac. This gonadal sac develops a series of lappetlike outgrowths, becoming eventually a racemose reproductive gland.

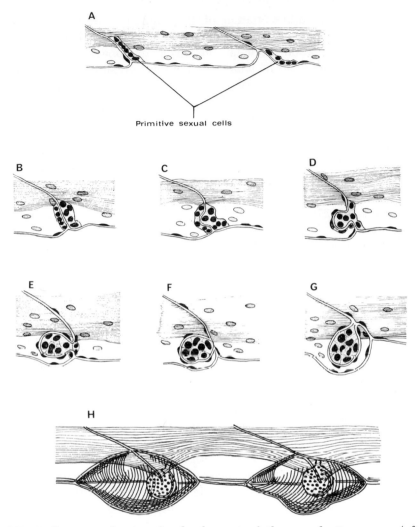

Fɪɢ. 3. Diagrams showing the development of the reproductive organs (after Boveri, 1892a). (A–G) Lateral view showing origin of primitive sexual cells from the myocoelic epithelium. (H) A lateral view of part of an individual ca. 13 mm long to show the young gonads within rhomboidal pouches of the myocoel projecting into the atrial cavity.

9.3.5 Gametogenesis

The formation of the gonadal pouches and the initial development of the sexual cells takes place at an early stage in the growth of amphioxus (see Section 9.3.4); but according to Langerhans (1875), in *B. lanceolatum,* differentiation of these sexual cells into male and female does not begin until the animal is ca. 17 mm long. I cannot find any information on the length at which the gonads are first visible in the whole animal of *B. lanceolatum;* it is probably about 35 mm. Compared with some other species, gonads can be detected in *B. belcheri* at 29 mm (Chin, 1941, p. 408). I have recorded a specimen of *Asymmetron lucayanum* of 10.0 mm total length in which there were 18 gonads, nearly mature (Wickstead, 1971). Also from my own records, gonads are visible in *Asymmetron cultellus* at a total length of 11 mm, and are virtually mature at slightly more than 13 mm.

9.3.5.1 SPERMATOGENESIS

Nogusa's (1957) observations on *Branchiostoma belcheri* during spermatogenesis showed a diploid number of 32 chromosomes and a haploid number of 16, all chromosomes being of the simple rod type. He reports ". . . 15 pairs of autosomes of isomorphic nature, and an XY pair forms a heteromorphic bivalent in the first meiosis and segregates into the X and Y in the first division, resulting in the production of two sorts of secondary spermatocytes." Howell and Boschung (1971), however, could not identify X and Y chromosomes in *B. floridae,* and they found a haploid range of 18 to 21 chromosomes with a mode of 19 and a diploid range of 37 to 40 with a mode of 38. Colombera (1974) affirmed a haploid number of 19 chromosomes in *B. lanceolatum,* and a diploid number of 38, all autosomes.

As far as is known there is nothing unusual in the development of spermatozoa in amphioxus. Growth and development is inward from the germinal epithelium lining the gonad. Spermatogonia originate from the germinal epithelium. These multiply by mitotic divisions until a certain number of primary spermatocytes are produced. These enlarge and each undergoes a primary meiotic division to form two secondary spermatocytes. Each of the latter, in turn, undergoes a secondary meiotic division to form two spermatids, each of which then develops into a spermatozoon.

Although, as stated above, there appears to be nothing unusual about the formation of the spermatozoa, there are conflicting ideas concerning the shape and structure of the spermatozoon.

According to Franzén (1956) the amphioxus sperm is of the primitive type. This follows from the work of Retzius (1905a,b), with which Franzén agrees, stating that: "This sperm consists of a small, conical to oviform head with a small, highly refracting particle at its anterior tip. The middle piece is short, and contains four to six globules of mitochondrial material. The tail is a thin filament beginning at the posterior pole of the head, and is about 50 μm long. Retzius considered this

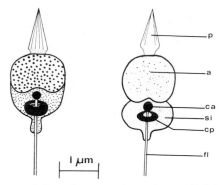

FIG. 4. Spermatozoon of *Branchiostoma lanceolatum* (left) according to Zarnik (1905). Interpretation by Drach (1948) of the spermatozoon (right) according to Zarnik. a, Acrosome; ca, cp, proximal and distal centrosomes; fl, flagella; p, perforator; si, intermediate segment.

sperm to be the primitive one within the Metazoa. . . ." Later, Franzén (1956, p. 453) describes some variations and modifications in different sperms, and says: "The only criterion which shows that we are after all in the presence of the primitive type is found in the four or five distinct mitochondrial spheres in the middle piece which is always short."

More specifically concerning the amphioxus sperm, Franzén (1956, p. 448) states that his results agree with those of Retzius (1905b, p. 103, plate 19). He goes on to say that the work of Retzius has been overlooked and it is the inaccurate and incorrect work of Zarnik (1905) which has been passed on, as for instance by Drach (1948, p. 1000, Fig. 442). This is perhaps a little unjust, since Drach's interpretation of the drawing of Zarnik is not quite as Zarnik expressed in his original paper in 1905. Figure 4 is from Zarnik's original paper and Drach's interpretation. There seems little justification for Drach's labeling as "perforateur" what Zarnik calls "eine Pars anterior'," or labeling as "acrosome" what Zarnik calls "eine Pars posterior." It is possible to see definite similarities between Zarnik's interpretation of a spermatozoon and the electron micrographs in Fig. 5, although the acrosome is clearly stylized. It is perhaps more a question of accurate observation followed by uncertain interpretation. Zarnik even notes the mitochondrial body in the form of a toroid, although he refers to it as "ringförmiges Centrosom." I would suggest then that the observations of Zarnik were in fact more accurate than the observations of Retzius.

In Fig. 5 are shown electron micrographs of a ripe testis of *B. lanceolatum*. It is important to stress that these micrographs show spermatozoa within the testis, and not after release. There is the possibility that

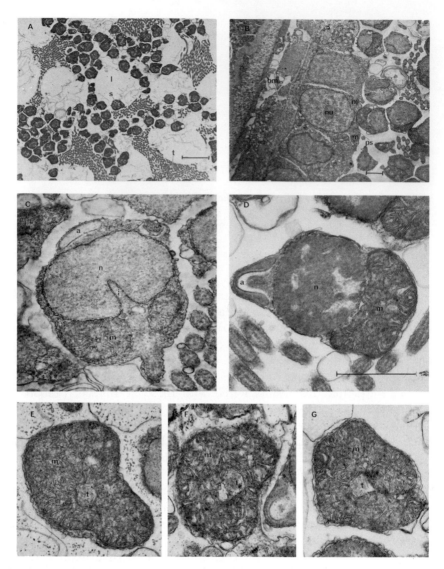

FIG. 5. Electron micrographs of amphioxus testes. (A) General view of testis showing groups of spermatozoa, heads, and tails in various sections, and lacunae. (B)
Wall of testes showing primary spermatocytes and basement membrane. (C) and
(D) Heads of spermatozoa showing insertion of tail (C) and prominent acrosome
(D). (E) and (F) Transverse sections of spermatozoa showing tail passing through
mitochondrial body. (G) Transverse section of spermatozoon showing where mitochondrial bodies are not completely fused. a, Acrosome; bm, basement membrane;
i, insertion of tail; j, junction of mitochondrial bodies; l, lacunae; m, mitochondrion;
n, nucleus; nu, nucleolus; ps, primary spermatocyte; s, spermatozoon; t, sperm tail.
Scale line in (A), 5 μm; (B) and (D), 1 μm; (C–G), all at the same magnification.

some further changes occur after discharge. Nevertheless the spermatozoa shown do appear to be reasonably mature, and probably undergo
little change, if any, after release. Figure 5A is a general view of a
section of the body of the testis. This shows the grouping of the spermatozoa interspersed with lacunae, and the groups of tails in various sections. Where the section happens to be in the right plane the acrosome
is conspicuous, as also are the nucleus and mitochondria. Figure 5B
shows, as I interpret it, the basement membrane and the peripheral
primary spermatocytes. It is possible to see the relatively huge nucleus
bounded by a distinct nuclear membrane and containing the nucleolus.
The mitochondria are relatively dispersed. Further in from the periphery
one can see the developing spermatozoa, with the prominent nucleus
and mitochondria, plus tails in cross section. Figure 5C and D are higher
magnifications of the heads of two spermatozoa. From examination of
the heads of spermatozoa in various sections it would seem that the
acrosome in Fig. 5D is a medial section. It is relatively larger and more
pointed than as depicted by Franzén. The nucleus is very large and,
while it approximates an oblate spheroid, Fig. 5C shows clearly the
insertion of the flagella into the nucleus, at the centriole.

A fundamental difference of interpretation concerns the mitochondria.
Retzius and Franzén consider the four or five mitochondrial spheres
in the middle piece to typify the primitive type of spermatozoa. Franzén
(1956, p. 449) states that there are usually four mitochondrial spheres
in the Acrania. From studies of electron micrographs, I have come to
the conclusion that the fusion of the mitochondrial elements in the developing spermatozoa is carried to the ultimate conclusion, and that
only a single mitochondrial body is present in the mature spermatozoon.
This is in the form of a toroid, with the tail passing from the centriole
through the center of the mitochondrial body. Figure 5E and F are
transverse sections across the mitochondrial body, and show clearly its
toroidal form. (Compare this with Fig. 5G, in which the mitochondrial
bodies are not completely fused.) Regarding primitive sperms in general,
Franzén (1956, p. 458) does state of the mitochondrial spheres: "They
form a regular ring around the base of the axial filament, and are usually
placed so closely together that the limits between them are visible only
under the highest magnification of the phase-contrast microscope." A
study of the electron micrographs does not reveal any separate mitochondrial spheres, and the conclusion must be that they have coalesced
to form a single body. Thus my interpretation of the mature spermatozoon of B. lanceolatum is as shown in Fig. 6.

Baccetti et al. (1972) agree with my conclusion about the mito-
chondrium and found that the mitochondria fuse into a single mass.

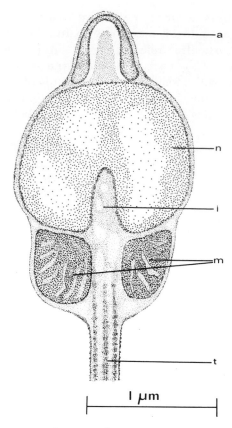

FIG. 6. Semidiagrammatic drawing of a spermatozoon of *Branchiostoma lanceo-latum* according to the author. a, Acrosome; n, nucleus; i, insertion of tail; m, single mitochondrial body in form of a toroid; t, tail.

They also state (p. 213) "The present ultrastructural picture of the spermatozoon of *Branchiostoma* fits the schematic description of Retzius (1905b) and Franzén (1956), who noted the characteristic asymmetry." However, Franzén's (1956) diagram of a sperm is symmetrical and he says (p. 449) "Retzius reports sperms with an asymmetrically placed middle piece to be fairly common. According to my own observations they constitute, however, exceptions."

9.3.5.2. OOGENESIS

The initiation of oogenesis follows the same pattern found in spermato-genesis. Cells of the germinal epithelium are probably the source of the oogonia. The oogonia which are small cells multiply to form the primary oocytes. The primary oocytes enlarge and become filled with the food reserves. Full-grown oocytes are often referred to as the "ovarian eggs." According to Cerfontaine (1906, p. 257) the full-grown oocytes pass from the germinal epithelium into the secondary ovarian cavity. Conklin (1932, p. 73) says that these, "ovarian eggs," are found in groups of a dozen or so attached to the membrane lining the "cavité ovarienne secondaire." Each of the full-grown oocytes is surrounded by a thin follicular epithelium (Conklin, 1932, p. 73; Raven, 1961, p. 26) and is about 0.1 mm diameter. In comparing the egg with an ascidian egg, Conklin (1932) states that underneath this epithelium is a peripheral layer of protoplasm free from yolk spherules. The works of van der Stricht (1895), Cerfontaine (1906), and Conklin (1932) give some details of the cytoplasmic structure of the egg, but the general appearance shows clearly the large, eccentrically positioned germinal vesicle and the granular cytoplasm containing yolk spherules. Recent work attempts to relate egg size with the species specific DNA content (Bier, 1970). Conklin (1932) attaches much importance to the positioning of the germinal vesicle, which is against the peripheral layer of cytoplasm on the side of the egg which is attached to the ovarian wall (see Fig. 2). This gives a clear indication of the axis and the positioning of the animal (attached) and vegetative (free) poles.

Oocyte development can be divided into three phases (Brachet, 1950; Raven, 1961), each being connected with a different metabolic process. Details of these stages are given by Cowden (1963) for B. caribaeum. Briefly, it may be said that the first phase is characterized by much RNA synthesis, the second by the accumulation of yolk granules, and the third by the development of a broad cortical region and a crenated nuclear membrane.

The secondary oocyte is formed after the first meiotic division and the first polar body thus formed lies outside the fertilization membrane (see the accounts of van der Stricht, 1895; Sobotta, 1897; Cerfontaine, 1906). Fertilization, as discussed below, is external, occurring at the metaphase of the second meiotic division, after which the second polar body is formed. The second polar body usually remains attached to the egg, and is within the fertilization membrane.

Although there are accurate times known for embryonic development, there appears to be no positive information on how long the gonads take to develop and ripen. In B. lanceolatum, when overwintering, the gonads can be detected as pale shadows along the sides of the body. Gametogenesis probably begins early in the year in this temperate species, and some 3 or 4 months are required for the gametes to become fully mature. In the Indian Ocean, from which B. lanceolatum has been recorded (see Wickstead, 1964a), gametogenesis probably takes only about 3 or 4 weeks, and individuals might be able to breed several times during the year.

From Chin (1941) it would appear that it takes about 6 to 7 weeks for the gonads of B. *belcheri* in the Amoy area to develop to ripeness. A similar period seems to apply to B. *nigeriense* (Webb, 1958).

9.3.6 Factors Influencing Gametogenesis

Acrania are normally benthic animals living in shallow water, i.e., <200, usually <50 m, or even intertidally at times, and are found generally in the warmer parts of the oceans. As such they are exposed to many variables, e.g., temperature, incident light, salinity, and tidal influences. Since acraniates appear to have definite breeding seasons, some environmental factor or factors clearly affect gametogenesis. Exactly what these factors are is not known with certainty, but sufficient information is available to make a reasonable assessment.

9.3.6.1 TEMPERATURE

One could go through the complete range of physiological conditions which might affect gametogenesis and breeding, but I think Kinne (1963, p. 314) sums up the situation neatly and concisely when he says: "Once certain prerequisites are given, such as the appropriate physiological conditions, food and space, the *time of reproduction* (breeding season) of most marine or brackish-water animals depends primarily on temperature." This statement almost certainly applies to acraniates. Thus when the amphioxus has reached that growth stage at which it is physiologically prepared for reproduction, and when sufficient food is available for the gonads to develop fully, the reproductive cycle is probably triggered by the environmental water attaining a temperature optimum. This temperature optimum may be sensed by Hatschek's pit.

Temperature cycles are present in tropical latitudes, although usually they are not as pronounced as those in the higher latitudes. However, a change of a few degrees is sufficient to result in associated breeding cycles (see Yonge, 1930; Stephenson, 1934; Gunter, 1957; Wickstead, 1961). In some animals it is possible to relate breeding to a definite temperature, a temperature minimum or a temperature maximum (see Orton, 1920; Thorson, 1946; Wickstead, 1963; Giese and Pearse, 1974). In the Acrania, such personal communications available to me indicate that the reproductive cycle is initiated by a rising temperature, there probably being a threshold temperature for any given area.

9.3.6.2 TIDES

Acrania, living in shallow water, will certainly be affected by the varying pressures resulting from different states of the tides. This effect could be important in such areas as the south coast of England, where tidal ranges vary between 4 and 6 m. On the other hand, acraniates do occur in the Mediterranean, Red Sea, and at various amphidromic points of the oceans. I think it reasonable, in general, to dismiss the idea of pressure fluctuations as a factor influencing gametogenesis of acraniates.

9.3.6.3 LIGHT

It is thought that light is the main factor concerning the breeding cycle of some animals, or at least the final part of the cycle (Clark, 1965). There is no doubt that amphioxus is light sensitive and reacts to flashes of light and high light intensities. It is common for the animal to leave the bottom at night and swim around in the plankton (Andrews, 1893; Wickstead, 1971). Reaction to light is by way of the pigment cells which are associated with the nerve cord; those along the body and tail are the most sensitive (Parker, 1908). Again the habitat of the acraniates is such that they could perceive long-term changes in light, e.g., length of day. It might be said of amphioxus in north temperate areas that gametogenesis and maturation of the gonads begins at about the Spring equinox, when the hours of daylight begin to exceed the hours of darkness, but no experiments have been performed on the effect of light. However, I have seen the reverse hold in the tropics, when gonads are ripe during that time of the year when hours of daylight are fewer than the hours of darkness, albeit that, being in the tropics, differences in daylight are small.

9.3.6.4 PARASITES

As it is well known that infestation with parasites can profoundly affect parts of the reproductive cycle and gametogenesis in some animals, it should be mentioned that the Acrania appear to be singularly free from parasites, and there is no record of any parasite having any effect on gametogenesis.

9.3.6.5 ENDOCRINE

The hormonal control of reproductive cycles in higher vertebrates, via an integrated endocrine system, is well known, the overriding control being provided by the pituitary gland. The lowest group of vertebrates, the Agnatha, possess a well-developed neurohypophysis and adenohypophysis (Adam, 1963, p. 459; Barrington, 1963, p. 303). While there does not yet appear to be positive proof that the tunicates, the group of animals which appears to be closest to the Acrania, possess a pituitary body (see Barrington, 1963, p. 306) there is nevertheless strong evidence to suggest that the neural gland, plus ciliated pit, of a tunicate is the homolog of the vertebrate pituitary gland, and plays an important part in their reproductive cycle (Carlisle, 1953).

Regarding a pituitary gland in amphioxus, there are two organs which can be considered; there are Hatschek's pit and the infundibular organ (Barrington, 1963, p. 303). Hatschek's pit is present at an early stage of the animal, being developed at about the 7 or 8 primary gill slit stage (Smith and Newth, 1917). Of a relationship with the neural gland of tunicates Young (1962, p. 65), speaking of the neural gland, says: "There is obvious similarity with Hatschek's pit of amphioxus. Both seem to be receptor organs, testing the water stream and also producing mucus." Carlisle (1953) states that: "The first function of the pituitary would appear to be one of chemoreception."

It does appear then that there is a distinct possibility that a pituitary-like gland exists in amphioxus, and that this could be the sensor unit for environmental conditions and the endocrine center that regulates gametogenesis.

9.3.7 Reproductive Behavior

So far as is known there is no set pattern or sequence of events which can justify the term reproductive behavior. No courtship has been observed; in fact there is no proof that the individuals emerge from their buried or semi-buried position to discharge the gametes, although one assumes that they do. As discussed below, the gametes are released at about sunset. Adult acraniates often emerge from the bottom at about sunset and swim around freely near the bottom, and one would assume that the gametes are released during this period of emergence. It is not known, however, whether the gametes are released while the animal is actually swimming or during the intermittent resting periods on the bottom; laboratory observations suggest the latter.

No swarming migration has been noted, the communities of acraniates being usually dense enough to ensure the mixing of the gametes. Smith (1932) indicates a concentration of probably more than $100/m^2$, and Webb (1958) comments on the high concentration of *B. nigeriense* which can be found in Lagos lagoon. Professor Goodbody has shown this author very high concentrations, up to 5,000 per m^2, of *Branchiostoma caribaeum* Sundevall in Kingston Harbour, Jamaica, in depths of only 0.25–0.5 m.

9.3.8 Spawning and Breeding Periods

9.3.8.1 SPAWNING MECHANISM

The gonads have no ducts; gametes are released by rupture of the atrial wall. Initially there was some confusion as to how the gametes reached the outside water. Kowalevsky (1867) asserted that the gametes were released through the mouth. Hatschek (1893), one of the most careful of observers, said: "I can fully confirm Kowalevsky's statement. . . ." As there is no doubt now that the gametes are discharged initially via the atriopore (see Bert, 1867), there must be an explanation of the observations of Kowalevsky and Hatschek. In a dense population of acraniates, which shows some degree of synchrony in spawning, it is certain that, at the actual time of spawning, there will be a concentration of gametes in the spawning area. Since there is a virtually continuous current of water passing through the mouth of individuals, and out through the atriopore, one can be certain of some gametes being taken in through the mouth, subsequent to release, and treated as food particles. Hence, Andrews (1893, p. 218) noted the fact that in sections of *A. lucayanum*: ". . . ova and sperm are found in all parts of the digestive tract, in the pre-oral chamber, stomach, intestine, and extreme end of the rectum." Since Andrews noted that some of the eggs had been fertilized, it is clear that they were taken in from the outside.

Another explanation lies in the habit of amphioxus to collapse the pharyngeal basket suddenly and energetically, ejecting water out of the mouth. This is a phenomenon very characteristic of the closely allied tunicates, giving rise to their popular name of sea squirts. Carlisle (1966), from his work with *Phallusia mammillata* (Cuvier), considers this sudden contraction of the gill basket to be a mechanism for cleaning off the gills when they become clogged. I think this is also the correct explanation for the behavior in amphioxus. Thus if there were to be this sudden collapse of the gill basket while the atrial cavity contained gametes, the gametes would be forced into the pharynx and, in some instances, out of the mouth.

When the gonadal wall ruptures it is not known with certainty if all the gametes are discharged at once and if so how long this takes, or whether they are discharged at intervals over hours, days or even weeks at varying time intervals. One point that appears definite is when an amphioxus is ripe all the gametes are at about the same stage of development, and the gonad is emptied on spawning. In some fish, for instance, when ripe, apart from the ripe eggs, a frequency distribution of all the eggs in an ovary often shows three or four different stages of development. Even so, while all the gametes in any gonad may be ripe, do all the gonads rupture at the same time?

Gonad development begins at about the center of the pharyngeal region, progressing anteriorly and posteriorly. A small individual will have, usually, fewer gonads than a larger individual of the same species, in all cases the center gonad of the row being the largest. Thus it may be that the center gonads mature and discharge before the anterior and posterior gonads are mature. Orton (1914) definitely states that some gonads were empty, the contents having been discharged, while other gonads remained full. Other authors are not as definite; but the impression is that the gonads ripen successively, being discharged when ripe and when weather conditions are suitable. Thus gametes are discharged at intervals over a period of several days to a week by the successive ripening and discharging of the gonads.

9.3.8.2. Spawning Time

Spawning appears to occur during a particular time of day. Regarding *B. lanceolatum*, Willey (1894b) wrote: "Spawning, when it occurs, invariably takes place at sundown—i.e. between 5 and 7 o'clock in the evening—and never, so far as is known, at any other time." Hatschek (1893) also found that spawning begins at sunset and: "When darkness begins, the spawning advances with great speed. . . ." Conklin (1932) says that ripe animals, if taken between 4 and 6 p.m., would often begin to shed gametes at once. These three observations were made in Mediterranean conditions. Orton (1914), at Plymouth, said: "Spawning apparently occurs usually overnight. . . ." When this observation was made sunset would have been at about 20.15 hours, so there probably was nobody present at this time to observe spawning. By extrapolation from the time of gastrulation, Orton considered that spawning had occurred at about midnight.

Hatschek (1893, p. 27) concluded that weather conditions were also important for spawning. No spawning occurred during "cool and stormy weather," but spawning did occur "on warm sunny days." Differences in weather conditions might be marked for acraniates living in only several meters depth of water, but one would not expect differences to be noticeable to, say, the Eddystone population off Plymouth, in deeper water (45–50 m).

9.3.8.3 Breeding Seasons

As stated above, temperature is considered an important factor for the initiation of gametogenesis. Thus any community of acraniates which is exposed to the same variations of temperature will show good synchrony regarding development of the gonads and release of the gametes. However, so far as is known, in nature there is not a rigid synchrony of gametogenesis, rather reproduction is restricted to part of the year. Thus, within a given population, there will be synchronous spawning of many, but not all, individuals at different times during the breeding season.

In the northern temperate area of Europe the breeding season of *B. lanceolatum* is in late Spring. Orton (1914) obtained ripe adults from Plymouth in June. From the stage of development of larvae taken off Plymouth in August and September (Wickstead, 1967), and knowing the approximate rate of development, it is possible to extrapolate backward and be confident that the main breeding period of *B. lanceolatum* off Plymouth is from mid-May to about the end of June. The Mediterranean populations around Italy appear to have a longer breeding season, from late March through the Summer according to Hatschek (1893), and from Spring to Autumn according to Willey (1894b). According to Hatschek and Conklin (1932) the peak of the breeding season appears to be late May–early June. In these areas there is only a single breeding season.

According to Azariah (1965a,b) there is a fairly extensive population of *B. lanceolatum* off the coast of India in the Madras area. From the appearance of the gonads in his samples, he deduces that this species has two breeding seasons in the area, one during July August, the other between December and February.

With *B. belcheri*, Chin (1941) concludes that this species, in the Amoy area, also has two breeding seasons, May–July and in December. A variety of this species, *Branchiostoma belcheri tsingtaoense*, described by Tchang and Koo in 1936, has been used for experimental embryological work in Shantung University. This is situated further north than Amoy, and Tung *et al.* (1958) state that the breeding season for this variety is from the middle of June to the middle of July.

Nelson (1968) concludes that *B. caribaeum* in Old Tampa Bay, Florida, spawns from late August into December. In addition he considers that: "some amount of spawning evidently occurs throughout the year." Andrews (1893) noted that some individuals of *A. lucayanum* taken at the Bahamas during early June had ripe gonads and discharged gametes.

Details from areas south of the equator are even fewer. Maximum numbers of the larvae of *A. cultellus* were taken in the Zanzibar area in January during the north-east monsoon (Wickstead, 1964b). Extrapolating backward from the stage of development of these larvae brings the breeding time to about October–December, probably mainly toward the end of November. Again it should be noted that this is during a rising temperature (Wickstead, 1963).

Blackburn (1956) took larvae of *Asymmetron bassanus* (Gunter) from the Bass Strait during April; extrapolation places the breeding time in about January–February. However, since Bone (1957) considered these larvae to be amphioxides larvae which might spend a much longer period in the plankton than the usual larva, this is a doubtful extrapolation.

With the species discussed above, it is assumed that the animals have a period of rest after the breeding season, before the gonads again begin to develop for the next breeding season. This is a reasonable assumption, but so far as I am aware there is no experimental proof that this is so since no one has bred the same individuals several years in succession. Chin (1941), however, concluded that individuals of *B. belcheri* live for between 2 and 3 years and breed when 1 year old, 1½ years old, and between 2 and 2½ years old. No similar data are available for other species.

In contrast to individuals breeding several times during their life span, Webb (1958) suggests that individuals of *B. nigeriense* breed but once and then die. He qualifies this by stating: "From the condition of lancelets which have recently spawned in the lagoon, it seems likely that spawning is followed by death, but this cannot be assumed with certainty as their moribund state may have been due to the salinity of the water." There is the possibility, therefore, that populations of *B. nigeriense* in the sea breed several times, and it is the physical conditions within the particular lagoon which restrict this population to a single spawning.

It may be that species in higher latitudes spawn once a year, while those in lower latitudes spawn twice a year. This generalization may also apply to populations of single species with wide distributions since *B. lanceolatum* spawns once a year in the Mediterranean and further north, and twice a year in the Indian Ocean (Azariah, 1965a).

9.3.8.4 SUMMARY

In summary, as I interpret the evidence, the sequence of events in general terms is as follows. A rising temperature, with a possible threshold temperature, provides the stimulation for gonad development. Thus spawning occurs at about late Spring, that is about May–June in the northern hemisphere, and about November–December in the southern hemisphere. Spawning occurs twice each year, or throughout the year, in tropical populations. There are no spawning migrations or aggregations. Spawning occurs at about, or soon after sunset, when the animals leave the substratum and swim about near the bottom. The gametes are probably released while the animal is laying on the bottom rather than actively swimming. There is some degree of control over discharging the gametes as it seems likely that appropriate weather conditions must prevail, these weather conditions being clear sunny weather. Discharge of gametes is intermittent and spread over several days; this is due to the successive ripening and discharge of the gonads, beginning at the center of the row.

9.4 Development

9.4.1 Embryonic Development

In view of the amount of original work undertaken by highly competent embryologists on the development of amphioxus, it would be presumptuous to embark here on a detailed description. Instead, reference should be made to the original works of Kowalevsky (1867, 1876), Hatschek (1893), and Conklin (1932), and, in addition, the works of Wilson (1892; 1893), Cerfontaine (1906), MacBride (1898, 1900, 1909), and Reverberi (1971).

A general description of the embryology, based on the literature quoted above, is as follows (see Fig. 7A–I). The fully formed egg (primary oocyte) is about 0.1 mm in diameter. The germinal vesicle is large, being about one-half that of the egg just before the first maturation division takes place. It is that part of the egg containing the germinal vesicle which is pressed against the peripheral layer of protoplasm which is attached to the ovarian wall. Conklin (1932) greatly stresses this orientation and the fact that the animal pole is the attached pole while the vegetative pole is free. The suggestion is that attachment by the animal pole can be correlated with the dorsal nerve cord in vertebrates, and attachment by the vegetative pole, as seen in invertebrates, can be correlated with a ventral nerve cord. Waddington (1956) makes the point that it is not clear as to whether most vertebrate eggs are, like amphioxus, attached to the ovary wall by the animal pole.

The first polar body is formed prior to the entry of the spermatozoon, consequently lying outside the fertilization membrane. Rothschild (1956; see also Kumé and Dan, 1968) states: "In all vertebrates and *Branchiostoma*, fertilization takes place at the second maturation division metaphase (Class 3 fertilization). . . . ." Entering at the vegetative pole, the sperm nucleus, plus centrosome and aster, moves toward the equator of the egg. The egg nucleus moves down to meet the sperm nucleus just above the equator, where the first-cleavage amphiaster is formed.

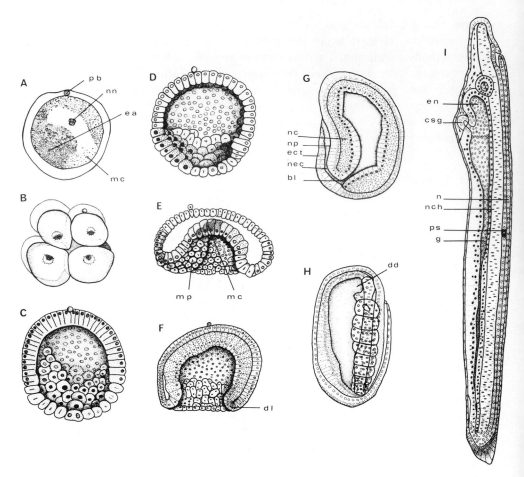

FIG. 7. Diagrams showing some developmental stages of *Branchiostoma lanceolatum* (after Conklin, 1932). (A) Egg 1 hour after fertilization. pb, Polar body; nn, egg and spermatozoon nuclei in contact; mc, mesodermal crescent; ea, endodermal area. (B) Eight-cell stage, 2½ hours after fertilization. (C) Section of blastula, 5½ hours after fertilization, showing large endoderm cells at bottom protruding into the blastocoel. (D) The beginning of gastrulation; note flattening in area of large endoderm cells. (E) Later gastrula of about 8 hours. The mesodermal cells, mc, are darkly stippled. Note bulging of first mesodermal pouch, mp. (F) Gastrula of about 11 hours. dl, Dorsal lip of blastopore. (G) Gastrula of about 15 hours, showing ectoderm growing over the neural plate. nc, Notochord cells; np, neural plate; ect, ectoderm; nec, neurenteric canal; bl, blastopore. (H) Embryo of about 18 hours, showing the somites of the left side overlying the notochord and neural plate. The dorasl diverticulum of the gut, dd, is apparent at the anterior end. (I) Medial optical section showing left view of larva of about 48 hours. en, Endostyle; csg, club-shaped gland; g, gut; n, nerve cord; nch, notochord; ps, pigment spot.

The second polar body is produced after fertilization, remaining, therefore, within the fertilization membrane.

After entry of the spermatozoon there is a general reorganization of the cytoplasm so that in the zygote, apart from being bilaterally symmetrical, ". . . the distribution of the different ooplasmic substances foreshadows the location of the three germinal layers of the embryo" (Conklin, 1932, p. 78). According to Hatschek (1893), working in Mediterranean conditions, as were the other authors, the first division occurs at about 1 hour after the release of the eggs. From there on Hatschek published a timetable, ending with the formation of the mouth and perforation of the first gill slit, which takes place at about 48 hours after entry of the sperm.

The first cleavage is in the plane of bilateral symmetry, dividing the mesodermal crescent and all other cytoplasmic substances into bilaterally symmetrical right and left halves. The second cleavage, also in a vertical plane, at virtually a right angle to the first, divides the egg into two anterior quarters and two posterior quarters, the former being somewhat larger than the latter. There was some confusion in earlier works as to which were anterior and which posterior; it was finally resolved by Conklin (1932). The third cleavage is horizontal, dividing each of the four blastomeres into an upper micromere and lower macromere. Conklin notes that the upper micromeres contains about one-third of the substance of the original ovarian egg and the macromeres about two-thirds.

Cleavage continues until the blastula stage is attained; this blastula, according to Conklin, is pear-shaped. Although the blastocoel (cleavage cavity), is present from the 4-cell stage, the blastula is completely formed with the eighth cleavage, and Conklin notes that it is also the end of the synchronous cleavages. The enlargement of the blastocoel, as seen in the blastula, is due partly to the swelling of the blastocoel jelly, a gelatinous substance formed between the early blastomeres. The second polar body is still present at this stage, enabling the correct orientation of the blastula to be determined.

Gastrulation begins with a slight flattening of the vegetative pole, where the cells are rather larger than the other cells of the blastula. This, which Conklin calls the endoderm plate, is surrounded on its lateral and posterior borders by the mesodermal cresent. As gastrulation proceeds the cavity thus formed remains in communication with the exterior by way of the blastopore. During gastrulation the blastopore changes shape; triangular, horse-shoe, becoming shaped like a pear in longitudinal section by the swelling of the lateral lips (to form the mesodermal pouches), then quadrilateral or oval by upward growth of the lower lip. During this phase, elongation of the embryo begins.

The neural plate can now be seen as a plate of rather larger cells on the flattened, dorsal side of the gastrula. This becomes covered by a growth of ectoderm cells in the ventral lip region of the blastopore. This growth takes initially the form of a shallow "**V**," the arms of which grow over the neural plate and join together, proceeding anteriorly, thus covering the neural plate. Meanwhile the neural plate becomes longer and narrower, rolling up on itself, eventually producing the neural tube much as is found in the vertebrates.

Meanwhile, small, rapidly dividing mesoderm cells, originating from the arms of the mesodermal crescent, extend forward to form grooves in the lateral wall of the gastrocoel, the cavity formed by gastrulation. These grooves are bounded on their dorsal and ventral side by notochordal and endodermal cells, respectively. Soon, constrictions in these mesodermal grooves, or bands, mark out the somites.

As the organs develop, the mouth and first gill slit appear after about 48 hours, the mouth connecting through into the gut at its anterior end. The anus forms an opening a little anterior to the site of the blastopore remnant. By the time the mouth has formed the reserve of yolk granules has been consumed, and the larva begins to feed.

Original work, particularly that of Conklin (1932), should be referred to for the complete details of embryological development and organogenesis. Waddington (1956) gives a useful summary and relates amphioxus to other animals.

9.4.1.1 INDIVIDUAL CELL DEVELOPMENT

Conklin (1933), after the work of Wilson (1892, 1893), followed the development of isolated blastomeres and parts of the developing egg of *B. lanceolatum*. He concluded that the egg of amphioxus is of the mosaic type, and that only the separated blastomeres of the first cleavage, if divided along the original cleavage axis, the left or right pair of blastomeres of the 4-cell stage would develop to give a normal larva. This is because, at this early stage, the "precursory" elements in the egg can be divided equally; this does not prove possible at the later stages of division. He makes the statement: "All axes and poles of the future larva are irreversibly determined at or before the first cleavage; complete regulation occurs only in the plane of bilateral symmetry; except for this bilateral regulation, development in Amphioxus, as also in Ascidians, is a mosaic work."

These conclusions are somewhat contradicted by the work of Tung, *et al.* (1958) who instead used *B. belcheri tsingtaoense* for their experiments; they found some differences in normal development including the fact that the blastomeres of the 4- and 8-cell stages were similar in size, and concluded: "Thus we may say that the potency of blasto-meres in the Amphioxus is neither totipotent nor mosaic as suggested by Wilson and Conklin."

Thus while the normal development of amphioxus is known in some detail, agreement has still to be reached as to what extent the cytoplasm retains a flexibility of development, and at what stage the flexibility ceases, and also what interspecific differences there are.

Recently a chapter "Amphioxus" in a book on experimental embryology (Reverberi, 1971) discusses particular developmental potentials of the egg and development of isolated blastomeres.

9.4.2 Larvae

The free-swimming ciliated stage is reached after about 8 hours, at which time gastrulation has formed the archenteron, and the neural plate is being covered. Willey (1894b, p. 113) says: "The fact that Amphioxus has a free-swimming, ciliated embryo is important as proving a general connecting link between the Vertebrates and the Invertebrates, since the possession of a ciliated ectoderm is very common among In-vertebrate embryos, but entirely unknown among the craniate Verte-brates." The ciliated epithelium persists, probably, throughout the larval life, disappearing at metamorphosis; certainly I have seen what I at-tribute to ciliary movement in a larva just prior to metamorphosis (Wickstead, 1967). However, Webb (1969, p. 58) notes: ". . . the larva of B. nigeriense from West Africa and B. lanceolatum from Helgoland, and probably of other species, too, lose the external ciliation of the embryo early in larval life."

Muscular movement is noticeable at an early stage, Bone (1958b) having noted twitching movements in neurulae of 27 hours. Muscular locomotion effectually replaces ciliary locomotion by the time the mouth has broken through.

At an early stage the larva develops its well-known asymmetry. Various explanations have been advanced in explanation of the asymmetry (van Wijhe, 1913; Garstang, 1928; Willey, 1891; Medawar, 1951; Bone, 1958a); none seems to be completely satisfactory. No doubt associated with this asymmetry is the fact that the larvae (and adults) rotate on the longitudinal axis when moving through the water. Bone (1958b) com-ments: "While swimming forward the neurulae usually rotate anti-clock-wise (viewed from behind), but occasionally reverse this rotation and proceed for two or three complete turns clockwise, before resuming their usual rotation." The later larvae rotate in a clockwise direction, when seen from behind (Wickstead, 1967) as do the adults (Franz, 1924). Various arguments have been proposed to explain an adaptive significance of this rotatory movement; I consider it probable that it is a reflection of the hydrodynamic properties of the larva. The asymmetric disposition of the smooth surfaces and orifices would impart a rotatory force as water passed over the larva. In addition water passing in through a laterally disposed mouth, water passing out through gill slits laterally disposed on the other side of the body from the mouth, and probably fecal matter plus water passed out from a laterally disposed anus would also help to produce a reaction imparting a rotatory movement to the larva. This is clearly not the complete answer, however, as the early larval stages, e.g., the neurula, are symmetrical, but still rotate.

The free-swimming larva now enters upon a feeding and growing phase, varying between about 1½ to 4 months, according to temperature, no doubt, and food availability. Feeding occurs, basically, from a current of water passing in through the mouth and out through the gills. Particulate matter in the water adheres to mucus, is passed by ciliary action to the roof of the branchial chamber, and back along the gut. Here the food string is rotated rather like a skipping rope while digestion takes place. Remains are then passed along through the hindgut and out the anus (see Bone, 1958b; Barrington, 1965; Wickstead, 1967). The view has been put forward recently by Webb (1969) that amphioxus larvae have the capability of securing large food particles with the mouth, e.g., a copepod or large diatom, and passing this through the gut for digestion.

One of the most important diagnostic characters of the Acrania is the number of myotomes in any particular species. A problem in identifying young acraniate larvae is the fact that they do not possess the full complement of myotomes. When active feeding begins I would assess from my observations that only about 75% of the myotomes have been developed. A feature of the young larva is a terminal "knob" of densely packed nuclei, easily revealed, if a simple nuclear stain is used. There can be little doubt that, as the larva grows, so are myotomes added posteriorly until the full complement has been reached, which stage is reached just a little before metamorphosis (Lankester and Willey, 1890; Wickstead, 1964a).

Of great interest is the possession in acraniates of the endostyle, which is homologous with the thyroid of vertebrates (see Barrington, 1963, 1965). To what extent the life of the adult acraniate, and the development of the larva, is subject to endocrine control is speculative, however. Mention has been made above regarding the possibility of the presence of a pituitary-like organ. I have suggested (Wickstead, 1967) that the club-shaped gland of the amphioxus larva up to, and including, metamorphosis, governs larval life. The club-shaped gland begins to form as an evagination of the gut about 25 hours after the beginning of development. At metamorphosis the club-shaped gland regresses, being absent in juvenile and adult, while the endostyle undergoes a marked reorientation and development.

During its planktonic life the larva is partly planktonic, exhibiting a pattern of diurnal vertical migration common to many planktonic organisms (Wickstead and Bone, 1959). Feeding appears to be intermittent, taking place during the daytime, when the larva is resting on the bottom. It is at sunset that the larva begins swimming actively and enters the plankton. Webb (1969), from a series of observations, considers that the larvae also feed in mid-water; the larvae of *B. lanceolatum* swim upward, mouth and gills closed, then periodically sink in a horizontal position with the mouth and gills open. Webb (1969) suggests that food is taken in during the sinking periods.

Hartmann and John (1971) note the distribution of 438 planktonic *Branchiostoma lanceolatum* larvae in the North Sea. The larvae at the surface were significantly smaller than those taken at greater depths. There did not appear to be any vertical migration pattern.

In *B. lanceolatum* metamorphosis occurs when the larva has, usually, 12–14 gill slits, and is between 3.5 and 5 mm long. At this stage the metamorphosed juvenile settles to the bottom and adopts the habits of the adult. To adapt to this change, great changes occur in the larva.

The larval mouth opens to the left side, while the gill slits open to the right. Secondary gill slits appear first as thickenings above the primary gill slits (Fig. 8A). These perforate and enlarge (Fig. 8B and C). The primary gill slits meanwhile are migrating slowly underneath the body, the final result being that primary and secondary gill slits are symmetrically disposed on the left and right sides, respectively (see Willey, 1894b, for an account of metamorphosis). At the same time the tongue bars develop on the gills and there is a closure and subsequent disappearance of the first one and last five or so gill slits (Fig. 8D). There is a noticeable decrease in length. In some of my experiments I found regression of the gill slits to continue (Wickstead, 1967) and suggested that this might be part of the normal course of events. The mouth moves to a ventral symmetrical position and the buccal cirri develop, as also does the caecum (Fig. 9).

With such profound changes occurring, metamorphosis is clearly a critical period in the life history of the animal; so far as I am aware nobody has been successful in rearing larvae through this stage and on to the adult.

One of the features of the larval stage is the coloration. Most of

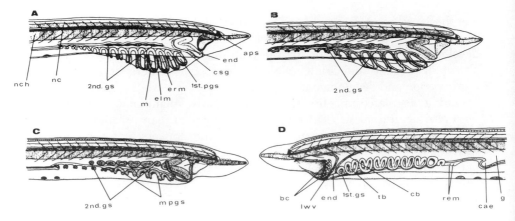

FIG. 8. Diagrams showing some stages in the metamorphosis of *Branchiostoma lanceolatum* larva (after Willey, 1891). (A) Thickenings initiating the secondary row of gill slits and the growing down of the metapleural folds to enclose the gills in the atrial chamber. aps, Anterior pigment spot; csg, club-shaped gland; elm, edge of left metapleure; end, endostyle; erm, edge of right metapleure; m, lower margin of mouth, on left side; nc, nerve cord; nch, notchord; 1st pgs, 1st primary gill slit; 2nd gs, thickenings indicating six secondary gill slits. (B) Thickenings of secondary gill slits have just perforated. 2nd gs, Secondary gill slits. (C) The enlarging secondary gill slits, a 7th gill slit having been added posteriorly, and the primary gill slits migrating under the body to the left side. The right and left metapleures have met and fused all the way along to form the atrium. 2nd gs, Secondary gill slits; mpgs, migrating primary gill slits. (D) Left side, gills now symmetrical, tongue bars have developed, the first primary gill slit has disappeared, the 1st gill slit in the diagram being, therefore, the 2nd primary gill slit. The remnants of the atrophied 11th and 12th primary gill slits can be seen. The mouth is moving to adopt a symmetrical position and the buccal cirri have begun to develop. bc, Buccal cirri; cae, caecum; cb, ciliated band; end, endostyle; g, gut; lwv, left wall of velum; rem, remnants of 11th and 12th primary gill slits; tb, tongue bar; 1st gs, 1st gill slit (=2nd primary gill slit).

the body is semitransparent when alive, but around the ilio-colonic ring, which is that part of the gut where digestion occurs, is a band of bright, fluorescent green; this green is also present, to a lesser extent, in the gill area. The purpose of this band of color is not clear; possibly it is not always present. Although the green band is a striking feature in all the larvae taken off Plymouth, it has been rarely noticed on larvae taken from other areas. Müller's original description (1851) made no comment on it, but it was noted by Leuckart and Pagenstecher (1858) and van Wijhe (1927). Webb (1958) made no mention of it in *B. nigeriense* larvae; I do not recall having noted it in *B. belcheri* and *Branchiostoma malayana* Webb larvae.

These metamorphic changes apply to *B. lanceolatum,* and have been

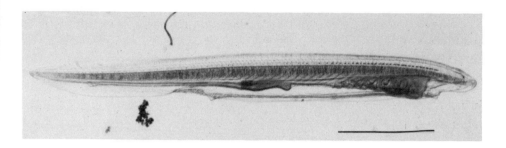

FIG. 9. Living specimen of a *Branchiostoma lanceolatum* larva at about the same stage as Fig. 8(D). Length before onset of metamorphosis 5.7 mm; length when photographed 4.6 mm; scale line, 1 mm.

well documented up to the assumption of symmetry by Willey (1894b). From observations in other species of acraniate larvae it can be reasonably assumed that this pattern of changes is general in acraniates, but the larvae are of varying sizes. A larger larvae requires more nourishment, and develops more gill slits. Thus Webb (1958) notes that *B. nigeriense* which has, usually, 21–22 gills when metamorphosis begins, averages 4.9–5.4 mm in length, while an individual with 25 gill slits averages 6.7 mm. Some idea of relationship between length and gill slit number is given in Figs. 10 and 11 (Wickstead, 1964b). These figures include data for the so-called amphioxides larva.

Amphioxides, described by Goldschmidt (1905) in a monograph as an adult animal of generic rank, was subsequently shown to be a larval form. Various theories were put forward to account for the amphioxides larva (Goldschmidt, 1933; Fuller, 1958; Bone, 1957, 1960). When there was even less information about acraniate larvae the larvae appeared to fall into two distinct size groups, the smaller, "normal", larva, and the larger, "amphioxides" group. Further information has filled in this gap and, as can be seen from Fig. 10, there is a continuous gradation of size.

Considering the latest information available, I have come to the conclusion (Wickstead, 1964b) that the amphioxides is in fact a "giant larva," a phenomenon familiar to oceanic planktonologists. I agree with Bone (1957) that the view of Bigelow and Farfante (1948) and supported by Berrill (1955) is unlikely to be correct. These giant larvae are found in various phyla and, although recognizable, are considerably larger than normal larvae. The key factor in the development of the amphioxides larva is considered to be lack of contact with the bottom. During the normal diurnal vertical migrations, when a larva is ready

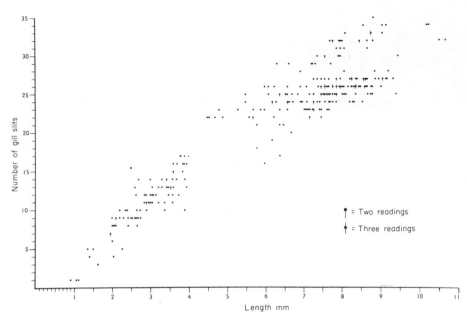

FIG. 10. Relationship between gill slit number and total length (in mm.) in 10 species of acraniate larvae.

to metamorphose, contact is made with the bottom during a migration, and this triggers off metamorphosis. If a larva is carried out to deep water, the larva cannot reach the bottom during its migrations, thus lacking the trigger to initiate metamorphosis. The great majority of amphioxides larvae are found near the surface of deep oceanic waters. If they are carried into shallow waters, contact is then made with the bottom and metamorphosis may be initiated (Wickstead, 1964a).

In order to survive the oceanic conditions for extended periods, the amphioxides has to become somewhat modified. Thus growth rate slows down considerably, the pharynx is divided into a "pars nutritoria" and a "pars respitoria" and, clearly, it must develop the capability to feed in mid-water. I have been fortunate enough to have the privilege of examining the acraniate larvae taken during the International Indian Ocean Expedition, and it seems that *Amphioxides pelagicus* and *Amphioxides valdiviae* can be found throughout the major current systems of the northern part of the Indian Ocean. This suggests the capability of living for a year, or possibly longer, as a larva.

When there is such a prolongation of larval life, one must always consider the possibilities of neoteny and paedogenesis. Some giant larvae develop sexual characteristics, e.g., the "appendix masculina" of male crustaceans (Gurney, 1942, p. 73). Goldschmidt (1905) reports, and

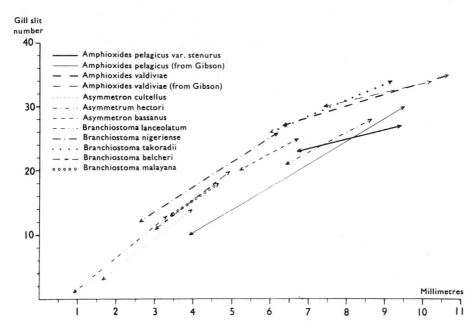

Gill slit number

——— Amphioxides pelagicus var. stenurus
——— Amphioxides pelagicus (from Gibson)
— — Amphioxides valdiviae
— — Amphioxides valdiviae (from Gibson)
········· Asymmetron cultellus
– · – · Asymmetrum hectori
– – – · Asymmetron bassanus
– – · · Branchiostoma lanceolatum
— · — Branchiostoma nigeriense
· · · · Branchiostoma takoradii
— – — Branchiostoma belcheri
o o o o o Branchiostoma malayana

Millimetres

Fɪɢ. 11. The range from the smallest gill slit number/smallest length to the greatest gill slit number/greatest length in 10 species of acraniate larvae (Figs. 10 and 11 from Wickstead, 1964b; published by permission of The Linnean Society of London).

illustrates, gonads in *Amphioxides pelagicus.* I have seen a metamorphosed individual of *Asymmetron lucayanum,* the accepted adult of *Amphioxides pelagicus,* of only 10.0 mm length with 18 gonads, on the one hand, and an *Amphioxides pelagicus* 13.8 mm long with no gonads, on the other (Wickstead, 1971). More recently (Wickstead, 1973), I found incipient gonads in about 100 specimens of amphioxides collected from the Indian Ocean, substantiating Goldschmidt's claim. The gonads occurred as lappet-like lower extensions of the myotomes, rather than intrusions into the myotomes as described by Goldschmidt (1905). Sections were made of a male 8.8 mm long and a female 9.1 mm long, with 27 and 28 gonads, respectively, and these showed spermatocytes and well developed oocytes.

It is interesting to note that in the view of Bone (1960) ". . . the Acrania are free-swimming neotenously produced adult forms. . . ." Thus it is, zoologically speaking, intriguing to speculate on a neotenously produced adult form which can produce a larva possibly with the capability of paedogenesis.

Acknowledgment

I am very pleased to acknowledge the electron microscope work done for me in this chapter by Mr. G. Best, of the Plymouth Laboratory.

9.5 Bibliography

Adam, H. (1963). The Pituitary Gland. In "The Biology of Myxine," pp. 459–476. Universitetsforlaget, Oslo.

Andrews, E. A. (1893). An Undescribed Acraniate; *Asymmetron lucayanum. Johns Hopkins Univ. Stud. Biol. Lab.* 5, 213–247.

Azariah, J. (1965a). Studies on the cephalochordates of Madras Coast. *J. Mar. Biol. Ass. India* 7, 348–363.

Azariah, J. (1965b). On the seasonal appearance of fin rays and their bearing on the reproductive cycle of *Branchiostoma lanceolatum. J. Mar. Biol. Ass. India* 7, 459–61.

Baccetti, B., Burrini, A. G., and Dallai, R. (1972). The spermatozoon of *Branchostoma lanceolatum* L. *J. Morphol.* 136, 211–226.

Barrington, E. J. W. (1963). "General and Comparative Endocrinology," 387 pp. Clarendon Press, Oxford.

Barrington, E. J. W. (1965). "The Biology of Hemichordata and Protochordata," 176 pp. Oliver and Boyd. London.

Benham, W. B. (1893). The structure of the pharyngeal bars of amphioxus. *Quart. J. Microsc. Sci.* 35, 97–118.

Berrill, N. J. (1955). "The Origin of Vertebrates," 257 pp. Oxford Univ. Press, Oxford.

Bert, M. P. (1867). On the anatomy and physiology of amphioxus. *Ann. Mag. Natur. Hist.* 20, (3rd Ser.), pp. 302–304.

Biberhofer, R. (1906). Über Regeneration bei *Amphioxus lanceolatus. Arch. Entwicklungsmech. Organismen* 22, 15–17.

Bier, K. (1970). Oögenestypen bei Insekten und vertebraten, ihre Bedeutung für die Embryogenese und Phylogenese. *Zool. Anz. Suppl.* 33, 7–29.

Bigelow, H. B. and Farfante, I. P. (1948). Lancelets, In "Fishes of the Atlantic Ocean," pp. 1–28. Mem. Sears Found. Mar. Res.

Biuw, L. W. and Hulting, G. (1971). Fine-grained secretory cells in the intestine of the lancelet, *Branchiostoma (Amphioxus) lanceolatum*, studied by light microscopy. *Z. Zellforsch.* 120, 546–554.

Blackburn, M. (1956). Sonic scattering layers of heteropods. *Nature (London)* 177, 374–377.

Bone, Q. (1957). The problem of the 'amphioxides' larva. *Nature (London)* 180, 1462–4.

Bone, Q. (1958a). The asymmetry of the larval amphioxus. *Proc. Zool. Soc. London* 130, 289–293.

Bone, Q. (1958b). Observations on the living larva of amphioxus. *Pubbl. Staz. Zool. Napoli* 30, 458–471.

Bone, Q. (1960). The origin of the chordates. *J. Linn. Soc. London Zool.* 44, 252–65.

Boveri, T. (1892a). Über die Bildungsstätte der Geschlechtsdrüsen und die Entstehung der Genitalkammern beim Amphioxus. *Anat. Anz.* 7, 170–181.

Boveri, T. (1892b). Die Nierenkanälchen des Amphioxus. Ein Beitrag zur phylogenie des Urogenitalsystems der Wirbelthiere. *Zool. Jahrb. Anat. Ontog. Tiere* 5, 429.

Brachet, J. (1950). "Chemical Embryology," 533 pp. Interscience, New York.

Carlisle, D. B. (1953). Origin of the pituitary body of chordates. *Nature (London)* 172, p. 1098.

Carlisle, D. E. (1966). The ciliary current of *Phallusia* (Ascidiacea) and the squirting of sea squirts. *J. Mar. Biol. Ass. U.K.* **46**, 125–127.

Casley-Smith, J. R. (1971). The fine structure of the vascular system of amphioxus: implications in the development of lymphatics and fenestrated blood capillaries. *Lymphology* **4**(3), 79–94.

Cerfontaine, P. (1906). Recherches sur le développement de l'Amphioxus. *Arch. Biol.* **22**, 229–418.

Chen, T. Y. (1931). On a hermaphrodite specimen of the Chinese amphioxus. *Peking Natur. Hist. Bull.* **5**, 11–16.

Chin, T. G. (1941). Studies on the biology of Amoy amphioxus *Branchiostoma belcheri* Gray. *Philipp. J. Sci.* **75**, 369–421.

Clark, R. B. (1965). Endocrinology and the reproductive biology of Polychaetes. *Oceanogr. Mar. Biol.* **3**, 211–255.

Colombera, D. (1974). Male chromosomes in two populations of *Branchiostoma lanceolatum*. *Experientia* **30**, 353–355.

Conklin, E. G. (1905). The early development of chordates in the light of the embryology of Ascidians. *Science* **21**, 264–5.

Conklin, E. G. 1932). The embryology of Amphioxus. *J. Morphol.* **54**, 69–151.

Conklin, E. G. (1933). The development of isolated and partially separated blastomeres of Amphioxus. *J. Exp. Zool.* **64**, 303–375.

Covelli, I., Salvatore, G., Sena, L., and Roche, J. (1960). Sur la formation d'hormones thyroidiennes et de leurs precurseurs par *Branchiostoma lanceolatus* Pallas (Amphioxus). *C. R. Soc. Biol.* **154**, 1165–1169.

Cowden, R. R. (1963). Cytochemical studies of oöcyte growth in the lancelet *Branchiostoma caribaeum*. *Z. Zeuforsch. Mikrosk. Anat.* **60**, 399–408.

de Beer, G. R. (1928). "Vertebrate Zoology," 505 pp. Sidgwick and Jackson, London.

Drach, P. (1948). Embranchement des Céphalocordés. *In:* "Traité de Zoologie" (P.-P. Grassé, ed.), Vol. 11, pp. 931–1040. Mason et Cie, Paris.

Eakin, R. M. (1968). Evolution of Photoreceptors. *In* "Evolutionary Biology" (T. Dobzhansky, M. K. Hecht, and W. C. Steere, eds.), Vol. 2, pp. 195–237. North Holland, Amsterdam.

Eismond, J. (1894). Zur Ontogenie des *Amphioxus lanceolatus*. *Biol. Zentrbl.* **14**, 353–360.

Fechter, H. (1971). "Manteltiere, Schaedellose, Rundmaeuler." (Tunicates, Acrania, Cyclostomata.), 206 pp. Walter de Gruyter and Company, Berlin.

Franz, V. (1922). Systematische Revision der Akranier. *Jena. Z. Naturwiss.* **58**, 309–451.

Franz, V. (1924), Lichtversuche am Lanzettfisch zur Ermittelung der Sinnesfunktion des Stirnoder Gehirnbläschens. *Wiss. Meeresunters. Abt. Helgoland* **1**5, 1 10.

Franz, V. (1927). Morphologie der Akranier. *Ergeb. Anat. Entwicklungsgesch.* **27**, 464–692.

Franzén, Å. (1956). On spermiogenesis, morphology of the spermatozoon, and biology of fertilization among Invertebrates. *Zool. Bidr. Uppsala* **31**, 355–480.

Fuller, A. S. (1958). Studies on some larval Cephalochordata, with special reference to the Amphioxides problem, 213 pp. Ph.D. Thesis, University of London.

Garstang, W. (1928). The morphology of the Tunicata, and its bearings on the phylogeny of the Chordata. *Quart. J. Microsc. Sci.* **72**, 147.

Gibson, H. O. S. (1909). The Cephalochorda: Amphioxides. *Trans. Linn. Soc. London Zool.* **13**, 213–56.

Giese, A. C. and J. S. Pearse (1974). Introduction: General principles. *In* "Reproduction of Marine Invertebrates" (A. C. Giese and J. S. Pearse, eds.), Vol. 1, pp. 1–49. Academic Press, New York.

Godeaux, J. (1967). Les Prochordés (Morphologie, Histologie, Embryologie). *Fortschr. Zool.* **18**, 350–59.

Goldschmidt, R. (1905). Amphioxides. *Wiss. Ergeb. Deut. Tiefsee-Exped. "Valdivia"* **12**, 1–92.

Goldschmidt, R. (1906). Amphioxides und Amphioxus. *Zool. Anz.* **30**, 443–448.

Goldschmidt, R. (1933). A note on Amphioxides from Bermuda based on Dr. W. Beebe's collections. *Biol. Bull.* **64**, 321–5.

Goodrich, E. S. (1912). A case of Hermaphroditism in Amphioxus. *Anat. Anz.* **42**, 318–320.

Goodrich, E. S. (1930). The development of the club-shaped gland in Amphioxus. *Quart. J. Microsc. Sci.* **74**, 155–164.

Goodrich, E. S. (1934). The early development of nephridia in Amphioxus. *Quart. J. Microsc. Sci.* **76**, 493–510; 655–674.

Gunter, G. (1957). Temperature. *In* "Treatise on Marine Ecology and Paleoecology" (J. W. Hedgpeth, ed.), Vol. 1, pp. 159–184. Geol. Soc. Amer. Mem.

Guraya, S. S. (1966). The origin and nature of cortical vacuoles in the amphioxus egg. *Z. Zellforsch. Mikrosk. Anat.* **79**, 326–331.

Gurney, R. (1942). Larvae of decapod crustacea. *Ray. Soc. Publ.* **129**, 306.

Hartmann, J. and John, H.-Ch. (1971). Planktische *Branchiostoma* nordwestlich der Doggerbank (Nordsee). *Ber. Deut. Wiss. Komm. Meeresforsch.* **22**, 80–84.

Hatschek, B. (1881). Studien zur Entwickelung des Amphioxus. *Arb. Zool. Inst. Univ. Wien* **4**, 1–88.

Hatschek, B. (1888). Ueber den Schichtenbau von Amphioxus. *Anat. Anz.* **3**, 23–25.

Hatschek, B. (1892). Die Metamerie des Amphioxus und des Ammocoetes. *Verh. Anat. Ges. Jena*, pp. 136–147.

Hatschek, B. (1893). "The Amphioxus and its Development" (Translated and Edited by James Tuckey), 183 pp. Swan Sonnenschein and Co., London.

Hewer, H. R. (1951). The early development of amphioxus. *Med. Biol. Illustr.* **1**, 172–5.

Howell, W. M. and Boschung, H. T., Jr. (1971). Chromosomes of the lancelet *Branchiostoma floridae* (Order Amphioxi). *Experienta* **27**, 1495–1496.

Huettner, A. F. (1941). "Fundamentals of Comparative Embryology of the Vertebrates," 416 pp. Macmillan, New York.

Huxley, J., Hardy, A. C., and Ford, E. B. (1954). Escape from Specialization. "Evolution as a Process," pp. 122–142. Allen & Unwin, London.

Kinne, O. (1963). The effects of temperature and salinity on marine and brackish water animals. I. Temperature. *Oceanogr. Mar. Biol.* **1**, 301–340.

Kowalewsky, A. (1867). Entwicklungsgeschichte des *Amphioxus lanceolatus*. *Zap. Imp. Akad. Nauk. VII Ser.* **11**.

Kowalewsky, A. (1876). Weitere Studien über die Entwicklungsgeschichte des *Amphioxus lanceolatus*, nebst einem Beitrage zur Homologie des Nervensystems der Würmer und Wirbelthiere. *Arch. Mikrosk. Anat. Entwicklungsmech.* **13**, 181–204.

Kumé, M. and Dan, K. (1968). Introduction. *In* "Invertebrate Embryology" (M. Kumé and K. Dan, eds.), pp. 1–70. Nolit, Belgrade.

Langerhans, P. (1875). Zur Anatomie des *Amphioxus lanceolatus*. *Arch. Mikrosk. Anat. Entwicklungsmech.* **12**, 334–335.

Lankester, E. R. (1889). Contributions to the knowledge of *Amphioxus lanceolatus*. *Quart. J. Microsc. Sci.* 31, 445–466.

Lankester, E. R. and Willey, A. (1890). The development of the atrial chamber in *Amphioxus. Quart. J. Microsc. Sci.* 31, 445–466.

Legros, R. (1895). Sur la Morphologie des glandes sexuelles de l'*Amphioxus lanceolatus. C. R. Séances 3rd Congr. Int. Zool.*, pp. 487–500.

Legros, R. (1898). Developpement de la cavité buccale de l'Amphioxus. *Arch. Anat. Microsc.* 1, 508–542; 2, 1–43.

Legros, R. (1910). Sur quelques points de l'anat. et du develop. de l'Amphioxus. *Anat. Anz.* 35, 561–587.

Leiber, A. (1903). Ueber Bau und Entwicklung der Weiblichen Geschlechtsorgane des *Amphioxus lanceolatus. Zool. Jahrb.*, pp. 1–41.

Leuckart, R. and Pagenstecher, A. (1858). Untersuchungen über niedere Seethiere. *Amphioxus lanceolatus. Arch. Anat. Physiol.*, pp. 558–569.

Lou, T. H. (1936). Note sur la larva d'Amphioxus recueillie à Tchefou. *Contr. Inst. Physiol. Nat. Acad. Peiping* 3, 115–124.

MacBride, E. W. (1898). The early development of Amphioxus. *Quart. J. Microsc. Sci.* 40, 589–612.

MacBride, E. W. (1900). Further remarks on the development of Amphioxus. *Quart. J. Microsc. Sci.* 43, 351–366.

MacBride, E. W. (1909). The formation of the layers in Amphioxus and its bearing on the interpretation of the early ontogenetic processes in other vertebrates. *Quart. J. Microsc. Sci.* 54, 279–345.

Makino, S. (1948). A review of the chromosome numbers in animals (1944), appended with recent additional data (1948). The Japan Society for the Promotion of Scientific Research. The fourth special committee, Rept. No. 3, pp. 1–26 and 136–137.

Marshall, A. M. (1893). "Vertebrate Embryology, 640 pp. Smith, Elder and Co, London.

Medawar, P. B. (1951). Asymmetry of larval amphioxus. *Nature (London)* 167, 852–853.

Morgan, T. H. (1896). The number of cells in larvae from isolated blastomers of amphioxus. *Arch. Entwicklungsmech. Organismen* 3, 269–294.

Müller, J. (1851). Über die Jugendzustände einiger Seethiere. *Mber. Akad. Wiss: Berlin* pp. 94–106.

Müller, W. (1875). Ueber das Urogenitalsystem des Amphioxus und der Cyclostomen. *Jena. Z. Naturwiss,* 9, 94.

Neidert, L. and Leiber, A. (1903). Ueber Bau und Endwickelung der Weiblichen Geschlechtsorgane des *Amphioxus lanceolatus. Zool. Jahrb. Anat. Onto. Tiere* 18, 187.

Nelsen, O. E. (1953). "Comparative Embryology of Vertebrates," 982 pp. Blakiston, London.

Nelson, G. E. (1968). Amphioxus in Old Tampa Bay, Florida. *Quart. J. Fla. Acad. Sci.* 31, 93–100.

Nogusa, S. (1957). The Chromosomes of the Japanese Lancelet, *Branchiostoma belcheri* (Gray), with special reference to the sex chromosomes. *Annot. Zool. Jap.* 30, 42–46.

Orton, J. H. (1914). On a hermaphrodite specimen of amphioxus with notes on experiments in rearing amphioxus. *J. Mar. Biol. Ass. U.K.* 10, 506–512.

Orton, J. H. (1920). Sea temperature, breeding and distribution in marine animals. *J. Mar. Biol. Ass. U.K.* **12**, 339–366.

Parker, G. H. (1908). The sensory reactions of Amphioxus. *Proc. Amer. Acad. Arts Sci.* **43**, 413–455.

Probst, G. (1930). Regenerationsstudien an Anneliden und *Branchiostoma lanceolatum* (Pallas). *Rev. Suisse Zool.* **37**, 343–351.

Raven, C. P. (1961). "Oögenesis—The Storage of Developmental Information," 274 pp. Pergamon Press, Oxford.

Reeves, C. D. (1931). Some observations of the behaviour of Amphioxus. Ginling College, Nanking, pp. 29–34.

Retzius, G. (1904). Zur Kenntnis der Spermien der Evertebraten, I. *Biol. Unters.* [N.F.] **11**, pp. 1–32.

Retzius, G. (1905a). Zur Kenntnis der Spermien der Evertebraten, II. *Biol. Unters.* **12**, 79–102.

Retzius, G. (1905b). Die Spermien der Leptokardier, Teleostier und Ganoider. *Biol. Unters.* **12**, 103–115.

Reverberi, G. (1971). Amphioxus. *In* "Experimental Embryology of Marine and Fresh-water Invertebrates" (G. Reverberi, ed.), pp. 551–572. North-Holland, Amsterdam.

Rice, H. J. (1880). Observations upon habits, structure and development of *Amphioxus lanceolatus. Amer. Natur.* **14**, 1–19, 73–95.

Riddell, W. (1922). On a hermaphrodite specimen of Amphioxus. *Ann. Mag. Natur. Hist.* **1**, 613–617.

Rolph, W. H. (1876). Untersuchungen über den Bau des *Amphioxus lanceolatus. Morphol. Jahrb.* **2**, 1–80.

Rothschild, L. (1956). "Fertilization," 170 pp. Methuen, London.

Sharp, L. W. (1926). "An Introduction to Cytology," 581 pp. McGraw-Hill, New York.

Smith, J. E. (1932). The shell gravel deposits and the infauna of the Eddystone grounds. *J. Mar. Biol. Ass. U.K.* **18**, 243–78.

Smith, K. M. and Newth, H. G. (1917). A note concerning the collar cavities of the larval Amphioxus. *Quart. J. Microsc. Sci.* **62**, 243–251.

Sobotta, J. (1895). Die Befruchtung des Eies von *Amphioxus lanceolatus. Anat. Anz.* **11**, 129–137.

Sobotta, J. (1897). Die Reifung und Befruchtung des Eies von *Amphioxus lanceolatus. Arch. Mikrosk. Anat. Entwicklungsmech.* **50**, 15.

Stephenson, A. (1934). The breeding of reef animals. Part II. Invertebrates other than corals. *Great Barrier Reef Exped.* 1928–29, *Sci. Rept.* **3**, 247–272.

Tambs-Lyche, H. (1967). *Branchiostoma lanceolatum* (Pallas) in Norway. *Sarsia* **29**, 177–182.

Tchang, S. and Koo, K. C. (1936). Description of a new variety of *Branchiostoma belcheri* Gray from Kiaochow Bay, Stangtung, China. *Contr. Inst. Physiol. Nat. Acad. Peiping* **3**, 77–132.

Thorson, G. (1946). Reproduction and larval development of Danish marine bottom invertebrates. *Medd. Komm. Danmarks Fiskeri-og Havunders., Ser. Plankton* **4**, 1–523.

Tung, T. C., Wu, S. C., and Tung, Y. F. Y. (1958). The development of isolated blastomeres of Amphioxus. *Sci. Sinica* **7**, 1280–1320.

van der Stricht, O. (1895). La maturation et la fécondation de l'oeuf d'*Amphioxus lanceolatus. Arch. Biol.* **14**, 469–491.

van Wijhe, J. W. (1913). On the metamorphosis of *Amphioxus lanceolatus*. *Proc. Sect. Sci. Kon. Ned. Akad. Wetensch.* **16**, 574–83.

van Wijhe, J. W. (1918). On the anatomy of the larva of Amphioxus and the explanation of the symmetry. *Verh. Kon. Akad. Wetensch.* **21**, 1013–1020.

van Wijhe, J. W. (1927). Observations on the adhesive apparatus and the function of ilio-colon ring in the living larva of amphioxus in the growth-period. *Proc. Sect. Sci. Kon. Ned. Akad. Wetensch.* **30**, 991–1003.

Waddington, C. H. (1956). "Principles of Embryology," 510 pp. George Allen & Unwin Ltd, London.

Webb, J. E. (1958). The ecology of Lagos Lagoon III. The life-history of *Branchiostoma nigeriense* Webb. *Phil. Trans. Roy. Soc. London* Ser. B, **241**, 393–419.

Webb, J. E. (1969). On the feeding and behaviour of the larva of *Branchiostoma lanceolatum*. *Mar. Biol.* **3**, 58–72.

Wickstead, J. H. (1961). A quantitative and qualitative survey of some Indo-West-Pacific plankton. *Colonial Office, Fish. Publ. London No.* **16**, 1–200.

Wickstead, J. H. (1963). The cladocera of the Zanzibar area of the Indian Ocean, with a note on the comparative catches of two plankton nets. *East Afr. Agr. Forest. J.* **29**, 164–172.

Wickstead, J. H. (1964a). Acraniate larvae from the Zanzibar area of the Indian Ocean. *J. Linn. Soc. London Zool.* **45**, 191–199.

Wickstead, J. H. (1964b). The status of the 'amphioxides' larva. *J. Linn. Soc. London Zool.* **45**, 201–7.

Wickstead, J. H. (1967). *Branchiostoma lanceolatum* larvae: some experiments on the effect of Thiouracil on metamorphosis. *J. Mar. Biol. Ass. U.K.* **47**, 49–59.

Wickstead, J. H. (1971). Report on the Acrania (Phylum Chordata) collected by the 'Dana' Expedition of 1928–1930. *Cah Pacif.*, No. 15, 163–168.

Wickstead, J. H. (1973). Report to the Council. *J. Mar. Biol. Assoc. U.K.* **53**, 1017.

Wickstead, J. H. and Bone, Q. (1959). Ecology of Acraniate larvae. *Nature (London)* **184**, 1849–1851.

Willey, A. (1891). The later larval development of Amphioxus. *Quart. J. Microsc. Sci.* **32**, 183–234.

Willey, A. (1894a). Report on a collection of Amphioxus made by Professor A. C. Haddon in Torres Straits, 1888–1889. *Quart. J. Microsc. Sci.* **35**, 361–371.

Willey, A. (1894b). "Amphioxus and the Ancestry of the Vertebrates," 316 pp. MacMillan, London.

Willey, A. (1906). A review of Dr. Richard Goldschmidt's monograph of Amphioxides. *Quart. J. Microsc. Sci.* **50**, 581–597.

Wilson, E. B. (1892). On the multiple and partial development in *Amphioxus*. *Anat. Anz.* **7**, 322–40.

Wilson, E. B. (1893). *Amphioxus* and the mosaic theory of development. *J. Morphol.* **8**, 579–638.

Yonge, C. M. (1930). "A Year on the Great Barrier Reef," 246 pp. Putnam, London.

Young, J. Z. (1962). "Life of the Vertebrates," 2nd. ed., 767 pp. Clarendon Press, Oxford.

Zapf, K. (1932). Beitrage zur Oogenese des *Branchiostoma lanceolatum. Jena Z. Naturwiss.* **66**, 223–262.

Zarnik, B. (1905). Über die Geschlechtsorgane bei Amphioxus. *Zool. Jahrb. Anat. Onto. Tiere* **21**, 253–338.

Zietz, A. (1908). A synopsis of the fishes of S. Australia, Part 1. *Trans. Roy Soc. S. Austr.* **32**, 288–299.

AUTHOR INDEX

Numbers in italics refer to the pages on which the complete references are listed.

321

SUBJECT INDEX

Entries in this index which are from a table are indicated by a "t" following the page number; page numbers for entries which are from a figure are set in boldface type.

TAXONOMIC INDEX

Entries in this index which are from a table are indicated by a "t" following the page number; page numbers for entries which are from a figure are set in boldface type.